T0257781

Senescence in Humans

Senescence in Humans

Edited by **Brandon Chesser**

New York

Published by Callisto Reference,
106 Park Avenue, Suite 200,
New York, NY 10016, USA
www.callistoreference.com

Senescence in Humans
Edited by Brandon Chesser

International Standard Book Number: 978-1-63239-557-3 (Hardback)

Contents

Preface

The phenomenon related to ageing in human beings is explained in this book. It covers various aspects relating to ageing and discusses the factors associated with it. Renowned experts and researchers in the field of senescence have made their valuable contributions to this book. It aims to benefit students and researchers interested in this field.

This book unites the global concepts and researches in an organized manner for a comprehensive understanding of the subject. It is a ripe text for all researchers, students, scientists or anyone else who is interested in acquiring a better knowledge of this dynamic field.

I extend my sincere thanks to the contributors for such eloquent research chapters. Finally, I thank my family for being a source of support and help.

Editor

Human

Parkinson's Disease: Insights from the Laboratory and Clinical Therapeutics

Jing-ye Zhou, Yong Yu, Xian-Lun Zhu,
Chi-Ping Ng, Gang Lu* and Wai-Sang Poon
*Division of Neurosurgery, Department of Surgery,
The Chinese University of Hong Kong, Hong Kong,
China*

1. Introduction

Parkinson's disease (PD), a neurodegenerative disorder that was first described by James Parkinson (1755-1824) in 1817, is characterized partly by a progressive loss of dopaminergic neurons in the substantia nigra pars compact. It affects approximately 1.5% of the global population over 65 years of age. PD is the type of Parkinsonism that is defined as any combination of six specific and independent motoric features: bradykinesia, resting tremor, rigidity, loss of postural reflexes, flexed posture and the freezing phenomenon. Current dopamine replacement strategies, which include levodopa (L-DOPA, the precursor of dopamine) and dopamine receptor agonists, as well as monoamine oxidase B and catechol O-methyltransferase inhibitors, can effectively improve these symptoms. Many reviews of this field are available elsewhere; therefore we focus here on the most recent outcomes regarding the identification of key biomedical progress in PD, describe the most promising biological research targets that are currently being assessed to find ideal treatments, and provide insights from progress in laboratory research and clinical therapeutics.

2. The pathogenesis of Parkinson's disease

Decades of research have not found a single cause for PD and therefore a single factor is unlikely to emerge. Current research is mainly carried out on animal models of PD induced by intoxication with 1-methyl-4-phenyl-1,2,3,6-tetrahydropyridine (MPTP) and models of postencephalitic parkinsonism, neither of which has fully reproduced the clinical and pathological features of true PD. However, it is believed that PD is a multifactorial disease caused by both environmental factors and genetic susceptibility. Aging is an obvious factor because PD mainly targets elderly people. Studies have shown that the incidence of PD is around 10-15 cases per 100 000 person–years (1), but this figure increased to 93.1 in people aged between 70 and 90 years (2). Male sex appeared to be another risk factor, because the incidence of PD in men was 1.5 times higher than that in women (3). Geographically, China has a similar prevalence of PD to western countries (4), whereas Africans have a lower rate compared with African Americans (5).

---

* Corresponding Author

2.1 Environmental factors

Many environmental factors may increase the risk of developing PD. Priyadarshi et al. (6-7) showed the association between PD and farming, professional pesticide use, and drinking well-water in a meta-analysis. Other environmental factors, such as metals, solvents, electromagnetic fields and lifestyles, have also been determined as possible risk factors (8). Studies over the past two to three decades have provided more supportive findings. Tanner and Goldman (9) linked the consumption of well-water to the occurrence of PD. Living in rural areas was associated with farming and pesticide use, which led to an increased incidence of PD patients (10-11). This was association was clarified by another study that demonstrated that the effect of pesticide use was independent from that of farming (12). A lifestyle study found similar evidence of increased herbicide exposure in patients with PD (13). The discovery that exposure to MPTP induced parkinsonian syndromes initiated a new field in PD research – the study of exposure to pesticides (14-15). Many different pesticides have been investigated. MPTP has a similar structure to paraquat, a herbicide that is widely used in many countries. Paraquat was found to be associated with PD based on a 20-year exposure study (16). In a study in Germany, organochlorine pesticides were identified as risk factors for PD (17). Dithiocarbamates, which have been shown to enhance MPTP toxicity (18), were considered to be another risk factor for PD (19). Manganese, a constituent of several pesticides and herbicides, induced parkinsonism in humans following chronic exposure (20). Pesticides and herbicides may be used in combination, which results in a higher level of toxicity. One study showed that exposure to paraquat plus manganese ethylenebis-dithiocarbamate (maneb) resulted in 4.17-fold greater risk for PD compared with unexposed populations (21).

In addition to the above agricultural risk factors, industrial factors also play an important role in the development of PD. Chronic exposure to copper, manganese, and lead was associated with the risk for PD (22) and PD patients who had worked in factories that used chemicals, iron or copper had higher death rates (23). A German study also reported an association between exposure to lead and PD (17). Furthermore, the relationship between PD and head trauma has been investigated: a history of head trauma was associated with onset of PD at an earlier age (24-25).

As discussed above, many risk factors are involved in the development of PD; however, two environmental factors could lower this risk: cigarette smoking (26) and coffee drinking (27), although their mechanisms are unknown. Studies of twins showed an inverse association between cigarette smoking and PD (28-29), and similar results were reported by a study that compared PD cases with their unaffected siblings (30). A meta-analysis reported an inverse association between PD and coffee drinking which was independent of smoking (31). However, this was seen in men but not in women (27).

2.2 Genetic susceptibility

Most cases of PD are sporadic, but some patients (10-15%) show a positive familial history of the disease (32). Although the cause of PD is still unknown, both environmental and genetic factors are considered to be important. The discovery of several causative mutations and genes (33) has allowed a better understanding of PD.

2.2.1 α-Synuclein (PARK1)

α-Synuclein, also called PARK1, was the first gene to be linked to PD (34), and mutations in α-synuclein gene have been linked to rare cases of familial PD (35-37). Genomic multiplications have been reported and both mRNA and protein levels of α-synuclein were increased in the brain (38). However, a large screening study has shown that α-synuclein multiplication is a rare cause of parkinsonism (39). Nevertheless, there is a link between α-synuclein level and age at onset and severity: when α-synuclein duplication causes the disease at an earlier age, then PD has a more aggressive form (39-40). α-Synuclein is a small neuronal protein that is involved in neurotransmitter release and synaptic vesicle recycling. Without genetic changes, α-synuclein is an abundant protein and a major component of Lewy bodies (LBs) in idiopathic, apparently sporadic PD (41-42). This supports the role of α-synuclein in the pathogenesis of PD.

2.2.2 Parkin (PARK2)

The PARK2 gene was identified as parkin in autosomal recessive forms of familial juvenileparkinsonism (AR-JP) (43). AR-JP is most commonly seen in Japanese populations and typically has an onset before the age of 40 years (44-45). Interestingly, no LBs have been found in parkin-positive brains. Parkin was reported to act as an E3-ubiquitin ligase that targets cytoplasmic proteins for proteasomal degradation and plays a role in receptor trafficking (46-47). A wide variety of parkin mutations have been found including large homozygous deletions in exons (43); frame-shift mutations, point mutations, duplications and triplications of exons (48); and deletions in the promoter (49). Parkin mutations were identified in nearly 50% of familial cases with disease onset before the age of 45 years (50) and in 15% of sporadic young-onset cases (51). In the subset of cases with onset before the age of 20 years, this proportion increased to 70% (51).

2.2.3 Ubiquitin carboxyl-terminal hydrolase L1 (UCH-L1; PARK5)

UCH-L1 is an enzyme that hydrolyzes the C-terminus of ubiquitin to generate ubiquitin monomers that can be recycled to clear other proteins. A single missense mutation in UCH-L1 was reported in two siblings with typical PD in a German family (52). A second rare mutation was reported in French families but was not restricted to PD (53). No other carriers of this mutation and no other mutations in UCH-L1 have been identified (54-55), which has raised doubts about the relevance of UCH-L1 to PD.

2.2.4 PTEN-induced kinase 1 (PINK1; PARK6)

PINK1 encodes a widely expressed protein kinase that is localized in mitochondria. PINK1 is the second most common cause of AR-JP (56) and may play an important role in sporadic PD (57). Several mutations have been identified including transitions (56), single heterozygous mutations (57), and heterozygous deletion of the PINK1 gene plus a splice site mutation on the remaining copy (58). One study suggested that heterozygous mutations are a significant risk factor in the development of PD (59). Briefly, PINK1 mutations may cause loss of function in patients with recessively inherited forms of PD because most mutations fall in the kinase domain (60).

2.2.5 Oncogene DJ-1 (PARK7)

The DJ-1 gene encodes a ubiquitous and highly conserved protein, and has been identified as a causative gene for early-onset autosomal recessive PD (61). A couple of mutations have been reported but these were found in only a few patients with early-onset PD (61-62). DJ-1 protein is not an essential component of LBs but is localized in mitochondria that protect against neuronal death (63).

2.2.6 Leucine-rich repeat kinase 2 (LRRK2; PARK8)

LRRK2 mutations are the most common mutations identified in either familial or sporadic PD. Although other LRRK2 mutations have been described, the G2019S mutation has been found to be the most common pathogenic cause of PD, and has been reported in 5–6.6% of cases of autosomal dominant PD (64-65) and 2–8% of sporadic cases (66-67). Penetrance in G2019S patients was age dependent, and increased from 17% at the age of 50 years to 85% at the age of 70 years (68). Nigral neuron loss and LB formation have been observed in the brains of sporadic PD patients with G2019S mutations (66).

2.2.7 Adenosine triphosphatase type 13A2 (ATP13A2; PARK9)

ATP13A2 has been identified as the causative gene in Kufor-Rakeb syndrome (69), a rare form of juvenile-onset parkinsonism caused by autosomal recessive neurodegeneration. Studies in PD patients have reported mutations of 22bp duplication in exon 16 (70) and missense mutation in exon 15 (71), indicating that they are possible causes of PD.

2.2.8 OMI/HTRA serine peptidase 2 (OMI/HTRA2; PARK13)

A missense mutation in the OMI/HTRA2 gene has been found in sporadic PD patients (72). The OMI/HTRA2 gene is located within the PARK3 linkage region, but its role in PD is unknown.

Other genetic factors, such as glucocerebrosidase (73), microtubule-associated protein tau (74) and progranulin (75), have shown an association with PD but their causality has yet to be elucidated.

3. Experimental models in PD research

3.1 Neurotoxin models

The development of experimental models is essential for a better understanding of the etiopathogenesis of PD and to provide effective therapeutic agents. Neurotoxins that target the dopamine (DA) system, such as 6-hydroxydopamine (6-OHDA) and MPTP were used in early animal models for PD research and are still widely in current use (76).

3.1.1 6-OHDA

6-OHDA was the first agent used in an animal model of PD (77). Because of its structural similarity to DA and norepinephrine, 6-OHDA can enter and accumulate in both dopaminergic and noradrenergic neurons. Catecholaminergic structures are destroyed by 6-

OHDA through reactive oxygen species (ROS) and quinines (78-79). Because 6-OHDA crosses the blood-brain barrier (BBB) poorly, it is usually injected directly into the brain stereotaxically. Intraventricular and intracisternal administrations of 6-OHDA to rats produce a bilateral loss of DA and motor abnormalities that can be partially corrected by dopaminergic receptor agonists (80). However, the motor deficits induced are caused by considerable depletion of DA that requires high doses of 6-OHDA. Thus, animals often die due to aphagia and adipsia from severe stress (77, 81). In contrast, a unilateral intracerebral injection is more practical and useful. This model provides an approach to measure asymmetrical turning behavior in response to DA agonists with an internal control – the unlesioned contralateral side of the brain. To induce unilateral lesions, 6-OHDA is typically injected into the striatum, substantia nigra or the median forebrain bundle. Striatal injection of 6-OHDA produces slow retrograde degeneration of the nigrostriatal system over 1 month (82) and apoptotic morphology in the neurons that die (83-84). After injection of 6-OHDA into the substantia nigra or the medial forebrain bundle, dopaminergic neurons die more quickly than after striatal injection and no apoptotic morphology is seen (85). It should be noted that no typical LB formation has been demonstrated in this model (86). Unilateral lesions produce typical asymmetric circling motor behavior, especially after injection into the substantia nigra or the medial forebrain bundle which leads to more readily detectable behavioral deficits. The quantification of this circling behavior has been applied widely to evaluate new anti-parkinsonian drugs, and stem-cell and gene therapies (86).

6-OHDA has mainly been used in small animals, such as rodents, but has also been administered to non-human primates (86) and has been applied *in vitro* in various different models (87). The unilateral lesion induced by 6-OHDA in rats is one of the most popular models of PD (88-89). This model has advantages for testing cell replacement therapies and investigating regenerative therapies (90). However, 6-OHDA models demonstrate only one dimension of a complex illness: one type of cell loss and cellular stress. Moreover, 6-OHDA causes an acute model and cannot replicate many features of PD (90).

3.1.2 MPTP

MPTP was discovered in the early 1980s (14). Unlike 6-OHDA, MPTP is highly lipophilic and can cross the BBB easily after systemic administration. In the brain, it is metabolized to the 1-methyl-4-phenylpyridinium ion (MPP$^+$), which can enter dopaminergic neurons via the DA transporter (DAT) (91), and results in mitochondrial complex I inhibition and ROS formation (92). No LBs have been observed in MPTP-induced parkinsonism in either human or animal models (93-94).

Systemic administration of MPTP to many different species satisfies most of the requirements for an ideal parkinsonian model (90). The most commonly used PD model is the MPTP mouse model cause of the relatively low cost and acceptable timing for research. Acute administration of MPTP caused depletion in striatal DA (95), whereas the subacute model showed striatal DA depletion and cell loss in the substantia nigra (96-97). However, the acute model showed a greater loss of striatal DA than the subacute model. Moreover, the mechanism of cell death appears to differ in these two models: non-apoptotic mechanisms after acute (98) versus apoptotic mechanisms after subacute administration of MPTP (99). In addition, Acute treatment of animals with MPTP induced clear microglial activation in the striatum and the substantia nigra (100-101) and over-expression of inducible nitric oxide synthase (102), whereas the subacute model showed only minimal microglial activation (101). Consequently, anti-inflammatory and anti-microglial compounds could be potential neuroprotective agents for PD and have been investigated using the acute MPTP model.

Although the mouse model has been used extensively in PD research, the non-human primate model is considered to be the "gold standard" for assessment before clinical trials because it reproduces the main pathological defect of PD, and the parkinsonian symptoms induced match many clinical features of PD. Acute and chronic non-human primate models are available, and both show similar motor symptoms of parkinsonism, although the mechanisms differ. Schneider et al. reported that the acute MPTP monkey model inuced increased binding to striatal DA D1 and D2 receptors and increased striatal prepronkephalin mRNA expression. In contrast to these findings, striatal preprotachykinin mRNA expression was decreased in both acute and chronic MPTP monkey models. Notablely, chronic administration of MPTP to animals with cognitive but no motor deficits induced no changes in preprotachykinin expression in the striaturn (103).

MPTP has been used to produce the best-characterized model of PD (104) in many different species, including non-human primates, small vertebrates, such as mice, and even invertebrates, such as worms (105-107). However, rats are relatively resistant to MPTP-induced neurotoxicity compared with mice (108-109) and primates are the most sensitive model (110). MPTP models develop pathological and neurochemical changes similar to those of PD patients (111) and produce an irreversible and severe parkinsonian syndrome that includes rigidity, tremor, bradykinesia, posture abnormalities and even freezing (86, 112). PD is a slowly progressive illness but the MPTP mouse model is an acute or subacute process. The chronic administration of MPTP to primates induced the slow development of a parkinsonian syndrome (113).

3.1.3 Rotenone

Rotenone is one of the most recent neurotoxins to be used in PD models (114). It is widely used as an insecticide throughout the world, and the rotenone model was the first to use an environmental toxin. Rotenone is highly lipophilic and can thus move freely and rapidly across cellular membranes without transporters. In mitochondria, it interferes with the electron transport chain, resulting in mitochondrial complex I inhibition (115). Furthermore, it also inhibits the formation of microtubules from tubulin (116) and the excess of tubulin monomers may be toxic to cells (117). The complex I inhibition induced by rotenone led to increased levels of oxidative stress which occurred predominantly in dopaminergic regions including the striatum, ventral midbrain and the olfactory bulb (118). Rotenone is a mitochondrial complex I inhibitor and acts evenly throughout the brain (119), which indicates that dopaminergic neurons are uniquely sensitive to mitochondrial complex I inhibition. Sherer et al (120) detected microglial activation in the striatum and substantia nigra of rotenone-infused rats. In the same model, the hallmark of PD pathology – LBs – were observed (114, 121) in the ventral midbrain regions (122). Exposure of animals to rotenone caused selective nigrostriatal dopaminergic neurodegeneration but had minimal effects on neurons of other brain regions (114, 121). It has been reported that parkinsonian symptoms in humans, such as rigidity and bradykinesia, are caused by reduced striatal dopaminergic activity (123); this change in motor behavior has also been reported in rotenone-exposed rats (114, 121). Rats treated with rotenone displayed a significant increase in abnormal motor behavior and decline in locomotor activities (124).

Rotenone-treated animal models reproduce all the pathological and behavioral features of typical human PD. Rotenone has been used successfully in a variety of species, including non-human primates, mice, and snails. However, there is some variation between the models and not all treated rats displayed these features. Briefly, this model provides very similar clinical features to typical PD but the low reproducibility and high mortality rate (88) may limit its practical use (117). Interestingly, rotenone was found to be involved in a multisystem disorder (125): enteric nervous system dysfunction (126) and loss of myenteric neurons in rats (127).

3.1.4 Other neurotoxins

Other neurotoxins used in PD models include paraquat and maneb. Paraquat is a common herbicide, has a similar chemical structure to MPP^+, the oxidized metabolite that mediates MPTP neurotoxicity, and has been suggested to be a risk factor for PD (128). Epidemiological studies have also indicated that exposure to paraquat may play an important role in the development of PD (16). After crossing the BBB, paraquat inhibits mitochondrial complex I in dopaminergic neurons (129). The treatment of mice with paraquat caused destruction of dopamine neurons in the substantia nigra (130). The paraquat-induced neurodegeneration is probably triggered by c-Jun N-terminal kinase signaling pathways (131). In this model, microglial activation has been identified and may act as a risk factor for dopaminergic cell death (132). Further investigations showed that the activated microglia produce potentially harmful molecules, such as superoxide anion and nitric oxide, resulting in redox cycling reactions and ROS formation which enhance tissue vulnerability in the paraquat model (133-135). In addition, oxidative stress plays an important role in nigrostriatal degeneration (136-137). The paraquat model demonstrated a-synuclein up-regulation and aggregation associated with dopaminergic cell death in the substantia nigra pars compacta in mice (138).

Maneb is a fungicide that is always used in combination with paraquat in agriculture. In PD research, maneb potentiates the DA toxicity of paraquat in mice (139). Maneb alone inhibits mitochondrial complex III and causes selective dopaminergic neurodegeneration (140). The combination of maneb and paraquat induced more pronounced behavioral and pathological changes than paraquat alone (141-142). Barlow et al. (143) explained that this effect could be due to the ability of maneb to modify the biodisposition and thus increase the concentration of paraquat. Findings in co-exposure models have important implications for the risk of PD in humans because they are liable to be exposed to the synergistic mixtures in agricultural or residential areas where both agents are applied jointly (139).

As a model of environmental exposure, administration of paraquat/maneb reproduces neurodegenerative changes and is useful for the investigation and understanding of the neurotoxic mechanisms of risk factors for PD. Although this model cannot achieve the severe nigrostriatal neurodegeneration induced by MPTP and 6-OHDA, it is a good complement for the comprehensive understanding of PD.

3.2 Genetic models

The majority of PD cases are sporadic but several causative genes and mutations have been discovered and have led to new approaches in the investigation of the mechanisms involve in PD. Several genetic animal models of PD reported in recent years are discussed below.

3.2.1 α-Synuclein

Many different transgenic mice models that over-express human α-synuclein have been generated and applied to pathogenesis and drug research. Transgenic mice induced by the tyrosine hydroxylase promoter expressed α-synuclein containing A30P and A53T mutations and showed a progressive decline in locomotor activity and loss of substantia nigra neurons and striatal DA content (144-145). When transgenic mice were induced by the neuron-specific platelet-derived growth factor β promoter, α-synuclein over-expression was observed, together with reduced tyrosine hydroxylase immunoreactivity and DA content in the striatum and impaired motor performance (146). Mice that over-expressed A53T mutant α-synuclein under the mouse prion promoter (PrP) developed an adult-onset progressive neurodegenerative disorder (147-148). Another neuron-specific promoter, thymocyte differentiation antigen 1 (Thy1), was used in mice to induce a high level of widespread expression of α-synuclein in most neuronal populations (149-150). Both Thy-1 and PrP mice are the only models that have intraneuronal inclusions, degeneration and mitochondrial DNA damage in the neurons (151).

A rat model that over-expresses wild-type or mutant α-synuclein induced by adenoassociated viruses in substantia nigra neurons, displayed progressive age-dependent loss of DA neurons, motor impairment, and α-synuclein-positive cytoplasmic inclusions (152). In Drosophila, α-synuclein over-expression led to age-dependent loss of dorsomedial dopaminergic neurons, accumulation of LB-like inclusions with α-synuclein immunoreactivity and compromised locomotor activity (153). α-Synuclein over-expression in Caenorhabditis elegans caused accelerated dopaminergic neuronal loss and motor impairment (154-155).

PC12 cells have been widely used in PD research to understand the regulation of the neuronal level of α-synuclein. α-Synuclein expression in PC12 cells was low but could be greatly increased by treatment with nerve growth factor (NGF) (156). NGF signal transduction was indicated via the MAP/ERK and PI3 kinase pathways (157). Another study with PC12 cells reported that wild-type α-synuclein was selectively translocated into lysosomes and degraded by the chaperone-mediated autophagy pathway (158). However, the mutant α-synuclein bound to the receptor on the lysosomal membrane inhibited both its own degradation and that of other substrates (158). PC12 cells over-expressing mutant α-synuclein showed impaired proteasomal activity and enhanced sensitivity to proteasomal inhibitors (159). Endoplasmic reticulum stress and mitochondrial dysfunction played important roles in increased cell death (160). The same model in another study showed impairment in both proteasomal and lysosomal functions, a high level of autophagic cell death and loss of chromaffin granules (161).

Yeast has been widely used to investigate the role of α-synuclein toxicity in human diseases, including PD. When expressed in yeast, α-synuclein became cytotoxic in a concentration-dependent manner (162). α-Synuclein in yeast was highly selectively associated with the plasma membrane and formed cytoplasmic inclusions (162). The growth inhibition induced by α-synuclein was accompanied by cellular consequences, such as proteasome impairment, heat-shock and oxidative stress, formation of ubiquitin-positive α-synuclein inclusion bodies, and emergence of apoptotic markers (162-164). Because the molecules that inhibit α-synuclein toxicity are potential therapeutic agents, α-synuclein toxicity in the yeast model has been used for genetic screening to identify genetic modifiers (165-166) and for small molecule or chemical screening to identify novel compounds.

Studies of α-synuclein in cell-free systems have focused on its aggregation pathway, post-translational modification, self-assembly and structure characterization. Test-tube models are critical for the investigation of PD-related protein α-synuclein and its related molecules as they provide more detailed information than any other approaches. However, some findings may not fully account for the biological complexity of α-synuclein *in vivo*. Therefore, further validation in cell cultures or *in vivo* is required.

3.2.2 Parkin

Parkin mutations have been found in a number of cases with recessive juvenile onset (167) and are the second genetic cause of PD. In a rat model, over-expression of parkin protected against the toxicity of mutant α-synuclein as demonstrated by a reduction in α-synuclein-induced neuropathology (168). In addition, parkin over-expression through viral transduction protected mice from mild MPTP-induced lesions (169). However, parkin knockout mice showed no impairment in the dopaminergic system (170-171). A parkin knockout Drosophila model exhibited locomotor defects and male sterility (172). These tissue-specific phenotypes were due to mitochondrial dysfunction. Further studies showed that oxidative stress components and genes involved in innate immunity were induced in parkin mutants, which indicated that oxidative stress and/or inflammation may play a fundamental role in the etiology of AR-JP (173). Another study showed that the expression of mutant human parkin in Drosophila caused age-dependent, selective degeneration of dopaminergic neurons accompanied by progressive motor impairment (174). Both the loss of function and toxicity of parkin have been demonstrated in the Drosophila model.

In experiments using cell-free models, mutant forms of parkin associated with AR-JP were reported to have reduced ubiquitin ligase activity (175-176). Catechol-modified parkin in the substantia nigra – a vulnerability of parkin to modification by dopamine – suggested a mechanism for the progressive loss of parkin function in dopaminergic neurons (177). Another modification – phosphorylation by cyclin-dependent kinase 5 – may contribute to the accumulation of toxic parkin substrates and decrease the ability of dopaminergic cells to cope with toxic insults in PD (178).

3.2.3 PINK1

PINK1 knockout mice had no dramatic abnormalities in the dopaminergic system (179). In Drosophila models, compared with the loss of parkin, the loss of PINK1 showed a strong similarity in phenotype: shortened lifespan, infertility and wing postural defect; in addition, identical loss of mitochondrial integrity was found in both cases (180-182). However, one difference was also observed: up-regulation of parkin rescued PINK1 mutants, whereas PINK1 up-regulation could not rescue parkin mutation (180-182). Cell-free models have been used to investigate the enzymatic function of PINK1 and the effects of mutations.

PINK1 was shown to phosphorylate downstream effector tumor necrosis factor receptor-associated protein 1 directly and prevent oxidative stress-induced apoptosis (183). It was also reported that the PINK1 kinase domain catalyzed the phosphorylation of artificial protein substrates, including α-casein (184) and histone H1 (60). Kinase assays suggested that multiple PINK1 mutants associated with autosomal recessive PD have reduced kinase activity (60, 183-185).

3.2.4 DJ-1

DJ-1 knockout mice exhibited a deficit in scavenging mitochondrial hydrogen peroxide due to its function of atypical peroxiredoxin-like peroxidase (186). Further studies have been carried out using Drosophila. DJ-1 knockout caused selective sensitivity to the oxidative toxins, paraquat and rotenone (187-189). DJ-1 protein undergoes oxidative modification on cysteine residue, which was also seen in Drosophila (189), and the oxidative modification occurred with aging and after exposure to paraquat (190). Cell-free models have helped researchers to characterize the crystal structure of DJ-1, investigate its function and activity, and reveal the effect of oxidative modifications on the stability and function of DJ-1.

3.2.5 LRRK2

In cell models, LRRK2 mutations significantly increased autophosphorylation activity (191-194). Over-expression of mutant LRRK2 caused condensed and fragmented nuclei, resulting in increased cellular toxicity (191). Because the cellular toxicity induced by mutant LRRK2 can be prevented by inactivation of the kinase domain in cell models, the kinase domain could be a therapeutic target for LRRK2-associated PD. Cell-free systems have been used to investigate the kinase function and how it is affected by pathogenic mutations. Reports indicated that LRRK2 catalyzed its autophosphorylation or the phosphorylation of artificial substrates (191-192, 195). This suggested that LRRK2-mediated phosphorylation was regulated by the binding of guanine triphosphate (GTP); in addition, both GTP binding and protein kinase activity are necessary for LRRK2 neurotoxicity (192, 196).

4. Clinical therapeutic insights

From the traditional view, PD is considered to be a single clinical entity, but this is currently under scrutiny (197-198). Clinically, the subtypes of this heterogeneous disease can be recognized on the basis of age at onset, predominant clinical features and progression rate. There are two major clinical subtypes: the tremor-predominant form which is often observed in younger people, and generally leads to a slow decline in motor function; in the other type, known as "postural imbalance and gait disorder" that is often observed in older people (>70 years old), motor function declines more rapidly, and is characterized by akinesia, rigidity, and gait and balance impairment. (198).

4.1 Dopamine replacement therapies

During the years of disease progression, the treatment of PD has to be adapted to alternating periods of reduced mobility and abnormal involuntary movements and is complicated by the onset of motor fluctuations and dyskinesia (199). PD was essentially an untreated motor disorder before L-DOPA was developed as a treatment. For the next two decades, the symptoms of hallucinations and delirium or other motor complications and psychiatric manifestations became the prevailing clinical problems in PD after treatment with L-DOPA. However, bradykinesia, resting tremors and rigidity which are the major symptoms of PD (200) can be controlled by long-term use of L-DOPA and other dopaminergic agents. Although the dopamine precursor, L-DOPA, and dopamine agonists are very effective in treating motor symptoms, they can cause substantial motor and behavioural adverse effects. Many reports have claimed that some patients treated with dopaminergic drugs develop impulse control disorders, a dopamine dysregulation syndrome or other abnormal behaviors (201). Because of these flaws, new treatments for PD should be developed to tackle two unresolved problems: the alternation between therapies that alleviate symptoms and those that modify the disease; and reduction of the real causes of disability in long-term PD, which include autonomic dysfunction, balance loss, cognitive impairment and the growing prevalence of other non-motor symptoms.

Peak-dose dyskinesia, diphasic dyskinesia and off-period dystonia are the three forms of dyskinesias that commonly occur with L-DOPA use and negatively affect the quality of life of patients in the advanced stages of the disease (202). Peak-dose dyskinesia occurs when plasma L-DOPA levels are highest; diphasic dyskinesia refers to the abnormal involuntary movements that occur transiently at the onset and end of L-DOPA efficacy; and off-period dystonia occurs when a patient receives subtherapeutic levels of L-DOPA. Recent advances in the treatment of severe disabling dyskinesias have lessened but not entirely eliminated their effects. Specific examples of such advances include deep brain stimulation (DBS) of the subthalamic nucleus, continuous subcutaneous infusion of apomorphine and continuous duodenal infusion of L-DOPA. Currently, a major focus of drug development is the identification of agents that can acutely suppress existing disabling dyskinesias and of agents that do not induce dyskinesias.

More than 80% of patients who have had PD for 20 years develop dementia. Once this occurs, irrespective of their age or the duration of the disease, death follows shortly (203). From an anatomopathological point of view, PD dementia is believed to be due to a combination of the extension of Lewy bodies into limbic and cortical structures with concomitant Alzheimer's disease(AD)-related neurofibrillary tangles and amyloid-β plaque pathology (203-204). The recent observation that lower levels of amyloid-β_{1-42} in the cerebral spinal fluid may predict a more rapid cognitive decline supports the contribution of AD-related pathologies to the cognitive impairment that is seen in patients with PD (205). Relief from neuropsychiatric cognitive and behavioral symptoms without worsening motor impairment or altering the relief of symptoms that is provided by L-DOPA are the goals of current treatment in PD dementia. To achieve these goals, reliance is placed on fine-tuning the balance between dopaminergic and non-dopaminergic (prominently cholinergic) neurotransmission strategies.

4.2 Surgical treatment and deep brain stimulation (DBS)

In recent years, DBS has become an established treatment for the advanced stages of PD. It is efficacious and is approved by the US Food and Drug Administration for the treatment of advanced, L-DOPA-responsive PD and medically refractory essential tremor. New anatomical targets for DBS, such as the pedunculopontine nucleus, are currently being explored in patients with PD who have gait disorders. In the search for new targets, smart DBS techniques such as coordinated reset stimulation are currently under development (206).

Many reports have enlarged described the long-term outcome of DBS in PD, but, as with L-DOPS treatment, flaws still remain. Subthalamic nucleus DBS (STN DBS) can substantially improve motor function and quality of life in some patients with PD; however, a minority of patients experience cognitive and emotional difficulties after surgery. Better controlled randomized trials that compared STN DBS with the best medical therapy failed to substantiate the findings of widespread or marked cognitive deterioration (207-208).

Smeding and colleagues (209) reported on predictors of the cognitive and psychosocial effects of STN DBS in patients with PD. Varied mood outcomes were observed: 16 patients treated with STN DBS (15%) showed improvements, but the same percentage showed deterioration. Strutt and colleagues (210) have shown that mood (depression) changes cannot be attributed solely to symptoms of somatic depression that overlap with those of PD.

The pre-operative selection of patients who are suitable for STN DBS is critical; response to L-DOPA is considered to be not only a predictor of motor outcomes, but perhaps also of neurocognitive and quality of life outcomes. As pre-operative impairments can predict neuropsychological outcomes after therapy, neuropsychological evaluation should be undertaken before surgery. Mood states should also be evaluated, but reliance on self-reported questionnaires should be discouraged (211).

Although aging is suggested to be a prognostic factor of neurosurgical outcome (212-213), studies that trace the long-term clinical evolution among subgroups of patients with early-onset versus late-onset PD after STN DBS are still lacking. The latest study of a cohort of 19 subjects treated with subthalamic nucleus DBS after more than 20 years of disease reported clinical and neuropsychological data up to a mean of 30 years after disease onset (214). A higher prevalence of axial and non-L-DOPA-responsive symptoms was observed during long-term evaluations compared with other STN DBS follow-up studies. This confirms that, even in patients with an early onset of disease and a previous long-lasting response to dopaminergic therapies, several complex aspects underlie the development of non-motor symptoms and other features of the progression of PD. Therefore, the surgical option of STN DBS should be proposed earlier, since the progression of PD might not follow a single direction, and it is possible that age might affect the development of non-motor features more than the duration of the disease.

4.3 Transplantation treatment

Pharmacological agents that increase DA can alleviate motor symptoms as mentioned above; however, patients develop severe effects with long-term use. Cell transplantation therapy has therefore been investigated as an alternative treatment in recent years. Since only one cell type is affected in a distinct location of the brain, cell replacement therapy is liable to be successful for PD, and has already been used in many other diseases. Transplantation treatment is considered to be an on-going alternative strategy for an effective cure for PD.

Stem-cell replacement therapy has been suggested as a treatment for neurodegenerative diseases caused by the degeneration of DA neurons in the substantia nigra of the brain, and especially for PD (215). Stem cell-derived DA neurons can replace endogenous degenerated neurons. Clinical studies using fetal midbrain tissue proved the principle that cell transplantation could be a feasible treatment for PD(216).

Although it has shown promise for the treatment of PD, the safety and efficacy of transplanted stem cells induced by different methods are variable. Fetal-tissue transplants have gained some success, but their availability is limited. Human induced pluripotent stem cells (hiPSCs) are a promising alternative for personalized therapy; many cells can be generated and the chances of immunorejection are low. Several reprograming methods can generate hiPSCs, the most common of which are lentiviral and retroviral methods, but these can generate mutations and lead to chromosomal aberrations.

Recently, Rhee and colleagues (217) compared the safety of several types of hiPSCs, and found that they were able to generate healthy DA neurons. Neural precursor cells from protein-based hiPSCs were transplanted into a rat model of PD. The transplanted tissue not only survived well but also was able to rescue motor deficits in the model animals. These findings suggest that protein-based hiPSCs can be considered as a safe, viable alternative to virus-induced cells; moreover, they could potentially be used for transplantation and treatment in patients with PD (218).

4.4 Neuroprotective effects

DA substitution therapy and DBS do not completely relieve the symptoms of PD. Hence, there is still a need to identify neuroprotective agents that can modify the progression of the underlying disease processes.

Due to its robust effects in preventing degeneration of the nigrostriatal system in commonly used neurotoxin-based pre-clinical models of the disease, glial cell line-derived neurotrophic factor (GDNF) has gained most attention as a candidate neuroprotective molecule in PD. GDNF may be used in two ways to afford substantial neuroprotection in rodent and primate models of PD induced by either 6-OHDA or MPTP: infusion and viral-mediated delivery of GDNF, and transplantation of GDNF-producing cells (219-222).

Because of these promising pre-clinical results, more clinical trials to evaluate the efficacy of GDNF and neurturin in patients with PD are now in progress. However, the results obtained from these trials to date remain inconclusive (223-225).

Another recent study demonstrated that viral vector-mediated delivery of GDNF is unable to prevent the degeneration of the nigrostriatal DA neurons induced by over-expression of human wild-type α-synuclein at levels that have been shown to be efficient in the toxin models; this highlights the importance of performing pre-clinical tests on potential therapeutic compounds in mechanistically different models of PD (226).

5. Conclusions

As is the case for many other diseases that humans have been fighting for decades, there is a common gap between laboratory research and the ideal clinical therapy: how to ensure that products derived from laboratory experiments are both efficacious and safe. Although various studies have made progress towards a definitive solution for PD, several unresolved areas still remain. A better understanding of its biochemical pathogenesis is the best method to develop new disease-modifying therapies. However, through novel therapies and the refinement of old treatments, the management of this disease has been considerably upgraded over the past 20 years. Clinical experience shows that most patients who have accepted treatment now have a relatively good quality of life despite having suffered the effects of PD for many years. We should be confident that all these new developments will provide advances for PD treatment, and give us a hope of a final triumph in fighting the disease.

6. References

[1] Twelves D, Perkins KS, Counsell C. Systematic review of incidence studies of Parkinson's disease. Movement disorders : official journal of the Movement Disorder Society. 2003;18(1):19-31. doi: 10.1002/mds.10305.

[2] Bower JH, Maraganore DM, McDonnell SK, Rocca WA. Incidence and distribution of parkinsonism in Olmsted County, Minnesota, 1976-1990. Neurology. 1999;52(6):1214-20.

[3] Elbaz A, Bower JH, Maraganore DM, McDonnell SK, Peterson BJ, Ahlskog JE, et al. Risk tables for parkinsonism and Parkinson's disease. Journal of clinical epidemiology. 2002;55(1):25-31.

[4] Zhang ZX, Roman GC, Hong Z, Wu CB, Qu QM, Huang JB, et al. Parkinson's disease in China: prevalence in Beijing, Xian, and Shanghai. Lancet. 2005;365(9459):595-7. doi: 10.1016/S0140-6736(05)17909-4.

[5] Schoenberg BS, Osuntokun BO, Adeuja AO, Bademosi O, Nottidge V, Anderson DW, et al. Comparison of the prevalence of Parkinson's disease in black populations in the rural United States and in rural Nigeria: door-to-door community studies. Neurology. 1988;38(4):645-6.

[6] Priyadarshi A, Khuder SA, Schaub EA, Priyadarshi SS. Environmental risk factors and Parkinson's disease: a metaanalysis. Environmental research. 2001;86(2):122-7. doi: 10.1006/enrs.2001.4264.

[7] Priyadarshi A, Khuder SA, Schaub EA, Shrivastava S. A meta-analysis of Parkinson's disease and exposure to pesticides. Neurotoxicology. 2000;21(4):435-40.

[8] Migliore L, Coppede F. Genetics, environmental factors and the emerging role of epigenetics in neurodegenerative diseases. Mutation research. 2009;667(1-2):82-97. doi: 10.1016/j.mrfmmm.2008.10.011.

[9] Tanner CM, Goldman SM. Epidemiology of Parkinson's disease. Neurologic clinics. 1996;14(2):317-35.

[10] Barbeau A, Roy M, Bernier G, Campanella G, Paris S. Ecogenetics of Parkinson's disease: prevalence and environmental aspects in rural areas. The Canadian journal of neurological sciences Le journal canadien des sciences neurologiques. 1987;14(1):36-41.

[11] Gorrell JM, DiMonte D, Graham D. The role of the environment in Parkinson's disease. Environmental health perspectives. 1996;104(6):652-4.

[12] Gorell JM, Johnson CC, Rybicki BA, Peterson EL, Richardson RJ. The risk of Parkinson's disease with exposure to pesticides, farming, well water, and rural living. Neurology. 1998;50(5):1346-50.

[13] Semchuk KM, Love EJ, Lee RG. Parkinson's disease and exposure to agricultural work and pesticide chemicals. Neurology. 1992;42(7):1328-35.

[14] Langston JW, Ballard P, Tetrud JW, Irwin I. Chronic Parkinsonism in humans due to a product of meperidine-analog synthesis. Science. 1983;219(4587):979-80.

[15] Langston JW. The etiology of Parkinson's disease with emphasis on the MPTP story. Neurology. 1996;47(6 Suppl 3):S153-60.

[16] Liou HH, Tsai MC, Chen CJ, Jeng JS, Chang YC, Chen SY, et al. Environmental risk factors and Parkinson's disease: a case-control study in Taiwan. Neurology. 1997;48(6):1583-8.

[17] Seidler A, Hellenbrand W, Robra BP, Vieregge P, Nischan P, Joerg J, et al. Possible environmental, occupational, and other etiologic factors for Parkinson's disease: a case-control study in Germany. Neurology. 1996;46(5):1275-84.

[18] Corsini GU, Pintus S, Chiueh CC, Weiss JF, Kopin IJ. 1-Methyl-4-phenyl-1,2,3,6-tetrahydropyridine (MPTP) neurotoxicity in mice is enhanced by pretreatment with diethyldithiocarbamate. European journal of pharmacology. 1985;119(1-2):127-8.

[19] Semchuk KM, Love EJ, Lee RG. Parkinson's disease and exposure to rural environmental factors: a population based case-control study. The Canadian journal of neurological sciences Le journal canadien des sciences neurologiques. 1991;18(3):279-86.

[20] Huang CC, Lu CS, Chu NS, Hochberg F, Lilienfeld D, Olanow W, et al. Progression after chronic manganese exposure. Neurology. 1993;43(8):1479-83.

[21] Costello S, Cockburn M, Bronstein J, Zhang X, Ritz B. Parkinson's disease and residential exposure to maneb and paraquat from agricultural applications in the central valley of California. American journal of epidemiology. 2009;169(8):919-26. doi: 10.1093/aje/kwp006.

[22] Gorell JM, Johnson CC, Rybicki BA, Peterson EL, Kortsha GX, Brown GG, et al. Occupational exposures to metals as risk factors for Parkinson's disease. Neurology. 1997;48(3):650-8.

[23] Rybicki BA, Johnson CC, Uman J, Gorell JM. Parkinson's disease mortality and the industrial use of heavy metals in Michigan. Movement disorders : official journal of the Movement Disorder Society. 1993;8(1):87-92. doi: 10.1002/mds.870080116.

[24] Maher NE, Golbe LI, Lazzarini AM, Mark MH, Currie LJ, Wooten GF, et al. Epidemiologic study of 203 sibling pairs with Parkinson's disease: the GenePD study. Neurology. 2002;58(1):79-84.

[25] Goldman SM, Tanner CM, Oakes D, Bhudhikanok GS, Gupta A, Langston JW. Head injury and Parkinson's disease risk in twins. Annals of neurology. 2006;60(1):65-72. doi: 10.1002/ana.20882.

[26] Baron JA. Cigarette smoking and Parkinson's disease. Neurology. 1986;36(11):1490-6.

[27] Ascherio A, Zhang SM, Hernan MA, Kawachi I, Colditz GA, Speizer FE, et al. Prospective study of caffeine consumption and risk of Parkinson's disease in men and women. Annals of neurology. 2001;50(1):56-63.

[28] Wirdefeldt K, Gatz M, Pawitan Y, Pedersen NL. Risk and protective factors for Parkinson's disease: a study in Swedish twins. Annals of neurology. 2005;57(1):27-33. doi: 10.1002/ana.20307.

[29] Tanner CM, Goldman SM, Aston DA, Ottman R, Ellenberg J, Mayeux R, et al. Smoking and Parkinson's disease in twins. Neurology. 2002;58(4):581-8.

[30] Scott WK, Zhang F, Stajich JM, Scott BL, Stacy MA, Vance JM. Family-based case-control study of cigarette smoking and Parkinson disease. Neurology. 2005;64(3):442-7. doi: 10.1212/01.WNL.0000150905.93241.B2.

[31] Hernan MA, Takkouche B, Caamano-Isorna F, Gestal-Otero JJ. A meta-analysis of coffee drinking, cigarette smoking, and the risk of Parkinson's disease. Annals of neurology. 2002;52(3):276-84. doi: 10.1002/ana.10277.

[32] Elbaz A, McDonnell SK, Maraganore DM, Strain KJ, Schaid DJ, Bower JH, et al. Validity of family history data on PD: evidence for a family information bias. Neurology. 2003;61(1):11-7.

[33] Farrer MJ. Genetics of Parkinson disease: paradigm shifts and future prospects. Nature reviews Genetics. 2006;7(4):306-18. doi: 10.1038/nrg1831.

[34] Polymeropoulos MH, Higgins JJ, Golbe LI, Johnson WG, Ide SE, Di Iorio G, et al. Mapping of a gene for Parkinson's disease to chromosome 4q21-q23. Science. 1996;274(5290):1197-9.

[35] Polymeropoulos MH, Lavedan C, Leroy E, Ide SE, Dehejia A, Dutra A, et al. Mutation in the alpha-synuclein gene identified in families with Parkinson's disease. Science. 1997;276(5321):2045-7.

[36] Kruger R, Kuhn W, Muller T, Woitalla D, Graeber M, Kosel S, et al. Ala30Pro mutation in the gene encoding alpha-synuclein in Parkinson's disease. Nature genetics. 1998;18(2):106-8. doi: 10.1038/ng0298-106.

[37] Zarranz JJ, Alegre J, Gomez-Esteban JC, Lezcano E, Ros R, Ampuero I, et al. The new mutation, E46K, of alpha-synuclein causes Parkinson and Lewy body dementia. Annals of neurology. 2004;55(2):164-73. doi: 10.1002/ana.10795.

[38] Farrer M, Kachergus J, Forno L, Lincoln S, Wang DS, Hulihan M, et al. Comparison of kindreds with parkinsonism and alpha-synuclein genomic multiplications. Annals of neurology. 2004;55(2):174-9. doi: 10.1002/ana.10846.

[39] Lockhart PJ, Kachergus J, Lincoln S, Hulihan M, Bisceglio G, Thomas N, et al. Multiplication of the alpha-synuclein gene is not a common disease mechanism in Lewy body disease. Journal of molecular neuroscience : MN. 2004;24(3):337-42. doi: 10.1385/JMN:24:3:337.

[40] Fuchs J, Nilsson C, Kachergus J, Munz M, Larsson EM, Schule B, et al. Phenotypic variation in a large Swedish pedigree due to SNCA duplication and triplication. Neurology. 2007;68(12):916-22. doi: 10.1212/01.wnl.0000254458.17630.c5.

[41] Spillantini MG, Schmidt ML, Lee VM, Trojanowski JQ, Jakes R, Goedert M. Alpha-synuclein in Lewy bodies. Nature. 1997;388(6645):839-40. doi: 10.1038/42166.

[42] Spillantini MG, Crowther RA, Jakes R, Hasegawa M, Goedert M. alpha-Synuclein in filamentous inclusions of Lewy bodies from Parkinson's disease and dementia with lewy bodies. Proceedings of the National Academy of Sciences of the United States of America. 1998;95(11):6469-73.

[43] Kitada T, Asakawa S, Hattori N, Matsumine H, Yamamura Y, Minoshima S, et al. Mutations in the parkin gene cause autosomal recessive juvenile parkinsonism. Nature. 1998;392(6676):605-8. doi: 10.1038/33416.

[44] Takahashi H, Ohama E, Suzuki S, Horikawa Y, Ishikawa A, Morita T, et al. Familial juvenile parkinsonism: clinical and pathologic study in a family. Neurology. 1994;44(3 Pt 1):437-41.

[45] Ishikawa A, Tsuji S. Clinical analysis of 17 patients in 12 Japanese families with autosomal-recessive type juvenile parkinsonism. Neurology. 1996;47(1):160-6.

[46] Shimura H, Hattori N, Kubo S, Mizuno Y, Asakawa S, Minoshima S, et al. Familial Parkinson disease gene product, parkin, is a ubiquitin-protein ligase. Nature genetics. 2000;25(3):302-5. doi: 10.1038/77060.

[47] Fallon L, Belanger CM, Corera AT, Kontogiannea M, Regan-Klapisz E, Moreau F, et al. A regulated interaction with the UIM protein Eps15 implicates parkin in EGF

receptor trafficking and PI(3)K-Akt signalling. Nature cell biology. 2006;8(8):834-42. doi: 10.1038/ncb1441.

[48] Mata IF, Lockhart PJ, Farrer MJ. Parkin genetics: one model for Parkinson's disease. Human molecular genetics. 2004;13 Spec No 1:R127-33. doi: 10.1093/hmg/ddh089.

[49] Lesage S, Magali P, Lohmann E, Lacomblez L, Teive H, Janin S, et al. Deletion of the parkin and PACRG gene promoter in early-onset parkinsonism. Human mutation. 2007;28(1):27-32. doi: 10.1002/humu.20436.

[50] Lucking CB, Durr A, Bonifati V, Vaughan J, De Michele G, Gasser T, et al. Association between early-onset Parkinson's disease and mutations in the parkin gene. The New England journal of medicine. 2000;342(21):1560-7. doi: 10.1056/NEJM200005253422103.

[51] Periquet M, Latouche M, Lohmann E, Rawal N, De Michele G, Ricard S, et al. Parkin mutations are frequent in patients with isolated early-onset parkinsonism. Brain : a journal of neurology. 2003;126(Pt 6):1271-8.

[52] Leroy E, Boyer R, Auburger G, Leube B, Ulm G, Mezey E, et al. The ubiquitin pathway in Parkinson's disease. Nature. 1998;395(6701):451-2. doi: 10.1038/26652.

[53] Farrer M, Destee T, Becquet E, Wavrant-De Vrieze F, Mouroux V, Richard F, et al. Linkage exclusion in French families with probable Parkinson' s disease. Movement disorders : official journal of the Movement Disorder Society. 2000;15(6):1075-83.

[54] Harhangi BS, Farrer MJ, Lincoln S, Bonifati V, Meco G, De Michele G, et al. The Ile93Met mutation in the ubiquitin carboxy-terminal-hydrolase-L1 gene is not observed in European cases with familial Parkinson's disease. Neuroscience letters. 1999;270(1):1-4.

[55] Lincoln S, Vaughan J, Wood N, Baker M, Adamson J, Gwinn-Hardy K, et al. Low frequency of pathogenic mutations in the ubiquitin carboxy-terminal hydrolase gene in familial Parkinson's disease. Neuroreport. 1999;10(2):427-9.

[56] Valente EM, Abou-Sleiman PM, Caputo V, Muqit MM, Harvey K, Gispert S, et al. Hereditary early-onset Parkinson's disease caused by mutations in PINK1. Science. 2004;304(5674):1158-60. doi: 10.1126/science.1096284.

[57] Valente EM, Salvi S, Ialongo T, Marongiu R, Elia AE, Caputo V, et al. PINK1 mutations are associated with sporadic early-onset parkinsonism. Annals of neurology. 2004;56(3):336-41. doi: 10.1002/ana.20256.

[58] Marongiu R, Brancati F, Antonini A, Ialongo T, Ceccarini C, Scarciolla O, et al. Whole gene deletion and splicing mutations expand the PINK1 genotypic spectrum. Human mutation. 2007;28(1):98. doi: 10.1002/humu.9472.

[59] Abou-Sleiman PM, Muqit MM, McDonald NQ, Yang YX, Gandhi S, Healy DG, et al. A heterozygous effect for PINK1 mutations in Parkinson's disease? Annals of neurology. 2006;60(4):414-9. doi: 10.1002/ana.20960.

[60] Sim CH, Lio DS, Mok SS, Masters CL, Hill AF, Culvenor JG, et al. C-terminal truncation and Parkinson's disease-associated mutations down-regulate the protein serine/threonine kinase activity of PTEN-induced kinase-1. Human molecular genetics. 2006;15(21):3251-62. doi: 10.1093/hmg/ddl398.

[61] Bonifati V, Rizzu P, van Baren MJ, Schaap O, Breedveld GJ, Krieger E, et al. Mutations in the DJ-1 gene associated with autosomal recessive early-onset parkinsonism. Science. 2003;299(5604):256-9. doi: 10.1126/science.1077209.

[62] Pankratz N, Pauciulo MW, Elsaesser VE, Marek DK, Halter CA, Wojcieszek J, et al. Mutations in DJ-1 are rare in familial Parkinson disease. Neuroscience letters. 2006;408(3):209-13. doi: 10.1016/j.neulet.2006.09.003.

[63] Miller DW, Ahmad R, Hague S, Baptista MJ, Canet-Aviles R, McLendon C, et al. L166P mutant DJ-1, causative for recessive Parkinson's disease, is degraded through the ubiquitin-proteasome system. The Journal of biological chemistry. 2003;278(38):36588-95. doi: 10.1074/jbc.M304272200.

[64] Di Fonzo A, Rohe CF, Ferreira J, Chien HF, Vacca L, Stocchi F, et al. A frequent LRRK2 gene mutation associated with autosomal dominant Parkinson's disease. Lancet. 2005;365(9457):412-5. doi: 10.1016/S0140-6736(05)17829-5.

[65] Nichols WC, Pankratz N, Hernandez D, Paisan-Ruiz C, Jain S, Halter CA, et al. Genetic screening for a single common LRRK2 mutation in familial Parkinson's disease. Lancet. 2005;365(9457):410-2. doi: 10.1016/S0140-6736(05)17828-3.

[66] Gilks WP, Abou-Sleiman PM, Gandhi S, Jain S, Singleton A, Lees AJ, et al. A common LRRK2 mutation in idiopathic Parkinson's disease. Lancet. 2005;365(9457):415-6. doi: 10.1016/S0140-6736(05)17830-1.

[67] Zabetian CP, Samii A, Mosley AD, Roberts JW, Leis BC, Yearout D, et al. A clinic-based study of the LRRK2 gene in Parkinson disease yields new mutations. Neurology. 2005;65(5):741-4. doi: 10.1212/01.wnl.0000172630.22804.73.

[68] Kachergus J, Mata IF, Hulihan M, Taylor JP, Lincoln S, Aasly J, et al. Identification of a novel LRRK2 mutation linked to autosomal dominant parkinsonism: evidence of a common founder across European populations. American journal of human genetics. 2005;76(4):672-80. doi: 10.1086/429256.

[69] Hampshire DJ, Roberts E, Crow Y, Bond J, Mubaidin A, Wriekat AL, et al. Kufor-Rakeb syndrome, pallido-pyramidal degeneration with supranuclear upgaze paresis and dementia, maps to 1p36. Journal of medical genetics. 2001;38(10):680-2.

[70] Ramirez A, Heimbach A, Grundemann J, Stiller B, Hampshire D, Cid LP, et al. Hereditary parkinsonism with dementia is caused by mutations in ATP13A2, encoding a lysosomal type 5 P-type ATPase. Nature genetics. 2006;38(10):1184-91. doi: 10.1038/ng1884.

[71] Di Fonzo A, Chien HF, Socal M, Giraudo S, Tassorelli C, Iliceto G, et al. ATP13A2 missense mutations in juvenile parkinsonism and young onset Parkinson disease. Neurology. 2007;68(19):1557-62. doi: 10.1212/01.wnl.0000260963.08711.08.

[72] Strauss KM, Martins LM, Plun-Favreau H, Marx FP, Kautzmann S, Berg D, et al. Loss of function mutations in the gene encoding Omi/HtrA2 in Parkinson's disease. Human molecular genetics. 2005;14(15):2099-111. doi: 10.1093/hmg/ddi215.

[73] Sidransky E, Nalls MA, Aasly JO, Aharon-Peretz J, Annesi G, Barbosa ER, et al. Multicenter analysis of glucocerebrosidase mutations in Parkinson's disease. The New England journal of medicine. 2009;361(17):1651-61. doi: 10.1056/NEJMoa0901281.

[74] Zabetian CP, Hutter CM, Factor SA, Nutt JG, Higgins DS, Griffith A, et al. Association analysis of MAPT H1 haplotype and subhaplotypes in Parkinson's disease. Annals of neurology. 2007;62(2):137-44. doi: 10.1002/ana.21157.

[75] Baker M, Mackenzie IR, Pickering-Brown SM, Gass J, Rademakers R, Lindholm C, et al. Mutations in progranulin cause tau-negative frontotemporal dementia linked to chromosome 17. Nature. 2006;442(7105):916-9. doi: 10.1038/nature05016.

[76] Schober A. Classic toxin-induced animal models of Parkinson's disease: 6-OHDA and MPTP. Cell and tissue research. 2004;318(1):215-24. doi: 10.1007/s00441-004-0938-y.

[77] Ungerstedt U. Postsynaptic supersensitivity after 6-hydroxy-dopamine induced degeneration of the nigro-striatal dopamine system. Acta physiologica Scandinavica Supplementum. 1971;367:69-93.

[78] Saner A, Thoenen H. Model experiments on the molecular mechanism of action of 6-hydroxydopamine. Molecular pharmacology. 1971;7(2):147-54.

[79] Cohen G. Oxy-radical toxicity in catecholamine neurons. Neurotoxicology. 1984;5(1):77-82.

[80] Rodriguez Diaz M, Abdala P, Barroso-Chinea P, Obeso J, Gonzalez-Hernandez T. Motor behavioural changes after intracerebroventricular injection of 6-hydroxydopamine in the rat: an animal model of Parkinson's disease. Behavioural brain research. 2001;122(1):79-92.

[81] Zigmond MJ, Stricker EM. Recovery of feeding and drinking by rats after intraventricular 6-hydroxydopamine or lateral hypothalamic lesions. Science. 1973;182(113):717-20.

[82] Sauer H, Oertel WH. Progressive degeneration of nigrostriatal dopamine neurons following intrastriatal terminal lesions with 6-hydroxydopamine: a combined retrograde tracing and immunocytochemical study in the rat. Neuroscience. 1994;59(2):401-15.

[83] Marti MJ, James CJ, Oo TF, Kelly WJ, Burke RE. Early developmental destruction of terminals in the striatal target induces apoptosis in dopamine neurons of the substantia nigra. The Journal of neuroscience : the official journal of the Society for Neuroscience. 1997;17(6):2030-9.

[84] Marti MJ, Saura J, Burke RE, Jackson-Lewis V, Jimenez A, Bonastre M, et al. Striatal 6-hydroxydopamine induces apoptosis of nigral neurons in the adult rat. Brain research. 2002;958(1):185-91.

[85] Jeon BS, Jackson-Lewis V, Burke RE. 6-Hydroxydopamine lesion of the rat substantia nigra: time course and morphology of cell death. Neurodegeneration : a journal for neurodegenerative disorders, neuroprotection, and neuroregeneration. 1995;4(2):131-7.

[86] Bove J, Prou D, Perier C, Przedborski S. Toxin-induced models of Parkinson's disease. NeuroRx : the journal of the American Society for Experimental NeuroTherapeutics. 2005;2(3):484-94.

[87] Falkenburger BH, Schulz JB. Limitations of cellular models in Parkinson's disease research. Journal of neural transmission Supplementum. 2006;(70):261-8.

[88] Terzioglu M, Galter D. Parkinson's disease: genetic versus toxin-induced rodent models. The FEBS journal. 2008;275(7):1384-91. doi: 10.1111/j.1742-4658.2008.06302.x.

[89] Hisahara S, Shimohama S. Toxin-induced and genetic animal models of Parkinson's disease. Parkinson's disease. 2010;2011:951709. doi: 10.4061/2011/951709.

[90] Dawson T, Mandir A, Lee M. Animal models of PD: pieces of the same puzzle? Neuron. 2002;35(2):219-22.

[91] Bezard E, Gross CE, Fournier MC, Dovero S, Bloch B, Jaber M. Absence of MPTP-induced neuronal death in mice lacking the dopamine transporter. Experimental neurology. 1999;155(2):268-73. doi: 10.1006/exnr.1998.6995.

[92] Ramsay RR, Singer TP. Energy-dependent uptake of N-methyl-4-phenylpyridinium, the neurotoxic metabolite of 1-methyl-4-phenyl-1,2,3,6-tetrahydropyridine, by mitochondria. The Journal of biological chemistry. 1986;261(17):7585-7.

[93] Langston JW, Forno LS, Tetrud J, Reeves AG, Kaplan JA, Karluk D. Evidence of active nerve cell degeneration in the substantia nigra of humans years after 1-methyl-4-phenyl-1,2,3,6-tetrahydropyridine exposure. Annals of neurology. 1999;46(4):598-605.

[94] Forno LS, DeLanney LE, Irwin I, Langston JW. Similarities and differences between MPTP-induced parkinsonsim and Parkinson's disease. Neuropathologic considerations. Advances in neurology. 1993;60:600-8.

[95] Sonsalla PK, Heikkila RE. The influence of dose and dosing interval on MPTP-induced dopaminergic neurotoxicity in mice. European journal of pharmacology. 1986;129(3):339-45.

[96] Ricaurte GA, Langston JW, Delanney LE, Irwin I, Peroutka SJ, Forno LS. Fate of nigrostriatal neurons in young mature mice given 1-methyl-4-phenyl-1,2,3,6-tetrahydropyridine: a neurochemical and morphological reassessment. Brain research. 1986;376(1):117-24.

[97] Schneider JS, Denaro FJ. Astrocytic responses to the dopaminergic neurotoxin 1-methyl-4-phenyl-1,2,3,6-tetrahydropyridine (MPTP) in cat and mouse brain. Journal of neuropathology and experimental neurology. 1988;47(4):452-8.

[98] Jackson-Lewis V, Jakowec M, Burke RE, Przedborski S. Time course and morphology of dopaminergic neuronal death caused by the neurotoxin 1-methyl-4-phenyl-1,2,3,6-tetrahydropyridine. Neurodegeneration : a journal for neurodegenerative disorders, neuroprotection, and neuroregeneration. 1995;4(3):257-69.

[99] Tatton NA, Kish SJ. In situ detection of apoptotic nuclei in the substantia nigra compacta of 1-methyl-4-phenyl-1,2,3,6-tetrahydropyridine-treated mice using terminal deoxynucleotidyl transferase labelling and acridine orange staining. Neuroscience. 1997;77(4):1037-48.

[100] Kurkowska-Jastrzebska I, Wronska A, Kohutnicka M, Czlonkowski A, Czlonkowska A. The inflammatory reaction following 1-methyl-4-phenyl-1,2,3, 6-tetrahydropyridine intoxication in mouse. Experimental neurology. 1999;156(1):50-61. doi: 10.1006/exnr.1998.6993.

[101] Furuya T, Hayakawa H, Yamada M, Yoshimi K, Hisahara S, Miura M, et al. Caspase-11 mediates inflammatory dopaminergic cell death in the 1-methyl-4-phenyl-1,2,3,6-tetrahydropyridine mouse model of Parkinson's disease. The Journal of neuroscience : the official journal of the Society for Neuroscience. 2004;24(8):1865-72. doi: 10.1523/JNEUROSCI.3309-03.2004.

[102] Liberatore GT, Jackson-Lewis V, Vukosavic S, Mandir AS, Vila M, McAuliffe WG, et al. Inducible nitric oxide synthase stimulates dopaminergic neurodegeneration in the MPTP model of Parkinson disease. Nature medicine. 1999;5(12):1403-9. doi: 10.1038/70978.

[103] Wade TV, Schneider JS. Expression of striatal preprotachykinin mRNA in symptomatic and asymptomatic 1-methyl-4-phenyl-1,2,3,6-tetrahydropyridine-exposed monkeys is related to parkinsonian motor signs. The Journal of neuroscience : the official journal of the Society for Neuroscience. 2001;21(13):4901-7.

[104] Bloem BR, Irwin I, Buruma OJ, Haan J, Roos RA, Tetrud JW, et al. The MPTP model: versatile contributions to the treatment of idiopathic Parkinson's disease. Journal of the neurological sciences. 1990;97(2-3):273-93.

[105] Zigmond MJ, Stricker EM. Animal models of parkinsonism using selective neurotoxins: clinical and basic implications. International review of neurobiology. 1989;31:1-79.

[106] Kopin IJ. MPTP: an industrial chemical and contaminant of illicit narcotics stimulates a new era in research on Parkinson's disease. Environmental health perspectives. 1987;75:45-51.

[107] Kitamura Y, Kakimura J, Taniguchi T. Protective effect of talipexole on MPTP-treated planarian, a unique parkinsonian worm model. Japanese journal of pharmacology. 1998;78(1):23-9.

[108] Giovanni A, Sieber BA, Heikkila RE, Sonsalla PK. Studies on species sensitivity to the dopaminergic neurotoxin 1-methyl-4-phenyl-1,2,3,6-tetrahydropyridine. Part 1: Systemic administration. The Journal of pharmacology and experimental therapeutics. 1994;270(3):1000-7.

[109] Giovanni A, Sonsalla PK, Heikkila RE. Studies on species sensitivity to the dopaminergic neurotoxin 1-methyl-4-phenyl-1,2,3,6-tetrahydropyridine. Part 2: Central administration of 1-methyl-4-phenylpyridinium. The Journal of pharmacology and experimental therapeutics. 1994;270(3):1008-14.

[110] Przedborski S, Jackson-Lewis V, Naini AB, Jakowec M, Petzinger G, Miller R, et al. The parkinsonian toxin 1-methyl-4-phenyl-1,2,3,6-tetrahydropyridine (MPTP): a technical review of its utility and safety. Journal of neurochemistry. 2001;76(5):1265-74.

[111] Tolwani RJ, Jakowec MW, Petzinger GM, Green S, Waggie K. Experimental models of Parkinson's disease: insights from many models. Laboratory animal science. 1999;49(4):363-71.

[112] Sedelis M, Hofele K, Auburger GW, Morgan S, Huston JP, Schwarting RK. MPTP susceptibility in the mouse: behavioral, neurochemical, and histological analysis of gender and strain differences. Behavior genetics. 2000;30(3):171-82.

[113] Jenner P. The contribution of the MPTP-treated primate model to the development of new treatment strategies for Parkinson's disease. Parkinsonism & related disorders. 2003;9(3):131-7.

[114] Betarbet R, Sherer TB, MacKenzie G, Garcia-Osuna M, Panov AV, Greenamyre JT. Chronic systemic pesticide exposure reproduces features of Parkinson's disease. Nature neuroscience. 2000;3(12):1301-6. doi: 10.1038/81834.

[115] Schuler F, Casida JE. Functional coupling of PSST and ND1 subunits in NADH:ubiquinone oxidoreductase established by photoaffinity labeling. Biochimica et biophysica acta. 2001;1506(1):79-87.

[116] Marshall LE, Himes RH. Rotenone inhibition of tubulin self-assembly. Biochimica et biophysica acta. 1978;543(4):590-4.

[117] Burke D, Gasdaska P, Hartwell L. Dominant effects of tubulin overexpression in Saccharomyces cerevisiae. Molecular and cellular biology. 1989;9(3):1049-59.

[118] Sherer TB, Betarbet R, Testa CM, Seo BB, Richardson JR, Kim JH, et al. Mechanism of toxicity in rotenone models of Parkinson's disease. The Journal of neuroscience : the official journal of the Society for Neuroscience. 2003;23(34):10756-64.

[119] Uversky VN. Neurotoxicant-induced animal models of Parkinson's disease: understanding the role of rotenone, maneb and paraquat in neurodegeneration. Cell and tissue research. 2004;318(1):225-41. doi: 10.1007/s00441-004-0937-z.

[120] Sherer TB, Betarbet R, Kim JH, Greenamyre JT. Selective microglial activation in the rat rotenone model of Parkinson's disease. Neuroscience letters. 2003;341(2):87-90.

[121] Sherer TB, Kim JH, Betarbet R, Greenamyre JT. Subcutaneous rotenone exposure causes highly selective dopaminergic degeneration and alpha-synuclein aggregation. Experimental neurology. 2003;179(1):9-16.

[122] Betarbet R, Canet-Aviles RM, Sherer TB, Mastroberardino PG, McLendon C, Kim JH, et al. Intersecting pathways to neurodegeneration in Parkinson's disease: effects of the pesticide rotenone on DJ-1, alpha-synuclein, and the ubiquitin-proteasome system. Neurobiology of disease. 2006;22(2):404-20. doi: 10.1016/j.nbd.2005.12.003.

[123] Klockgether T. Parkinson's disease: clinical aspects. Cell and tissue research. 2004;318(1):115-20. doi: 10.1007/s00441-004-0975-6.

[124] Alam M, Schmidt WJ. Rotenone destroys dopaminergic neurons and induces parkinsonian symptoms in rats. Behavioural brain research. 2002;136(1):317-24.

[125] Hely MA, Morris JG, Reid WG, Trafficante R. Sydney Multicenter Study of Parkinson's disease: non-L-dopa-responsive problems dominate at 15 years. Movement disorders : official journal of the Movement Disorder Society. 2005;20(2):190-9. doi: 10.1002/mds.20324.

[126] Greene JG, Noorian AR, Srinivasan S. Delayed gastric emptying and enteric nervous system dysfunction in the rotenone model of Parkinson's disease. Experimental neurology. 2009;218(1):154-61. doi: 10.1016/j.expneurol.2009.04.023.

[127] Drolet RE, Cannon JR, Montero L, Greenamyre JT. Chronic rotenone exposure reproduces Parkinson's disease gastrointestinal neuropathology. Neurobiology of disease. 2009;36(1):96-102. doi: 10.1016/j.nbd.2009.06.017.

[128] Di Monte D, Sandy MS, Ekstrom G, Smith MT. Comparative studies on the mechanisms of paraquat and 1-methyl-4-phenylpyridine (MPP+) cytotoxicity. Biochemical and biophysical research communications. 1986;137(1):303-9.

[129] Shimizu K, Matsubara K, Ohtaki K, Fujimaru S, Saito O, Shiono H. Paraquat induces long-lasting dopamine overflow through the excitotoxic pathway in the striatum of freely moving rats. Brain research. 2003;976(2):243-52.

[130] Brooks AI, Chadwick CA, Gelbard HA, Cory-Slechta DA, Federoff HJ. Paraquat elicited neurobehavioral syndrome caused by dopaminergic neuron loss. Brain research. 1999;823(1-2):1-10.

[131] Peng J, Mao XO, Stevenson FF, Hsu M, Andersen JK. The herbicide paraquat induces dopaminergic nigral apoptosis through sustained activation of the JNK pathway. The Journal of biological chemistry. 2004;279(31):32626-32. doi: 10.1074/jbc.M404596200.

[132] Purisai MG, McCormack AL, Cumine S, Li J, Isla MZ, Di Monte DA. Microglial activation as a priming event leading to paraquat-induced dopaminergic cell degeneration. Neurobiology of disease. 2007;25(2):392-400. doi: 10.1016/j.nbd.2006.10.008.

[133] Dringen R. Oxidative and antioxidative potential of brain microglial cells. Antioxidants & redox signaling. 2005;7(9-10):1223-33. doi: 10.1089/ars.2005.7.1223.

[134] Bonneh-Barkay D, Langston WJ, Di Monte DA. Toxicity of redox cycling pesticides in primary mesencephalic cultures. Antioxidants & redox signaling. 2005;7(5-6):649-53. doi: 10.1089/ars.2005.7.649.

[135] Bonneh-Barkay D, Reaney SH, Langston WJ, Di Monte DA. Redox cycling of the herbicide paraquat in microglial cultures. Brain research Molecular brain research. 2005;134(1):52-6. doi: 10.1016/j.molbrainres.2004.11.005.

[136] McCormack AL, Atienza JG, Langston JW, Di Monte DA. Decreased susceptibility to oxidative stress underlies the resistance of specific dopaminergic cell populations to paraquat-induced degeneration. Neuroscience. 2006;141(2):929-37. doi: 10.1016/j.neuroscience.2006.03.069.

[137] Peng J, Stevenson FF, Doctrow SR, Andersen JK. Superoxide dismutase/catalase mimetics are neuroprotective against selective paraquat-mediated dopaminergic neuron death in the substantial nigra: implications for Parkinson disease. The Journal of biological chemistry. 2005;280(32):29194-8. doi: 10.1074/jbc.M500984200.

[138] Manning-Bog AB, McCormack AL, Li J, Uversky VN, Fink AL, Di Monte DA. The herbicide paraquat causes up-regulation and aggregation of alpha-synuclein in mice: paraquat and alpha-synuclein. The Journal of biological chemistry. 2002;277(3):1641-4. doi: 10.1074/jbc.C100560200.

[139] Thiruchelvam M, Brockel BJ, Richfield EK, Baggs RB, Cory-Slechta DA. Potentiated and preferential effects of combined paraquat and maneb on nigrostriatal dopamine systems: environmental risk factors for Parkinson's disease? Brain research. 2000;873(2):225-34.

[140] Zhang J, Fitsanakis VA, Gu G, Jing D, Ao M, Amarnath V, et al. Manganese ethylene-bis-dithiocarbamate and selective dopaminergic neurodegeneration in rat: a link through mitochondrial dysfunction. Journal of neurochemistry. 2003;84(2):336-46.

[141] Thiruchelvam M, Richfield EK, Baggs RB, Tank AW, Cory-Slechta DA. The nigrostriatal dopaminergic system as a preferential target of repeated exposures to combined paraquat and maneb: implications for Parkinson's disease. The Journal of neuroscience : the official journal of the Society for Neuroscience. 2000;20(24):9207-14.

[142] Thiruchelvam M, McCormack A, Richfield EK, Baggs RB, Tank AW, Di Monte DA, et al. Age-related irreversible progressive nigrostriatal dopaminergic neurotoxicity in the paraquat and maneb model of the Parkinson's disease phenotype. The European journal of neuroscience. 2003;18(3):589-600.

[143] Barlow BK, Thiruchelvam MJ, Bennice L, Cory-Slechta DA, Ballatori N, Richfield EK. Increased synaptosomal dopamine content and brain concentration of paraquat produced by selective dithiocarbamates. Journal of neurochemistry. 2003;85(4):1075-86.

[144] Richfield EK, Thiruchelvam MJ, Cory-Slechta DA, Wuertzer C, Gainetdinov RR, Caron MG, et al. Behavioral and neurochemical effects of wild-type and mutated human alpha-synuclein in transgenic mice. Experimental neurology. 2002;175(1):35-48. doi: 10.1006/exnr.2002.7882.

[145] Thiruchelvam MJ, Powers JM, Cory-Slechta DA, Richfield EK. Risk factors for dopaminergic neuron loss in human alpha-synuclein transgenic mice. The European journal of neuroscience. 2004;19(4):845-54.

[146] Masliah E, Rockenstein E, Veinbergs I, Mallory M, Hashimoto M, Takeda A, et al. Dopaminergic loss and inclusion body formation in alpha-synuclein mice: implications for neurodegenerative disorders. Science. 2000;287(5456):1265-9.

[147] Giasson BI, Duda JE, Quinn SM, Zhang B, Trojanowski JQ, Lee VM. Neuronal alpha-synucleinopathy with severe movement disorder in mice expressing A53T human alpha-synuclein. Neuron. 2002;34(4):521-33.

[148] Lee MK, Stirling W, Xu Y, Xu X, Qui D, Mandir AS, et al. Human alpha-synuclein-harboring familial Parkinson's disease-linked Ala-53 --> Thr mutation causes neurodegenerative disease with alpha-synuclein aggregation in transgenic mice. Proceedings of the National Academy of Sciences of the United States of America. 2002;99(13):8968-73. doi: 10.1073/pnas.132197599.

[149] Rockenstein E, Mallory M, Hashimoto M, Song D, Shults CW, Lang I, et al. Differential neuropathological alterations in transgenic mice expressing alpha-synuclein from the platelet-derived growth factor and Thy-1 promoters. Journal of neuroscience research. 2002;68(5):568-78. doi: 10.1002/jnr.10231.

[150] Song DD, Shults CW, Sisk A, Rockenstein E, Masliah E. Enhanced substantia nigra mitochondrial pathology in human alpha-synuclein transgenic mice after treatment with MPTP. Experimental neurology. 2004;186(2):158-72. doi: 10.1016/S0014-4886(03)00342-X.

[151] Martin LJ, Pan Y, Price AC, Sterling W, Copeland NG, Jenkins NA, et al. Parkinson's disease alpha-synuclein transgenic mice develop neuronal mitochondrial degeneration and cell death. The Journal of neuroscience : the official journal of the Society for Neuroscience. 2006;26(1):41-50. doi: 10.1523/JNEUROSCI.4308-05.2006.

[152] Kirik D, Rosenblad C, Burger C, Lundberg C, Johansen TE, Muzyczka N, et al. Parkinson-like neurodegeneration induced by targeted overexpression of alpha-synuclein in the nigrostriatal system. The Journal of neuroscience : the official journal of the Society for Neuroscience. 2002;22(7):2780-91. doi: 20026246.

[153] Feany MB, Bender WW. A Drosophila model of Parkinson's disease. Nature. 2000;404(6776):394-8. doi: 10.1038/35006074.

[154] Lakso M, Vartiainen S, Moilanen AM, Sirvio J, Thomas JH, Nass R, et al. Dopaminergic neuronal loss and motor deficits in Caenorhabditis elegans overexpressing human alpha-synuclein. Journal of neurochemistry. 2003;86(1):165-72.

[155] Kuwahara T, Koyama A, Gengyo-Ando K, Masuda M, Kowa H, Tsunoda M, et al. Familial Parkinson mutant alpha-synuclein causes dopamine neuron dysfunction in transgenic Caenorhabditis elegans. The Journal of biological chemistry. 2006;281(1):334-40. doi: 10.1074/jbc.M504860200.

[156] Stefanis L, Kholodilov N, Rideout HJ, Burke RE, Greene LA. Synuclein-1 is selectively up-regulated in response to nerve growth factor treatment in PC12 cells. Journal of neurochemistry. 2001;76(4):1165-76.

[157] Clough RL, Stefanis L. A novel pathway for transcriptional regulation of alpha-synuclein. The FASEB journal : official publication of the Federation of American Societies for Experimental Biology. 2007;21(2):596-607. doi: 10.1096/fj.06-7111com.

[158] Cuervo AM, Stefanis L, Fredenburg R, Lansbury PT, Sulzer D. Impaired degradation of mutant alpha-synuclein by chaperone-mediated autophagy. Science. 2004;305(5688):1292-5. doi: 10.1126/science.1101738.

[159] Tanaka Y, Engelender S, Igarashi S, Rao RK, Wanner T, Tanzi RE, et al. Inducible expression of mutant alpha-synuclein decreases proteasome activity and increases sensitivity to mitochondria-dependent apoptosis. Human molecular genetics. 2001;10(9):919-26.

[160] Smith WW, Jiang H, Pei Z, Tanaka Y, Morita H, Sawa A, et al. Endoplasmic reticulum stress and mitochondrial cell death pathways mediate A53T mutant alpha-synuclein-induced toxicity. Human molecular genetics. 2005;14(24):3801-11. doi: 10.1093/hmg/ddi396.

[161] Stefanis L, Larsen KE, Rideout HJ, Sulzer D, Greene LA. Expression of A53T mutant but not wild-type alpha-synuclein in PC12 cells induces alterations of the ubiquitin-dependent degradation system, loss of dopamine release, and autophagic cell death. The Journal of neuroscience : the official journal of the Society for Neuroscience. 2001;21(24):9549-60.

[162] Outeiro TF, Lindquist S. Yeast cells provide insight into alpha-synuclein biology and pathobiology. Science. 2003;302(5651):1772-5. doi: 10.1126/science.1090439.

[163] Dixon C, Mathias N, Zweig RM, Davis DA, Gross DS. Alpha-synuclein targets the plasma membrane via the secretory pathway and induces toxicity in yeast. Genetics. 2005;170(1):47-59. doi: 10.1534/genetics.104.035493.

[164] Flower TR, Chesnokova LS, Froelich CA, Dixon C, Witt SN. Heat shock prevents alpha-synuclein-induced apoptosis in a yeast model of Parkinson's disease. Journal of molecular biology. 2005;351(5):1081-100. doi: 10.1016/j.jmb.2005.06.060.

[165] Cooper AA, Gitler AD, Cashikar A, Haynes CM, Hill KJ, Bhullar B, et al. Alpha-synuclein blocks ER-Golgi traffic and Rab1 rescues neuron loss in Parkinson's models. Science. 2006;313(5785):324-8. doi: 10.1126/science.1129462.

[166] Willingham S, Outeiro TF, DeVit MJ, Lindquist SL, Muchowski PJ. Yeast genes that enhance the toxicity of a mutant huntingtin fragment or alpha-synuclein. Science. 2003;302(5651):1769-72. doi: 10.1126/science.1090389.

[167] Douglas MR, Lewthwaite AJ, Nicholl DJ. Genetics of Parkinson's disease and parkinsonism. Expert review of neurotherapeutics. 2007;7(6):657-66. doi: 10.1586/14737175.7.6.657.

[168] Lo Bianco C, Schneider BL, Bauer M, Sajadi A, Brice A, Iwatsubo T, et al. Lentiviral vector delivery of parkin prevents dopaminergic degeneration in an alpha-synuclein rat model of Parkinson's disease. Proceedings of the National Academy of Sciences of the United States of America. 2004;101(50):17510-5. doi: 10.1073/pnas.0405313101.

[169] Paterna JC, Leng A, Weber E, Feldon J, Bueler H. DJ-1 and Parkin modulate dopamine-dependent behavior and inhibit MPTP-induced nigral dopamine neuron loss in mice. Molecular therapy : the journal of the American Society of Gene Therapy. 2007;15(4):698-704. doi: 10.1038/sj.mt.6300067.

[170] Von Coelln R, Thomas B, Savitt JM, Lim KL, Sasaki M, Hess EJ, et al. Loss of locus coeruleus neurons and reduced startle in parkin null mice. Proceedings of the National Academy of Sciences of the United States of America. 2004;101(29):10744-9. doi: 10.1073/pnas.0401297101.

[171] Goldberg MS, Fleming SM, Palacino JJ, Cepeda C, Lam HA, Bhatnagar A, et al. Parkin-deficient mice exhibit nigrostriatal deficits but not loss of dopaminergic neurons.

The Journal of biological chemistry. 2003;278(44):43628-35. doi: 10.1074/jbc.M308947200.

[172] Greene JC, Whitworth AJ, Kuo I, Andrews LA, Feany MB, Pallanck LJ. Mitochondrial pathology and apoptotic muscle degeneration in Drosophila parkin mutants. Proceedings of the National Academy of Sciences of the United States of America. 2003;100(7):4078-83. doi: 10.1073/pnas.0737556100.

[173] Greene JC, Whitworth AJ, Andrews LA, Parker TJ, Pallanck LJ. Genetic and genomic studies of Drosophila parkin mutants implicate oxidative stress and innate immune responses in pathogenesis. Human molecular genetics. 2005;14(6):799-811. doi: 10.1093/hmg/ddi074.

[174] Sang TK, Chang HY, Lawless GM, Ratnaparkhi A, Mee L, Ackerson LC, et al. A Drosophila model of mutant human parkin-induced toxicity demonstrates selective loss of dopaminergic neurons and dependence on cellular dopamine. The Journal of neuroscience : the official journal of the Society for Neuroscience. 2007;27(5):981-92. doi: 10.1523/JNEUROSCI.4810-06.2007.

[175] Imai Y, Soda M, Inoue H, Hattori N, Mizuno Y, Takahashi R. An unfolded putative transmembrane polypeptide, which can lead to endoplasmic reticulum stress, is a substrate of Parkin. Cell. 2001;105(7):891-902.

[176] Shimura H, Schlossmacher MG, Hattori N, Frosch MP, Trockenbacher A, Schneider R, et al. Ubiquitination of a new form of alpha-synuclein by parkin from human brain: implications for Parkinson's disease. Science. 2001;293(5528):263-9. doi: 10.1126/science.1060627.

[177] LaVoie MJ, Ostaszewski BL, Weihofen A, Schlossmacher MG, Selkoe DJ. Dopamine covalently modifies and functionally inactivates parkin. Nature medicine. 2005;11(11):1214-21. doi: 10.1038/nm1314.

[178] Avraham E, Rott R, Liani E, Szargel R, Engelender S. Phosphorylation of Parkin by the cyclin-dependent kinase 5 at the linker region modulates its ubiquitin-ligase activity and aggregation. The Journal of biological chemistry. 2007;282(17):12842-50. doi: 10.1074/jbc.M608243200.

[179] Kitada T, Pisani A, Porter DR, Yamaguchi H, Tscherter A, Martella G, et al. Impaired dopamine release and synaptic plasticity in the striatum of PINK1-deficient mice. Proceedings of the National Academy of Sciences of the United States of America. 2007;104(27):11441-6. doi: 10.1073/pnas.0702717104.

[180] Clark IE, Dodson MW, Jiang C, Cao JH, Huh JR, Seol JH, et al. Drosophila pink1 is required for mitochondrial function and interacts genetically with parkin. Nature. 2006;441(7097):1162-6. doi: 10.1038/nature04779.

[181] Park J, Lee SB, Lee S, Kim Y, Song S, Kim S, et al. Mitochondrial dysfunction in Drosophila PINK1 mutants is complemented by parkin. Nature. 2006;441(7097):1157-61. doi: 10.1038/nature04788.

[182] Yang Y, Gehrke S, Imai Y, Huang Z, Ouyang Y, Wang JW, et al. Mitochondrial pathology and muscle and dopaminergic neuron degeneration caused by inactivation of Drosophila Pink1 is rescued by Parkin. Proceedings of the National Academy of Sciences of the United States of America. 2006;103(28):10793-8. doi: 10.1073/pnas.0602493103.

[183] Pridgeon JW, Olzmann JA, Chin LS, Li L. PINK1 protects against oxidative stress by phosphorylating mitochondrial chaperone TRAP1. PLoS biology. 2007;5(7):e172. doi: 10.1371/journal.pbio.0050172.

[184] Silvestri L, Caputo V, Bellacchio E, Atorino L, Dallapiccola B, Valente EM, et al. Mitochondrial import and enzymatic activity of PINK1 mutants associated to recessive parkinsonism. Human molecular genetics. 2005;14(22):3477-92. doi: 10.1093/hmg/ddi377.

[185] Beilina A, Van Der Brug M, Ahmad R, Kesavapany S, Miller DW, Petsko GA, et al. Mutations in PTEN-induced putative kinase 1 associated with recessive parkinsonism have differential effects on protein stability. Proceedings of the National Academy of Sciences of the United States of America. 2005;102(16):5703-8. doi: 10.1073/pnas.0500617102.

[186] Andres-Mateos E, Perier C, Zhang L, Blanchard-Fillion B, Greco TM, Thomas B, et al. DJ-1 gene deletion reveals that DJ-1 is an atypical peroxiredoxin-like peroxidase. Proceedings of the National Academy of Sciences of the United States of America. 2007;104(37):14807-12. doi: 10.1073/pnas.0703219104.

[187] Park J, Kim SY, Cha GH, Lee SB, Kim S, Chung J. Drosophila DJ-1 mutants show oxidative stress-sensitive locomotive dysfunction. Gene. 2005;361:133-9. doi: 10.1016/j.gene.2005.06.040.

[188] Menzies FM, Yenisetti SC, Min KT. Roles of Drosophila DJ-1 in survival of dopaminergic neurons and oxidative stress. Current biology : CB. 2005;15(17):1578-82. doi: 10.1016/j.cub.2005.07.036.

[189] Meulener M, Whitworth AJ, Armstrong-Gold CE, Rizzu P, Heutink P, Wes PD, et al. Drosophila DJ-1 mutants are selectively sensitive to environmental toxins associated with Parkinson's disease. Current biology : CB. 2005;15(17):1572-7. doi: 10.1016/j.cub.2005.07.064.

[190] Meulener MC, Xu K, Thomson L, Ischiropoulos H, Bonini NM. Mutational analysis of DJ-1 in Drosophila implicates functional inactivation by oxidative damage and aging. Proceedings of the National Academy of Sciences of the United States of America. 2006;103(33):12517-22. doi: 10.1073/pnas.0601891103.

[191] West AB, Moore DJ, Biskup S, Bugayenko A, Smith WW, Ross CA, et al. Parkinson's disease-associated mutations in leucine-rich repeat kinase 2 augment kinase activity. Proceedings of the National Academy of Sciences of the United States of America. 2005;102(46):16842-7. doi: 10.1073/pnas.0507360102.

[192] West AB, Moore DJ, Choi C, Andrabi SA, Li X, Dikeman D, et al. Parkinson's disease-associated mutations in LRRK2 link enhanced GTP-binding and kinase activities to neuronal toxicity. Human molecular genetics. 2007;16(2):223-32. doi: 10.1093/hmg/ddl471.

[193] Greggio E, Lewis PA, van der Brug MP, Ahmad R, Kaganovich A, Ding J, et al. Mutations in LRRK2/dardarin associated with Parkinson disease are more toxic than equivalent mutations in the homologous kinase LRRK1. Journal of neurochemistry. 2007;102(1):93-102. doi: 10.1111/j.1471-4159.2007.04523.x.

[194] MacLeod D, Dowman J, Hammond R, Leete T, Inoue K, Abeliovich A. The familial Parkinsonism gene LRRK2 regulates neurite process morphology. Neuron. 2006;52(4):587-93. doi: 10.1016/j.neuron.2006.10.008.

[195] Ito G, Okai T, Fujino G, Takeda K, Ichijo H, Katada T, et al. GTP binding is essential to the protein kinase activity of LRRK2, a causative gene product for familial Parkinson's disease. Biochemistry. 2007;46(5):1380-8. doi: 10.1021/bi061960m.

[196] Smith WW, Pei Z, Jiang H, Dawson VL, Dawson TM, Ross CA. Kinase activity of mutant LRRK2 mediates neuronal toxicity. Nature neuroscience. 2006;9(10):1231-3. doi: 10.1038/nn1776.

[197] Langston JW. The Parkinson's complex: parkinsonism is just the tip of the iceberg. Ann Neurol. 2006;59(4):591-6. doi: 10.1002/ana.20834.

[198] Selikhova M, Williams DR, Kempster PA, Holton JL, Revesz T, Lees AJ. A clinico-pathological study of subtypes in Parkinson's disease. Brain. 2009;132(Pt 11):2947-57. doi: awp234 [pii] 10.1093/brain/awp234.

[199] Marsden CD. Studies on the normal and disordered human motor cortex. Electroencephalogr Clin Neurophysiol Suppl. 1982;36:430-4.

[200] Jankovic J. Parkinson's disease and movement disorders: moving forward. Lancet Neurol. 2008;7(1):9-11. doi: S1474-4422(07)70302-2 [pii] 10.1016/S1474-4422(07)70302-2.

[201] Voon V, Krack P, Lang AE, Lozano AM, Dujardin K, Schupbach M, et al. A multicentre study on suicide outcomes following subthalamic stimulation for Parkinson's disease. Brain. 2008;131(Pt 10):2720-8. doi: awn214 [pii] 10.1093/brain/awn214.

[202] Pechevis M, Clarke CE, Vieregge P, Khoshnood B, Deschaseaux-Voinet C, Berdeaux G, et al. Effects of dyskinesias in Parkinson's disease on quality of life and health-related costs: a prospective European study. Eur J Neurol. 2005;12(12):956-63. doi: ENE1096 [pii] 10.1111/j.1468-1331.2005.01096.x.

[203] Kempster PA, O'Sullivan SS, Holton JL, Revesz T, Lees AJ. Relationships between age and late progression of Parkinson's disease: a clinico-pathological study. Brain. 2010;133(Pt 6):1755-62. doi: awq059 [pii] 10.1093/brain/awq059.

[204] Emre M, Aarsland D, Brown R, Burn DJ, Duyckaerts C, Mizuno Y, et al. Clinical diagnostic criteria for dementia associated with Parkinson's disease. Mov Disord. 2007;22(12):1689-707; quiz 837. doi: 10.1002/mds.21507.

[205] Siderowf A, Xie SX, Hurtig H, Weintraub D, Duda J, Chen-Plotkin A, et al. CSF amyloid {beta} 1-42 predicts cognitive decline in Parkinson disease. Neurology. 2010;75(12):1055-61.   doi: WNL.0b013e3181f39a78   [pii] 10.1212/WNL.0b013e3181f39a78.

[206] Hauptmann C, Tass PA. Therapeutic rewiring by means of desynchronizing brain stimulation. Biosystems. 2007;89(1-3):173-81. doi: S0303-2647(06)00256-5 [pii] 10.1016/j.biosystems.2006.04.015.

[207] Weaver FM, Follett K, Stern M, Hur K, Harris C, Marks WJ, Jr., et al. Bilateral deep brain stimulation vs best medical therapy for patients with advanced Parkinson disease: a randomized controlled trial. JAMA. 2009;301(1):63-73. doi: 301/1/63 [pii] 10.1001/jama.2008.929.

[208] Williams A, Gill S, Varma T, Jenkinson C, Quinn N, Mitchell R, et al. Deep brain stimulation plus best medical therapy versus best medical therapy alone for advanced Parkinson's disease (PD SURG trial): a randomised, open-label trial. Lancet Neurol. 2010;9(6):581-91. doi: S1474-4422(10)70093-4 [pii] 10.1016/S1474-4422(10)70093-4.

[209] Smeding HM, Speelman JD, Huizenga HM, Schuurman PR, Schmand B. Predictors of cognitive and psychosocial outcome after STN DBS in Parkinson's Disease. J Neurol Neurosurg Psychiatry. 2011;82(7):754-60. doi: jnnp.2007.140012 [pii] 10.1136/jnnp.2007.140012.

[210] Strutt AM, Simpson R, Jankovic J, York MK. Changes in cognitive-emotional and physiological symptoms of depression following STN-DBS for the treatment of Parkinson's disease. Eur J Neurol. 2011. doi: 10.1111/j.1468-1331.2011.03447.x.

[211] Voon V, Saint-Cyr J, Lozano AM, Moro E, Poon YY, Lang AE. Psychiatric symptoms in patients with Parkinson disease presenting for deep brain stimulation surgery. J Neurosurg. 2005;103(2):246-51. doi: 10.3171/jns.2005.103.2.0246.

[212] Ory-Magne F, Brefel-Courbon C, Simonetta-Moreau M, Fabre N, Lotterie JA, Chaynes P, et al. Does ageing influence deep brain stimulation outcomes in Parkinson's disease? Mov Disord. 2007;22(10):1457-63. doi: 10.1002/mds.21547.

[213] Parent BA, Cho SW, Buck DG, Nalesnik MA, Gamblin TC. Spontaneous rupture of hepatic artery aneurysm associated with polyarteritis nodosa. Am Surg. 2010;76(12):1416-9.

[214] Merola A, Zibetti M, Angrisano S, Rizzi L, Lanotte M, Lopiano L. Comparison of subthalamic nucleus deep brain stimulation and Duodopa in the treatment of advanced Parkinson's disease. Mov Disord. 2011;26(4):664-70. doi: 10.1002/mds.23524.

[215] Lindvall O, Kokaia Z, Martinez-Serrano A. Stem cell therapy for human neurodegenerative disorders-how to make it work. Nat Med. 2004;10 Suppl:S42-50. doi: 10.1038/nm1064 nm1064 [pii].

[216] Freed WJ, Vawter MP. Microarrays: applications in neuroscience to disease, development, and repair. Restor Neurol Neurosci. 2001;18(2-3):53-6.

[217] Rhee YH, Ko JY, Chang MY, Yi SH, Kim D, Kim CH, et al. Protein-based human iPS cells efficiently generate functional dopamine neurons and can treat a rat model of Parkinson disease. J Clin Invest. 2011;121(6):2326-35. doi: 45794 [pii] 10.1172/JCI45794.

[218] Merola A, Zibetti M, Angrisano S, Rizzi L, Ricchi V, Artusi CA, et al. Parkinson's disease progression at 30 years: a study of subthalamic deep brain-stimulated patients. Brain. 2011;134(Pt 7):2074-84. doi: awr121 [pii] 10.1093/brain/awr121.

[219] Bjorklund A, Lindvall O. Parkinson disease gene therapy moves toward the clinic. Nat Med. 2000;6(11):1207-8. doi: 10.1038/81291.

[220] Kirik D, Georgievska B, Bjorklund A. Localized striatal delivery of GDNF as a treatment for Parkinson disease. Nat Neurosci. 2004;7(2):105-10. doi: 10.1038/nn1175 nn1175 [pii].

[221] Ramaswamy S, Soderstrom KE, Kordower JH. Trophic factors therapy in Parkinson's disease. Prog Brain Res. 2009;175:201-16. doi: S0079-6123(09)17514-3 [pii] 10.1016/S0079-6123(09)17514-3.

[222] Klein AA, Arrowsmith JE. Should routine pre-operative testing be abandoned? Anaesthesia. 2010;65(10):974-6. doi: 10.1111/j.1365-2044.2010.06503.x.

[223] Gill SS, Patel NK, Hotton GR, O'Sullivan K, McCarter R, Bunnage M, et al. Direct brain infusion of glial cell line-derived neurotrophic factor in Parkinson disease. Nat Med. 2003;9(5):589-95. doi: 10.1038/nm850 nm850 [pii].

[224] Lang AE, Gill S, Patel NK, Lozano A, Nutt JG, Penn R, et al. Randomized controlled trial of intraputamenal glial cell line-derived neurotrophic factor infusion in Parkinson disease. Ann Neurol. 2006;59(3):459-66. doi: 10.1002/ana.20737.

[225] Bartus RT, Herzog CD, Chu Y, Wilson A, Brown L, Siffert J, et al. Bioactivity of AAV2-neurturin gene therapy (CERE-120): differences between Parkinson's disease and nonhuman primate brains. Mov Disord. 2011;26(1):27-36. doi: 10.1002/mds.23442.

[226] Decressac M, Ulusoy A, Mattsson B, Georgievska B, Romero-Ramos M, Kirik D, et al. GDNF fails to exert neuroprotection in a rat {alpha}-synuclein model of Parkinson's disease. Brain. 2011;134(Pt 8):2302-11. doi: awr149 [pii] 10.1093/brain/awr149.

2

The Functioning of "Aged" Heterochromatin

Teimuraz A. Lezhava, Tinatin A. Jokhadze and Jamlet R. Monaselidze
*Department of Genetics, Iv.Javakhishvili Tbilisi State University, Tbilisi,
Georgia*

1. Introduction

1.1 Heterochromatin – Substratum of aging

The aging process is programmed in the genome of each organism and is manifested late in life. Any change in normal homeostasis, particularly any further loss of the cell function with aging, occurs in the functional units of the chromatin domains.

Modification of the chromatin structure and function by hetero- or deheterochromatinization occurs throughout life and plays a pivotal role in the irreversible process in aging by affecting gene expression, replication, recombination, mutation, repair, and programming (Gilson and Magdinier, 2009; Elcock and Bridger, 2010). Among chromatin modifications, methylation and acetilation of lysine residues in histones H3 and H4 are critical to the regulation of chromatin structure and gene expression. Compacted heterochromatin regions are generally hypoacetylated and methylated in a discrete combination of lysine methylated marks such as H3K9me2 and 3 (its recognition by specific structural proteins such as HP1 is required for heterochromatin assembly and spreading) and H4K20me1 (Trojer and Reinberg, 2007; Vaquero, 2009). Hypermethylation may cause heterochromatinization and thus would result in gene silencing (Mazin, 1994, 2009). It was found that HP1 is associated with transcripts of more than one hundred euchromatic genes. All these proteins are in fact involved both in RNA transcript processing and in heterochromatin formation. Loss of HP1 proteins causes chromosome segregation defects and lethality in some organisms; a reduction in levels of HP1 family members is associated with cancer progression in humans (Dialynas et al., 2008). This suggests that, in general, similar epigenetic mechanisms have a significant role on both RNA and heterochromatin metabolisms (Piacentini et al., 2009).

Current evidence suggests that SirT1-7 (NAD-dependent deacetylase activity proteins), now called "sirtuins," have been emerging as a critical epigenetic regulator for aging (Imai, 2009). The first event, arrival of and SirT1 at chromatin, results in deacetylation of H4K 16 and H3K9Ac, and direct recruitment of the linker histone H1, in the formation of heterochromatin, a key factor in the formation of the 30 nm fiber (Vaquero, 2004; Michishita *et al.*, 2005). The fact that such histones modifications are reversible (Dialynas *et al.*, 2008; Kouzarides, 2007) offers the potential for therapy (Dialynas *et al.*, 2008). The first level of chromatin organization, the 10 nm fiber, corresponds to a nucleosome array. This fiber is accessible to the transcriptional machinery and is associated with transcriptionally active regions, which are also known as active chromatin or euchromatin (Trojer and Reinberg, 2007).

Heterochromatin is divided into two main forms according to their distinct structural functional dynamics: constitutive heterochromatin (CH) and facultative heterochromatin (FH). CH refers to the regions that are always maintained as heterochromatin; these span large portions of the chromosome and have a structural role. CH regions contain few genes and are located primarily in pericentromeric regions and telomeres. FH refers to those regions that can be formed as heterochromatin in a certain situation but can revert to euchromatin once required. FH can span from a few kilobases to a whole chromosome and generally includes regions with a high density of genes. SirT1 contains both forms of heterochromatin (Prokofieva-Belgobskaya, 1986; Vaquero, 2004, 2009). Heterochromatin composed of distinct life-important functional domains, includes: 1. constitutive heterochromatin, almost entirely composed of non-coding sequences (satellite DNA) that are mostly localized at or are adjacent to the centromeric and telomeric regions; 2. NOR-satellite stalk heterochromatin reflecting the activity of synthetic processes (Ag-positive - coding chromatin and Ag-negative – non-coding chromatin) and 3. facultative heterochromatin (heterochromatinization - condensed euchromatic regions) that mainly consist of "closed" transcribe genes.

According to this view, we discuss of the levels of: 1) total heterochromatin; 2) constitutive (structural) heterochromatin; 3) nucleolus organizer regions (NORs) heterochromatin and 4) facultative heterochromatin in lymphocytes cultured from individuals at the age of 80 and over.

2. Facultative heterochromatin (condensation of eu- and heterochromatin regions)

We have used differential scanning microcalorimetry to produce a calorimetric curve in cultured human lymphocytes over the temperature range 38–130°C. It was determined that the clearly expressed shoulder of the heat absorption curve in the temperature interval from 40°C to 50°C with $T_m(I)=45+1$°C corresponds to melting of membranes and some cytoplasm proteins, maxima at $T_m(II)=55+1$°C correspond to melting (denaturation) of non-histone nuclei proteins, maxima at $T_m(IV)70+1$°C corresponds to the ribonucleoprotein complex, and maxima at $T_m(III)=63+1$°C and $T_m(V)=83+1$°C correspond to cytoplasm proteins. Other clearly expressed peaks at $T_m(VI)=96+1$°C and $T_m(VII)=104+1$°C correspond to the chromatin denaturation (Monaselidz et al.,2006,2008). The heating process produced clear and reproducible endothermic heat absorption peaks. We found that an endothermic peak at $T_m=104\pm1$°C corresponds to melting of 30 nm-thick fibers, which represents the most condensed state of chromatin in interphase nuclei (heterochromatin), and that an endothermic peak at 96 ± 1°C corresponds to melting of 11 nm-thick filaments.

The chromatin heat absorption peaks VI and VII changed significantly with age. In particular, in the shifted endotherms VI and VII, the temperatures increased by 2°C and 3°C accordingly in old age (80-86 years). Additional condensation of the eu- and heterochromatin was demonstrated by an increase in T_m by 2°C and 3°C in comparison with the meddle age (25-40 years) (Fig.1). These prominent changes in chromatin stability indicated transformation of eu- and heterochromatin in condensed chromatin (heterochromatinization).

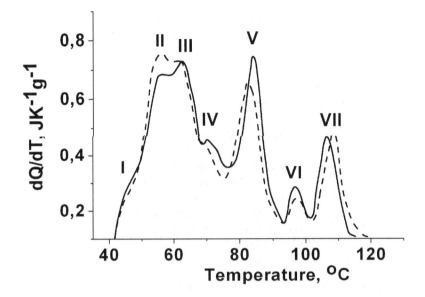

Fig. 1. The excess of heat capacity ($\Delta C_p=dQ/dT$) as function of temperature for lymphocytes cultures from young donors (------) and old donors (– –) (48 –hour cell culture), dry biomass (------) - 8.5 mg and 87 µg DNA, dry biomass (– –) 8.8mg and 90 µg DNA

One of the potential epigenetic mechanisms is heterochromatinization of chromatin within the region of the genome containing a gene sequence, which inhibits any further molecular interactions with that underlying gene sequence and effectively inactivates that gene (Ellen *et al.*, 2009). The chromatin peak behavior described above shows progressive heterochromatinization of lymphocyte chromosomes from old individuals and confirms previously reported data (Lezhava, 1984, 2001, 2006; Vaquero, 2004).

These significant changes in chromatin stability in old age indicate that the aging process involves transformation of the eu- and heterochromatin into condensed forms and that further compaction or progressive heterochromatinization occurs during aging.

3. Constitutive heterochromatin (pericentromeric and telomeric heterochromatin)

Centromeric and telomeric heterochromatin differs from each other by structure and sensitivity to exogenous factors. Centromeric heterochromatin showed increased H3-K27 trimethilation in the absence of SUV39h1 and Suv39h2HMTases. Such modification was not detectable at telomeric heterochromatin. Despite the differences between the two heterochromatin domains and the distinction of functions, they have much in common (Blasco, 2004; Lam *et al.*, 2006).

3.1 Pericentromeric heterochromatin

The heterochromatin regions of human chromosomes near the centromere vary and the degree of variability is related to the amount and molecular organization of DNA, which contains only a fraction of satellite DNA. The amount and function of heterochromatin regions have a close relationship with the organization and functioning of the entire genome.

Satellite DNA (tandemly repeated noncoding DNA sequences) stretch over almost all native centromeres and surrounding pericentromeric heterochromatin. Satellite DNA was considered to be an inert by-product of genome dynamics in heterochromatic regions. However, recent studies have shown that the evolution of satellite DNA involved an interplay of stochastic events and selective pressure. This points to the functional significance of satellite sequences, which in (peri) centromeres may play some fundamental roles. First, specific interactions between satellite sequences and DNA-binding proteins are proposed to complement sequence-independent epigenetic processes. Second, transcripts of satellite DNA sequences initialize heterochromatin formation through an RNAi mechanism. In addition, satellite DNAs in (peri)centromeric regions affect chromosomal dynamics and genome plasticity (Mehta *et al.*, 2007; Plohl *et al.*, 2008). Satellite DNA is localized in human (peri) centromeres heterochromatin chromosomes 1,9, 16 and Y.

The data on comparative of (peri) centromeric heterochromatin (C-segment) were provided for all three chromosome pairs (1, 9 and 16) indicating that the variants of large C-segments (d and e) were registered more often in old individuals than in the cells of the younger ones: for chromosome 1 – X^2_4 =21.9, (p<0.001); for chromosome 9 – X^2_4 =10,6 (p<0.001); for chromosome 16 – X^2_4 =18.7, (p<0.001). The increased size of the C-segments were also found in the Y- chromosomes of the family : the father and the grandfather (59 and 88 years, respectively), compared with the 30 year old son (Lezhava, 2006).

Thus, the (peri) centromeric heterochromatin on three chromosome pairs (1, 9 and 16) and the C-segments of the Y chromosome increase in size in old age, pointing to the heterochromatinization of these heterochromatin regions of chromosomes.

In some cases, without pretreatment metaphases from old individuals, blocks of centromeric heterochromatin were common on homologous chromosomes 1qh C-band locations were similar to those seen after an alkaline or thermal pretreatment or staining with buffered Giemsa.

In a percentage without pretreatment of metaphases, the heterochromatin-positive 1qh chromosomes displayed some packing impairment. Sizes and distribution of centromeric heterochromatin on the 1qh homologous varied in some metaphases of 6 from 24 individuals aged 81 to 114 years and was absents in control group ranging in age from 13 to 34 years.

Of interest was a sample from a 114-year-old man whose 1qh showed dark-stained heterochromatin sites sized 1.5-fold greater than counterpart sites in other individuals samples. However, intrahomologous variability was often related to sizes and the absents of heterochromatin blocks in one of the homologous chromosome 1 (Fig.2).

The control of cellular senescence by specific human chromosomes was examined in interspecies cell hybrids between diploid human fibroblasts and an immortal, Syrian hamster cell line. Most such hybrids exhibited a limited life span comparable to that of the human fibroblasts, indicating that cellular senescence is dominant in these hybrids. Karyotypic analyses of the hybrid clones that did not senesce revealed that all these clones had lost both copies of human chromosome 1, whereas all other human chromosomes were observed in at least some of the immortal hybrids. The application of selective pressure for retention of human chromosome 1 to the cell hybrids resulted in an increased percentage of hybrids that senesced. Further, the introduction of a single copy of human chromosome 1 to the hamster cells by microcell fusion caused typical signs of cellular senescence. These findings indicate that human chromosome 1 may participate in the control of cellular senescence and further support a genetic basis for cellular senescence (Sugawara et al., 1990).

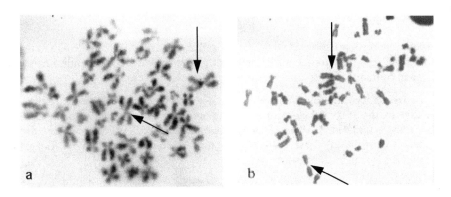

Fig. 2. Distribution of C-bands on one of the homologous of the 1qh chromosomes without preparation pretreatment and unbuffered Unna blue staining. Metaphases: from 114-year-old man (a) and from 83-year-old man (b). Arrows indicate: homologous chromosomes 1 with and without bands

3.2 Telomeric heterochromatin

Telomeres are specialized DNA–protein structures that form loops at the ends of chromosomes (Boukamp et al., 2005). In human cells they contain short DNA repeat sequences $(TTAGGG)_n$ added to the ends of chromosomes by telomerase. Telomere heterochromatin in most human somatic cells loses 50–200 bp per cell division (lansdorp, 2000; Geserick and Blasco, 2006). Telomeres serve multiple functions, including the protection of chromosome ends and prevention of chromosome fusions. They are essential for maintaining individuality and genome stability (Lo *et al.*, 2002; Murnane, 2006). A major mechanism of cellular senescence involves telomere shortening (Horikawa and Barrett, 2003; Opresko et al., 2005), which is directly associated with many DNA damage–response proteins that induce a response similar to that observed with DNA breaks (Bradshow et al., 2005; Wright and Shay, 2005).

Terminal telomere structures consist of tandemly repeated DNA sequences, which vary in length from 5 to 15 kb in humans. Several proteins are attached to this telomeric DNA, including PARP-1, Ku70/80, DNA-PKcs, Mre11, XRCC4, ATM, NBS and BLM, some of which are also involved in different DNA damage response (repair) pathways. Mutations in the genes coding for these proteins cause a number of rare genetic syndromes characterized by chromosome and/or genetic instability and cancer predisposition (Callen and Surralles, 2004; Hande, 2004; Bradshow et al., 2005).

Based on the presented data, we concluded that telomeric chromatin undergoes progressive heterochromatinization (condensation) with aging that determines: (a) inactivation of the gene coding for the catalytic subunit of telomerase, hTERT; and (b) switching off the genes for Ku80, Mre11, NBS, BLM, etc causing chromosome disorders related to chromosome syndromes. Telomere shortening is another consequence of age-related.

Heterochromatinization that is reportedly due to unrepaired single-strand breaks of DNA in telomere regions resulting in unequal interchromatid and interchromosome exchanges and inactivation of the telomerase-coding gene-determining telomere length (Golubev,2001; Gonzalo et al., 2006).

Our experimental data showed that the number of cell with end-to-and telomere associations and the total frequency of aberrant telomeres were considerably increased at the old age in comparison with those at middle age (Iezhava,2006).

The higher frequency of chromosome end-to-and telomere associations in extreme old age may be due to the loss of heterochromatin telomere regions (Fig.3). Mouse embryonic fibroblast cells lacking Suv39h1 and Suv39h2 exhibit reduced levels of H3K9me and HP1 (deheterochromatinization).These alterations in chromatin correlate with telomere elongation (Garcia-Cao et al., 2004).

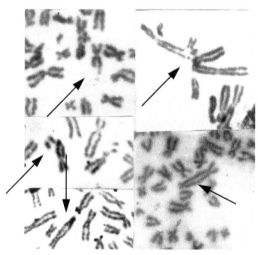

Fig. 3. Telomeres aberrations and end-to-end associations of chromosomes from elderly are shown by arrows.

According to previous publications (Prokofieva-Belgovskaya, 1986; Hawley and Arbe, 1993) sister chromosome exchanges (SCEs) do not occur or are less frequent in heterochromatin or heterochromatinized regions. The evaluation of SCE in individuals aged 80 years and more has revealed that single-cell SCE counts appear to be lower than in middle age (lezhava, 2006), that is, exchanges between sister chromatids mostly take place in euchromatic regions.

In old age, $CoCl_2$ alone and in combination with the tetrapeptide bioregulator Livagen enhanced the distribution of SCE; that is, pericentromeric heterochromatin appeared to be more sensitive to the $CoCl_2$ effect alone (15.4 ± 1.8% SCE), whereas SCE was mostly observed in telomere heterochromatin when $CoCl_2$ in combination with livagen was used (12 ± 1.2% SCE) (control, 2.8 ± 0.5% SCE, respectively).Because exchanges occur in euchromatic uncondensed regions, the obvious effect of $CoCl_2$ alone and in combination with Livagen could be attributed to its decondensing deheterochromatinization effect on pericentromeric and telomeric heterochromatin, which would elevate the possibility of SCE ((lezhava and Jokhadze, 2007). At the same time, the deheterochromatinization of telomeric heterochromatin contributes to activation of DNA repair. That is, the intensity of unscheduled DNA synthesis increases (lezhava and Jokhadze, 2004) and creates a basis for activation of inactivated genes during aging and development of diseases.

4. Nucleolus Organizer Regions (NOR) heterochromatin

The heterochromatic regions of secondary constrictions (NORs) in human D (13, 14, 15) and G (21, 22) group acrocentric chromosomes contain genes coding for 18S and 28 ribosomal RNA. It has been established that genetically active NORs can appear with nucleolar form of DNA-dependent RNA-polymerase and selectively stain with silver (Ag-stained). It has also been found that association between Ag-stained satellite stalks of acrocentric chromosomes in metaphase cells (Fig.4) are determined primarily by their function as nucleolar organizers.

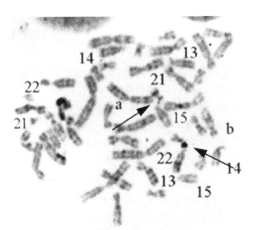

Fig. 4. Metaphases with variable sizes of Ag-positive nucleolar organizer regions. Arrows indicate a - "open" satellite stalks association; b - "closed" satellite stalks association.

The acrocentric association phenomenon may induce acrocentric nondisjunction during the meiosis or early zygote division, and chromosome rearrangements. Chromosomes can associate when two chromatid satellites are available, and so they are defined as associated, when their satellites make up a pair. Therefore, prematurely condensed silver-stained acrocentrics have similar rates of interphase and metaphase association. It was shown (Lezhava,1984; Verma and Rodriguez, 1985) that the likelihood of acrocentric chromosome associations is related to an extent of satellite stalk heterochromatinization.

Heterochromatinization of stalks – NORs has been studied by association frequencies in lymphocytes. In humans of a very old age (80–93 years), the estimated number of Ag-positive nucleolus organizer regions (NORs) for all chromosomes per cell, both associated and nonassociated, was significantly lower (6.10 in individuals 80–93 years old) in comparison with that in young individuals (7.05; $p < 0.01$). The frequency of acrocentric chromatid association in individuals aged 80 years and over was significantly decreased in comparison with those in a control group.

Increase of associations frequency was parallel to the growth of Ag segment size. At the same time, chromosomes containing NORs of grade 2 frequently formed associations among the middle-aged individuals, rather than in the older group.

Moreover, the transcriptional activity of ribosomal cistrons, which determine activity of a nucleolar form of DNA-dependent RNA polymerase - were from 668–721 imp/min in old individuals. They were significantly decreased in comparison with the control: from 1020 to 1120 imp/min.

The above considerations imply that a decreased number of chromosomes with Ag-positive NORs, a lower frequency of association of acrocentric chromatids, and a decrease in endogenic RNA-polymerase activity of ribosomal cistrons, result in alterations in the length of chromosomal satellite stalks that is caused by heterochromatinization in the process of aging (Lezhava and Dvalishvili,1992).

4.1 Cis- and trans-types of chromatid association

Most of acrocentric chromosome associations (85 percent) are formed by single chromatid satellite stalks (Lezhava et al., 1972; Verma et al., 1983). The exposure of lymphocyte cultures to 5-bromodeoxyuridine (BrDU) during two replication cycles revealed two-acrocentric associations that were either at a cis-position (differentially stained acrocentric chromatids with a dark-to-dark or light-to-light association) or a trans-position (chromatids with a dark-to-light or light-to-dark association) (Chemitiganti et al., 1984).

Frequencies of the cis- and trans-orientation of acrocentric chromatid association have been studied in old individuals. Lymphocyte cultures were prepared with a conventional methodology. The study examined 173 metaphases from 9 individuals aged 80 to 89 years and 124 metaphases from 6 individuals aged 20 to 48 years. For differential staining of sister chromatids BrDU (7.7 μg/ml) was added to the cultures immediately on their initiation. The lymphocytes were incubated in darkness for 96 h at 37°C. Giemsa stain was employed after DNA thymine was substituted by BrDU. In DNA thymine was totally substituted in one of second-mitosis sister chromatids which stained light and was denoted chromatid 1; only half of DNA thymine was substituted in the other chromatid which stained dark and was defined as chromatid 2 (Fig. 5). According to association criteria of cis-1 position was the term adopted for the light-to-light association, cis-2 position for the dark-to-dark association, and trans-position for the light-to-dark association (Fig. 5).

Statistical analysis of association frequencies proceeded from the assumption that the cis-1 and cis-2 associations have similar chances to occur, and the chances make half of the probability of the trans-oriented association, that is

$$P_{\text{cis-1}}(DD) = P_{\text{cis-2}}(DD) = 1/2\, P_{\text{trans}}(DD) \qquad (1)$$

$$P_{\text{cis-1}}(GG) = P_{\text{cis-2}}(GG) = 1/2\, P_{\text{trans}}(GG) \qquad (2)$$

$$P_{\text{cis-1}}(DG) = P_{\text{cis-2}}(DG) = 1/2\, P_{\text{trans}}(DG) \qquad (3)$$

These equalities represent the hypothesis that chromatids-1 and chromatids-2 participated in the association with the same probability.

The data of the middle-aged group fitted the hypotheses (2) and (3). The statistics

$$X^2(GG) = \frac{\left(V_{cis-1}(GG) - V(GG)/4\right)^2}{(1/4)V(GG)} + \frac{\left(V_{cis-2}(GG)\right) - V(GG)/4\right)^2}{(1/4)V(GG)} + \frac{\left(V_{trans}GG\right) - V\left(GG/2\right)^2}{(1/2)V(GG)}$$

should be almost $X^2(2)$-distributed if (3) is true; they yielded the value of 0.69. Similar statistics $X^2(DG)$ for testing (3) gave the value of 1.54. Equalities (1) proved less supportive: the verifying statistics $X^2(DD)$ gave 5.14 while the presumptive value was 0.08.

A different pattern was seen in the old individuals group. While the data fitted equalities (2), (1) and (3) had to be rejected since the statistics were $X^2(DD) = 5.76$ and $X^2(DG) = 18$.

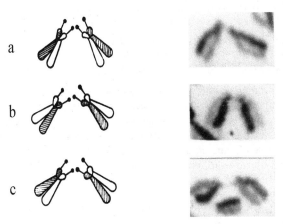

Fig. 5. Associations of acrocentric chromatid satellite stalks. a - cis-1 position (light -to-light chromatid association); b - cis -2 position (dark-to dark association); c - trans-position (dark-to-light association).

An important consideration is deviation of the data from the hypotheses (1)-(3). The deviation suggested that chromatids 1 and 2 of D chromosomes had different associative activities, unlike G-chromosome chromatids. Indeed, if D-chromosome chromatid 2 were more active than chromatid 1, probabilities should be

$$P_{\text{cis-1}}(DD) < 1/2\, P_{\text{trans}}(DD) < P_{\text{cis-2}}(DD) \tag{4}$$

$$P_{\text{cis-1}}(DG) < 1/2\, P_{\text{trans}}(DG) < P_{\text{cis-2}}(DG) \tag{5}$$

and these agreed well with the actual findings.

In conclusion, sister chromatids of acrocentric chromosomes show a functional heterogeneity in very old individuals (Lezhava, 1987, 2006).

5. Correlation between mutation, repair and hetheterochromatinization of chromosomes in aging

Progressive heterochromatinization of chromosome regions observed during aging correlates with the greater frequency of chromosome aberrations and the reduced intensity of reparative events. Chromosome alterations have been studied in 70 individuals aged 80–114 years (30 women and 40 men). In these samples, the percentages of aberrant metaphase and chromosomal aberrations were 4.08±0.41% in women and 5.15±0.45% in men; these values are significantly higher than the published control levels (aged 25–40 years)of 1.8±0.42% and 2.15±0.35%, respectively (Lezhava, 2001, 2006).

The incidence of cell with chromosome aberrations in 80- to 90-year-old individuals was 4.75±0.71% for 25 women and 3.06±0.54% for 31 men; these means were also above those of 20-to 48-year-old individuals. The incidence of aberrant cells in men aged 91 to 114 years (5.62±1.45%) was higher than that in women aged 91 to 108 years and control individuals (Fig. 6, 7).

Our studies have also demonstrated a marked decline in the unscheduled DNA synthesis (repair) rates in 80-90 year- old individuals in response to UV irradiation at a dose of 10-15 J/mm² compared with the middle-aged individuals (P < 0.03, P < 0.01 respectively). These data suggest that human lymphocytes from older people have a significantly reduced capacity for unscheduled DNA synthesis–excision repair (Lezhava, 1984, 2001).

Progressive heterochromatinization of chromosome regions observed during senescence correlates with the lowered intensity of reparative events and the increases frequency of chromosome aberrations. To explain the prevalence of the accumulation of damage in heterochromatin and in the heterochromatinization regions, it has been assumed that the repair of lesions capable of causing aberrations is possible only in those areas of DNA that are actively involved in transcription and that are within physically accessible of reparative enzymes, i.e. in euchromatin areas (Yeilding, 1971). Assuming that heterochromatinized regions are inaccessible to reparative enzymes and therefore number of cells with chromosome aberrations profoundly affects the functioning of the genome in old age (Fig.8).

Our results indicate that decreases in the repair processes and increases in the frequency of chromosomal aberrations in aging are secondary to the progressive heterochromatinization and that chromosome heterochromatinization is a key factor in aging.

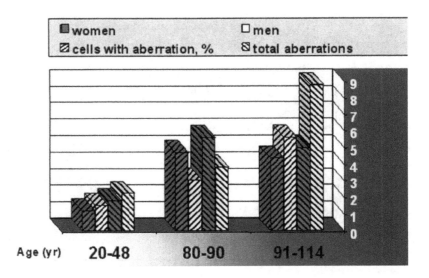

Fig. 6. Spontaneously structural chromosome aberration at 80 years and over

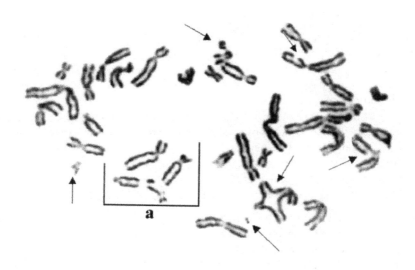

Fig. 7. 114-year-old man's metaphase with aberrant chromosomes. a – association of telomeric regions

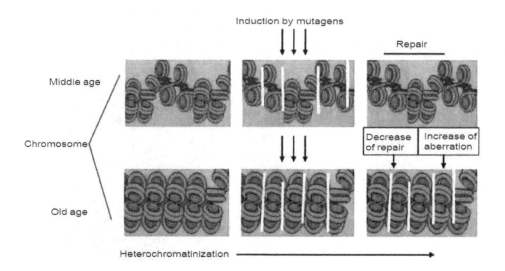

Fig. 8. Heterochromatinized regions inaccessible to reparative enzymes and therefore the number of cells with chromosome aberrations profoundly affects the functioning of the genome in old age.

6. Heterochromatin and pathology

Heterochromatinization progresses with aging and can deactivate many previously functioning active genes. It blocks certain stages of normal metabolic processes of the cell, which inhibits many specific enzymes and leads to aging pathologies. The action of genetic systems reveals general rules in the behavior of such systems, such as the connection between the structural and functional interrelationships between the "directing" and "directed" structures. In the respect, it should be noted that heterochromatinized regions in chromosomes can reverse. Many physical and chemical agents, hormones and peptide bioregulators (Epitalon - Ala-Glu-Asp-Gly; Livagen - Lys-Glu-Asp-Ala; Vilon - Lys-Glu) (Khavinson *et al.*, 2003; Lezhava and Bablishvili, 2003; Lezhava *et al.*, 2004, 2008) cause deheterochromatinization (decondensation) releasing the inactive (once being active)genes that seems to favour purposive treatment of diseases of aging.

We have demonstrated also that Co^{2+} ions alone and in combination with the bioregulator Livagen can reverse the deheterochromatinization of precentromeric and telomeric heterochromatin (Fig.9), to normalize the telomere length in cells from old individuals (Lezhava and Jokhadze, 2007; Lezhava *et al.*,2008). Blood cholesterol levels in an animal model (rabbit) for atherosclerosis was reduced (41% on the average) by pretreatment with combination of livagen and $CoCl_2$ - normalization of telomere length (unpublished data of research – STCU 4307- grants in 2007-2009) (Lezhava *et al.*, 2007-2009).

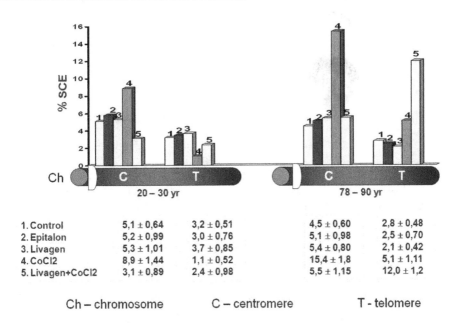

1. Control	5,1 ± 0,64	3,2 ± 0,51	4,5 ± 0,60	2,8 ± 0,48
2. Epitalon	5,2 ± 0,99	3,0 ± 0,76	5,1 ± 0,98	2,5 ± 0,70
3. Livagen	5,3 ± 1,01	3,7 ± 0,85	5,4 ± 0,80	2,1 ± 0,42
4. CoCl2	8,9 ± 1,44	1,1 ± 0,52	15,4 ± 1,8	5,1 ± 1,11
5. Livagen+CoCl2	3,1 ± 0,89	2,4 ± 0,98	5,5 ± 1,15	12,0 ± 1,2

Ch – chromosome   C – centromere   T - telomere

Fig. 9. The effect of Co^{2+} ions separate and with peptide bioregulators Epitalon (Ala-Glu-Asp-Gly) and Livagen (Lys -Glu-Asp-Ala) distribution of SCE among centromer and telomer heterochromatin regions.

7. Conclusion

In the present investigation, we assessed the modification of total, constitutive (pericentromeric, telomeric and nucleolus organizer region (NOR) heterochromatin) and facultative heterochromatin in cultured lymphocytes exposed to the influence of heavy metal and bioregulators from individuals aged 80 years and over.

The results showed that: (1) progressive heterochromatinization of total, constitutive (pericentromeric, telomeric and NOR heterochromatin) and facultative heterochromatin occurred with aging; (2) a decrease in repair processes and an increase in frequency of chromosome aberrations with aging is secondary to the progressive heterochromatinization of chromosomes; (3) peptide bioregulators induce deheterochromatinization of chromosomes in old age and (4) Co^{2+} ions alone and in combination with the tetrapeptide bioregulator, Livagen (Lys-Glu-Asp-Ala), have different chromosomal target regions; that is, deheterochromatinization of pericentromeric (Co^{2+} ions) and telomeric (Co^{2+} ions in combination with livagen) heterochromatin regions in lymphocytes of olderaged individuals.

The proposed genetic mechanism responsible for constitutive (pericentromeric, telomeric and nucleolus organizer region (NOR) heterochromatin) and facultative heterochromatin remodeling (hetero- and deheterochromatinization) of senile pathogenesis highlights the importance of external and internal factors in the development of diseases and may lead to the development of therapeutic treat.

8. Acknowledgements

This article is dedicated to the memory of professor **A.A. Prokofieva-Belgovskaya** from her greatful students.

9. References

Blasco, M. (2004). Telomere epigenetics: a higher-order control of telomere length in mammalian cells. *Carcinogenesis*, Vol.25, pp. 1083-1087

Boukamp, P.; Popp, S. & Krunic, D. (2005). Telomere-dependent chromosomal instability. *J Investig Dermatol Sym Proc*, Vol.10, pp. 89-94

Bradshow, P.; Stavropoulos, P. & Mey, M. (2005). Human telomeric protein TRF2 associates with genomic double-strand breaks as an early response to DNA damage. *Nat Genet*, Vol.37, pp. 116-118

Callen, E. & Surralles, J. (2004). Telomere dysfunction in genome instability syndromes. *Mutat Res*, Vol.567, pp. 85-104

Chemitiganti, S.; Verma, R.; Ved Brat, S.; Dosik, H.(1984).Random single chromatid type segregation of human acrocentric chromosomes in BrdU-labeled mitosis. *Can J Genet and Cytol*, Vol.26, pp. 137-140

Dialynas, G.; Vitalini, M. & Wallrath, L. (2008). Linking heterovhromatin protein 1 (HP1) to cancer progression. *Mutat Res*, Vol. 647, pp. 13-28

Elcock, L. & Bridger, J. (2010). Exploring the relationship between interphase gene positioning, transcriptional regulation and the nuclear matrix.*Biochem Soc Trans*, Vol. 38, pp. 263-267

Ellen, T.; Kluz, T.; Harder, M.; Xiong, J. & Costa, M. (2009). Heterochromatinization as a potential mechanism on nickel-induced carcinogenesis. *Biochemistry*, Vol.48, pp. 4626-4632

Garcia-Cao, M.; O'Sullivan, R.; Peters, A. et al. (2004). Epigenetic regulation of telomere length in mammalian cells by the Suv39h1 and Suv39h2 histone methyltransferases. *Nat Genet*, Vol.36, pp. 94-99

Geserick, C. & Blasco, M. (2006). Novel roles for telomerase in aging. *Mech Ageing Dev*, Vol.127, pp. 579-583

Gilson, E. & Magdinier, F. (2009). Chromosomal position effect and aging. In Epigenetics and aging. *Springer New York*, Vol.2, pp. 151-175

Golubev, A. (2001). The natural history of telomeres. *Adv.Gerontol* , Vol.7, pp. 95-104

Gonzalo, S.; Jaco, I.; Fraga, M. et al. (2006). DNA methyltransferases control telomere length and telomere recombination in mammalian cells. *Nat Cell Biol*, Vol.8, pp. 416-424

Hande,M.(2004).DNA repair factors and telomere-chromosome intergrity in mammalian cells. *Cytogenet Genome Res*. Vol.104, pp.116-122.

Hawley, R. & Arbel, T. (1993). Yeast genetics and the fall of classical view of meiosis. *Cell*, Vol. 72, pp. 301-303

Horikawa, I. & Barrett, J. (2003). Transcriptional regulation of the telomeraze hTERT gene as a target for cellular and viral oncogenic mechanisms. *Carcinogenesis*, Vol.24, pp. 1167-1176

Imai, S. (2009). From heterochromatin islands to the NAD World: a hierarchical view of aging through the functions of mammalian Sirt1 and systemic NAD biosynthesis. *Biochim Biophys Acta*, Vol.1790, pp. 997-1004

Khavinson, V.; Lezhava, T.; Monaselidze, J. et al. (2003). Peptide Epitalon activates chromatin at the old age. *Neuroendocrinol letters*, Vol.24, pp. 329-333

Kouzarides, U. (2007). Chromatin modifications and their function. *Cell*, Vol.128, pp. 693-705

Lam, A.; Bovin, C.; Bonney, C. et al. (2006). Human centromeric is a dynamic chromosomal domain that can spread over noncentromeric DNA. *Proc Natl Acad Sci USA*, Vol.103, pp. 4186-4191

Lansdorp, P. (2000). Repair of telomeric DNA prior to replicative. *Mech Ageing Dev*, Vol.118, pp. 23-34

Lezhava, T. hitashvili, R.; Khmaladze E.(1972).Use of mathematical"satellite model" for association of acrocentric chromosomes depending on human age. *Bio-medical Computing*, Vol.3, pp. 101-199

Lezhava, T. (1984). Heterochromatinization as a key factor of aging. *Mech Ageing and Dev*, Vol.28, pp. 279-288

Lezhava, T. (1987).Sister chromatidexchange in human lymphocyte in extreme age. *Proc Japan Acad*, Vol.63, pp. 369-372

Lezhava, T. (2001). Chromosome and aging:genetic conception of aging. *Biogerontology*, Vol.2, pp. 253-260

Lezhava, T. (2006). Human chromosomes and aging. From 80 to 114 years. *Nova biomedical*, ISBN 1-60021-043-0, New York, USA

Lezhava, T. & Bablishvili, N. (2003). Reactivation of heterochromatin induced by sodium hydrophospate at the old age. *Proc Georg Acad Sci, Biol Ser B* Vol.1, pp. 1-5

Lezhava, T. & Dvalishvili, N. (1992). Cytogenetic and biochemical studies on the nucleolus organizing regions of chromosomes in vivo and in vitro aging. *Age*, Vol.15, pp. 41-43

Lezhava, T. & Jokhadze, T. (2004). Variability of unscheduled DNA synthesis induced by nikel ions and peptide bioregulator epitalon in old people. *Proc Georg Acad Sci*, Vol.2, pp. 65-70

Lezhava, T. & Jokhadze, T. (2007). Activation of pericentromeric and telomeric heterochromatin in cultured lymphocytes from old individuals. *Ann N Y Acad Sci*, Vol.1100, pp. 387-399

Lezhava, T.; Khavinson, V.; Monaselidze, J. et al. (2004). Bioregulator Vion-induced reactivation of chromatin in cultured lymphocytes from old people. *Biogerontology*, Vol.4, pp. 73-79

Lezhava, T.; Monaselidze, J. & Jokhadze, T. (2008). Decondensation of chromosmes heterochromatinization regions by effect of heavy metals and bioregulators in cultured lymphocytes from old individuals. *Proceeding of the 10th International Symposium of Metal Ions in Biology and Medicine, Bastia France May 19-22 Edited by Philippe Collery*, 10, pp. 569- 576

Lezhava, T.; Monaselidze, J.; Jokhadze, T.; Kakauridze, N. & Kordeli, N. (2007-2009). Decondensation of telomeric heterochromatin as a protective means from Atherosclerosis. *Project Proposal, STCU 4307*

Lo, A.; Sprung, C.; Fouladi, B.; Pedram, M. et al. (2002). Chromosome instability as a result of double – strend breaks near telomeres in mouse embryonic stem cells. *Mol Cell Biol*, Vol.22, pp. 4836-4850

Kouzarides, U. (2007). Chromatin modifications and their function. *Cell*, Vol.128, pp. 693-705

Mazin, A. (1994). Enzimatic DNA methylation as an aging mechanism. *Mol Biol Mosc,* Vol.28, pp. 21-51

Mazin, A. (2009). Suicidal function of DNA methylation in age-related genome disintegration. *Ageing Res Rev,* Vol. 8, pp. 314-327

Mehta, I.; Figgitt, M.; Clements, C. et al. (2007). Alterations to nuclear architecture and genome and genome behavior in senescent cells. *Ann NY Acad Sci,* Vol.1100 pp. 250-263

Michishita, E.; Park, J.; Burneskis, J. et al. (2005). Etvolutionarily conserved and nonconserved cellular localizations and functions of human SIRT proteins. *Mol Biol,* Vol.16, pp. 4623-4635

Monaselidze, J.; Abuladze, M.; Asatiani, N. et al. (2006). Characterization of Chromium-induced Apoptosis in Cultured Mammalian Cells. A Different Scanning Calolorimetry Study. *Thermochemia Acta,* Vol.441, pp. 8–15

Monaselidze, J.; Bregadze, V.; Barbakadze, Sh. et al. (2008). Influence of metal ions of thermodina stability of leukemic DNA in vivo. Microcalorimetri investigation. *Proceeding of the 10th International Symposium of Metal Ions in Biology and Medicine, Bastia France May 19-22 Edited by Philippe Collery,* 10, pp. 451-457

Murnane, J. (2006). Telomeres and chromosome instability. *DNA Repair, (Amst)* Vol.8, pp. 1082-1092

Opresko, P.; Fan, J.; Danzy, S. et al. (2005). Oxidative damage in telomeric DNA disrupts recognition by TRF1 and TRF2. *Nucleic Acids Res,* Vol.33, pp. 1230-1239

Piacentini, L.; Fanti, L.; Negri, R. at al. (2009). Heterochromatin protein 1 (HP1a) positively regulates euchromatic gene expression through RNA transcript association and interaction with hnRNPs in Drosophila. *PloS Genet,* 10, e1000670

Plohl, M.; Luchetti, A.; Metrovic, N.; Mantovani, B. (2008). Satellite DNAs between selfishness andfunctionality: structure, genomics and evolution of tandem repeats in centromer (hetero) chromatin. *Gene,* Vol.409, pp. 72-82

Prokofieva-Belgovskaya, A. (1986). *Heterochromatin regions of chromosomes.* M Nauka, ISBN 575.113+576.316

Sugawara, O.; Oshimura, M.; Koi, M. et al. (1990) Induction of cellular senescence in immortalized cells by human chromosome. *Science,* Vol.247, pp. 707-710

Trojer, P. & Reinberg, D. (2007). Facultative Heterochromatin. Is There a Distinctive Molecular Signature? *Mol Cell,* Vol.28, pp. 1-13

Vaquero, A. (2009). The conserved role of sirtuims in chromatin regulation. *Int J De Biol,* Vol.53, pp. 303-322

Vaquero, A.; Scher, M.; Lee, D. et al. (2004). Human SirT1interacts with histone H1 and promotes formation of facultative heterochromatin. *Mol Cell,* Vol.16, pp. 93-105

Verma, R.; Shah ,J.; Dosic H. (1983). Frequencies of chromosome and chromatid types of associations of nucleolar human chromosomes demonstrated by the N-banding technique. *Cytobios,* Vol.36, pp. 25-29

Verma, R. & Rodriguez, J. (1985) Structural organization of ribosomal cistrons in human nucleolar organizing chromosomes. Cytobios, Vol.44, pp. 25-28

Wright, W. & Shay, J. (2005). Telomera-binding factors and general DNA repair. *Nat Genet,* Vol.37, pp. 193-197

Yelding, K. (1974). Model for aging based on differential of somatic mutational damage. *Perspect Biol Med,* Vol.17, pp. 201-208

3

Female Vascular Senescence

Susana Novella[1], Ana Paula Dantas[2], Gloria Segarra[1],
Carlos Hermenegildo[1] and Pascual Medina[1]

[1]*Departamento de Fisiología, Universitat de València,
Instituto de Investigación Sanitaria INCLIVA, Hospital Clínico Universitario, Valencia,*
[2]*Institut d'Investigacions Biomèdiques August Pi i Sunyer (IDIBAPS)
Institut Clinic de Tòrax, Hospital Clinic Barcelona
Spain*

1. Introduction

Long before the existence of cardiovascular imaging, Sir William Osler axiom that *"man is as old as his arteries"*. Followed by several physicians for decades, this aphorism has been widely confirmed by studies demonstrating that risk factors for cardiovascular disease increase as we age (Cooper et al., 1994; Lakatta & Levy, 2003). Nevertheless, a flaw in this statement is the generalization that men and women age similarly. Much data from clinical and basic research have established that vascular aging in women does not follow the same chronology as in men (Shaw et al., 2006; Pereira et al., 2010; Takenouchi et al., 2009). If known risk factors that influence cardiovascular aging are excluded (e.g. smoking, cholesterol, hypertension), men display a pattern of progressive vascular aging, while timing for vascular aging in women presents a clear hallmark, i.e. menopause (Taddei et al., 1996; Bucciarelli & Mannucci, 2009). Until menopause women are considered "hemodynamically younger" than men, based on epidemiological studies showing that the incidence of cardiovascular diseases in premenopausal women is markedly low compared to age-matched men (Messerli et al., 1987; Bairey Merz et al., 2006; Shaw et al., 2006). After menopause, however, these numbers rise to values that are close, or even higher, to those found in men (Lerner & Kannel, 1986; Eaker et al., 1993; Eaker et al., 1994). And so it one could say that *"man is as old as his arteries, although the arteries of a woman are as young as her hormones"*.

Cardiovascular disease is the primary cause of death among women after menopause (55%), compared to men (43%) even above all cancers combined (Rosamond et al., 2008). With increasing recognition of the importance of cardiovascular disease in women, the interest and emphasis on research concerning women and cardiovascular disease have grown substantially (Bairey Merz et al., 2006; Shaw et al., 2006). Despite this, there is still a concerning gap in the knowledge, understanding, and general awareness of mechanisms for cardiovascular aging in women. In this review, we will discuss clinical and experimental data that document the effects of aging, estrogens and hormonal replacement therapy on vascular function of females.

2. Effects of aging on vascular function

Vascular aging is a natural phenomenon that could be simply described as a consequence of physical stress. Arteries are elastic tissues, and as such are predisposed to fatigue and fracture with time, as a consequence of extension-relaxation cycles during heartbeats (Avolio et al., 1983; Avolio et al., 1985; O'Rourke & Hashimoto, 2007). In fact, fracture of elastic lamellae is observed with aging in aorta, and can account for the major physical changes seen in elder: dilation (after fracture of load-bearing material) and stiffening (by transfer of stress to the more rigid collagenous component of the arterial wall) (Lakatta, 2003).

There is growing evidence that vascular aging begins early in life, with evidence for alteration in vascular matrix proteins as early as the third decade in health individuals (Wallace, 2005; Tracy, 2006; Redheuil et al., 2010). This theory is mathematically supported by engineering studies establishing that fatigue and fracture of 10% of natural rubber occurs at 8×10^8 extension-relaxation cycles, which is equivalent to 30 years at a heart rate of 70 beats/min (O'Rourke & Hashimoto, 2007). Biologically, a combination of imaging and histology studies have described age-associated increase in arterial thickening and a progressive reduction in aortic strain and distensibility, and have linked those changes to increased risk for cardiovascular disease (Lakatta & Levy, 2003; Lakatta, 2003; O'Rourke & Nichols, 2005; Redheuil et al., 2010). Although age-associated remodeling of arterial wall has been mostly described in patients with established risk for cardiovascular disease, few recent studies have shown similar age-related changes in healthy asymptomatic individuals (Redheuil et al., 2010). Similar age-related effects on arterial remodeling have been described in rodents and non-human primate without risk factors for cardiovascular disease, strengthening the hypothesis that aging *per se* can cause a series of alterations on mechanical properties that affect vascular function and lead to subsequent increased risk of cardiovascular disease.

Besides mechanical modifications, aging is also associated with several biochemical changes that are also implicated on the development and progression of cardiovascular disease. Dysfunction of both endothelial and smooth muscle molecular signaling appear to occur during aging process and favors vasospasm, thrombosis, inflammation and abnormal cell migration and proliferation (Lakatta, 2003; Briones et al., 2005; Barton, 2010; Herrera et al., 2010). The presence of endothelial dysfunction in the elder has been largely associated with malfunctioning of vascular tissue resulting, in turn, into cardiovascular disease (including atherosclerosis, hypertension or coronary artery disease) (Lakatta, 2003; Herrera et al., 2010), as well as renal dysfunction (Schmidt et al., 2001; Erdely et al., 2003), Alzheimer (Price et al., 2004) and erectile dysfunction (Burnett, 2006).

The mechanisms for age-associated endothelial dysfunction are multiple, though they are mostly associated to a decrease on nitric oxide (NO) bioavailability (Hayashi et al., 2008; Santhanam et al., 2008; Erusalimsky, 2009; Kim et al., 2009). NO is the major vascular messenger molecule involved in many physiological processes, including vasodilation and inhibition of thrombosis, cell migration and proliferation (Dudzinski & Michel, 2007; Lamas et al., 2007; Michel & Vanhoutte, 2010). Reduced endothelium-dependent and NO-mediated vasodilation has been described during aging in both human and animal models (Kim et al., 2009; Virdis et al., 2010).

A lower NO production in elderly may be based in either decreased NO synthesis or increased NO degradation. Several mechanisms to explain a reduction on NO production have been pointed out and include: 1) a decrease on the expression of endothelial NO synthase (eNOS) (Briones et al., 2005; Yoon et al., 2010); 2) a deficiency on NO precursor (L-arginine) (Santhanam et al., 2008) and eNOS cofactor (tetrahydrobiopterin - BH_4) (Yoshida et al., 2000; Eskurza et al., 2005); or 3) an increase of endogenous eNOS inhibitors (asymmetric dimethylarginine – ADMA) (Xiong et al., 2001; Kielstein et al., 2003). On the other hand, strong evidences support the hypothesis that age-associated increase in oxidative stress, and consequent production of superoxide anion (O_2^-) is a potent contributor to lowering NO bioavailability and increasing endothelial dysfunction (Jacobson et al., 2007; Rodriguez-Manas et al., 2009).

Despite the decline in NO bioavailability could sufficiently explain most of the changes in the functioning of vascular cells, other molecules that are crucial to control vascular function have also been described to be modified by aging. In the regulation of vasomotion, cyclooxygenase (COX)-derived factors are of particular importance as they control both vascular relaxation and contraction. Under normal condition, COX-derived relaxing (PGI_2) and contracting (TXA_2 and PGH_2) are in perfect balance, and few studies have reported a prevalence in the production of relaxing COX factors in the vasculature of young and healthy individuals. During aging, however, a swap in this balance favoring to the release of contracting factors occurs, leading to an increase of vascular contraction. Moreover, activation of inflammatory pathways in the vascular wall plays a central role in the process of vascular aging. Several studies have created an important link between arterial aging and a pro-inflammatory endothelial phenotype, even in the absence of traditional risk factors for atherosclerosis. An age–associated shift to a pro-inflammatory gene expression profile, known as endothelial activation, induces up-regulation of cellular adhesion molecules and cytokines which increases endothelial–leukocyte interactions and permeability, mechanisms considered crucial on the initial steps for the development of atherosclerosis (Herrera et al., 2010; Seals et al., 2011).

Even though endothelial function is undoubtedly impaired in the elderly, how aging affects molecular biochemistry of vascular cells is largely unknown. Going back to the observation that vascular aging is a consequence of mechanical fatigue, one might speculate that the mechanical forces on the vascular wall could contribute to the damage on endothelial cell functioning. In fact, it is well known that blood vessels are under constant mechanical loading from flowing blood which cause internal stresses, known as endothelial shear stress (caused by flow) and circumferential stretch (caused by pressure). These mechanical forces not only cause morphological changes of endothelium and blood vessel wall, but also trigger a myriad of intracellular events in endothelial cells and activate biochemical and biological events (Lu & Kassab, 2011). The triggering of endothelial signaling by mechanic forces seems to be mostly determined by the cytoskeleton, which represents a highly dynamic network that constantly assembles and disassembles, playing an active role in responding to mechanical stimuli (Wong et al., 1983). The cytoskeleton rearranges upon changes on stress and stretch and activates signaling molecules, such as NO production, that are capable to regulate vascular tone in order to keep homeostasis (Su et al., 2005; Su et al., 2007). An increase in arterial wall stiffening by aging could alter the impact of a mechanical stimulus, and therefore induce a significant reduction or dysfunction in the signaling pathways activated by shear stress (Kliche et al., 2011). In this regard, the chronically stiffed cells will lead to a decrease of NO, which will eventually lead to endothelial dysfunction.

Continuous damage to the endothelium from the daily pounding of the cycling pressure can also activate maintenance repair systems. When maintenance system is efficient (as in young individuals), endothelial cells likely correct the defect and keep going. On the other hand, when an irreversible damage occur or when endothelial cells are senescent, those inefficient cells are eventually eliminated by a mechanism yet to be described, while a "sister" circulating progenitor endothelial cells assume some repair function and will divide to fill up the gap (Thorin & Thorin-Trescases, 2009). Recent findings on progenitor stem cell research suggest that continuous division of progenitor endothelial cells for maintenance is likely the main response of an injured endothelium (Hill et al., 2003; Van Craenenbroeck & Conraads, 2010). Continuous cell division during life causes shortening in telomeres, a region of repetitive DNA sequences at the end of a chromosome, which protects the chromosomes from deterioration (Allsopp et al., 1995). Increasing evidence have support a role for reduction on telomere length with changes on cellular function and cellular senescence that may contribute to increased risk of vascular damage. In the long term, therefore, the regenerated endothelium may become dysfunctional as senescent endothelial cells start to express a pro-inflammatory, pro-oxidative, and pro-atherogenic phenotype (Chang & Harley, 1995; Bekaert et al., 2007; De Meyer T. et al., 2011).

In addition to mechanical fatigue, the vascular endothelium also undergoes important oxidative damage. The free-radical theory of aging states that organisms age because cells accumulate oxidative stress damage over time (de Grey, 2006; Camici et al., 2011). In other words, one can say that the body literally *"rusts"* with time. Growing evidence from research studies have supported this theory and have described an intimate relationship of increased oxidative stress with vascular dysfunction and increased risk for cardiovascular disease (Touyz, 2003; Griendling & Alexander, 1997; Harrison, 1997). Numerous studies underscore the importance of dysregulated oxidant and antioxidant balance in advancing age (Moon et al., 2001) and in the development and progression of atherosclerosis (Wassmann et al., 2004). Aging-associated increase in reactive oxygen species (ROS) are common to many species and despite decades of investigation, the mechanisms for the aging-related increase in ROS and how they affect vascular function have yet to be defined.

The main ROS proposed to be implicated on vascular aging process is the O_2^-. Increased O_2^- in the vessel wall has been well associated with decrease of NO bioavailability due to its rapid interaction and inactivation by O_2^-. In this regard, an increase of oxidative stress, and more specifically O_2^-, during aging could cause vascular damage simply by reducing the protective effect of NO in the vessel wall (Squadrito & Pryor, 1998; Harrison, 1997). However, increased oxidative stress has been implicated in more complex modulatory mechanisms that may affect vascular function by aging. Numerous studies have demonstrated that increase of oxidative stress contributes to the activation of transcriptional factors (such as NF-κB) that are key regulators of endothelial activation. By this way, aging-associated increase of ROS could favor endothelial cells to express a pro-inflammatory phenotype and increase the risk for cardiovascular disease (Herrera et al., 2010).

But proper vascular function does not lean on endothelium only. Vascular smooth muscle cells comprised by medial layer of blood vessels represent a dynamic component of the vasculature, and thus may also be affected by aging. In fact, vascular smooth muscle cells degenerate and decrease in number when subjects reach middle or advanced age. Smooth muscle cells are intercalated between the elastic lamina and the elastic fibers that also undergo a process of degeneration, thinning, sectioning, fracture and decrease in volume with aging. In parallel, there is a marked increase on collagen fibers, mucinous substrate, and calcification of the intercellular substrates begins (Toda et al., 1980).

Biochemical studies have shown that the content of elastin in human aorta decreases with age (Spina et al., 1983). Large amounts of elastin are produced during the fetal or neonatal period but not later (Godfrey et al., 1993). An age-related decrease in the cross-links in elastin contributes significantly to the reduction in arterial elasticity (Watanabe et al., 1996). As the turnover of elastin and collagen requires a very long period of time (lasting more than 10 years), these molecules are likely to undergo the addition of a sugar or a glycooxidative reaction. Thus, advanced glycation end-products accumulate in the arteries with age and partially contribute to age-related arterial stiffness (Konova et al., 2004; Semba et al., 2009). Type I, III, and V collagens are the major components of the collagen fibers of large conductance vessels such as aorta. During infancy or early childhood, collagen fibers are absent in the aorta and begin to accumulate with age; this process is known as fibrosis or sclerosis. Most studies have shown an age-related increase in the collagen content in the aorta (Spina et al., 1983) and increase in the number of collagen cross-links (Watanabe et al., 1996). Both an increase in the collagen content and the number of cross-links contributes significantly to the stiffening of the elastic arteries, namely atherosclerosis.

Senescent vascular smooth muscle cells have been shown to exhibit a pro-calcificatory/osteoblastic phenotype (Reid & Andersen, 1993; Burton, 2009; Nakano-Kurimoto et al., 2009), that could play a major role in the pathophysiology of age-related vascular calcification, a well-known major risk factor for the development of cardiovascular diseases (Adragao et al., 2004; Thompson & Partridge, 2004). Calcification in tunica media (medial calcification) increases throughout ageing, and accumulation of calcium in the elastin-rich layer of the media is ≥30-times more in the thoracic aorta at 90 years of age than that at 20 years of age (Elliott & McGrath, 1994). The underlying mechanisms that lead to the development of vascular calcification currently remain elusive. Calcification in the media usually occurs in the absence of macrophages and lipids, and is associated with α-smooth muscle actin-positive vascular smooth muscle cells, suggesting that vascular smooth muscle cells are the main key player in medial calcification (Luo et al., 1997). Alternatively, ROS may have some involvement in the osteoblastic transition of vascular smooth muscle cells (Byon et al., 2008).

Researchers have examined the role of the redox state in vascular smooth muscle cells in the pathogenesis of vascular disease (Clempus & Griendling, 2006; Lyle & Griendling, 2006). Vascular smooth muscle cells present in atherosclerotic lesions proliferate more rapidly and show increased expression of genes for growth factors and other molecules involved in extracellular matrix remodeling (Schwartz, 1997; Newby, 2006). Proliferation of vascular smooth muscle cells is part of the initiation and the progression of atherosclerosis (Ross, 1993) and may occur in response to injury or as a result of aberrant apoptosis (Clarke et al., 2006). Besides, vascular smooth muscle cells appear to undergo an age-associated phenotypic modulation toward a dedifferentiated and synthetic state. Smooth muscle cell migration from the medial to the intimal compartment is a plausible mechanism for the increased number of vascular smooth muscle cells within the diffusely thickened intima of central arteries as animals age (Miller et al., 2007).

In general, growth factors and hormones are the most potent activators that stimulate vascular smooth muscle growth, migration, and extracellular matrix synthesis. For instance, angiotensin II (Ang II) signaling has been widely linked to an age-associated increase in the migratory capacity of vascular smooth muscle cells and to the proinflammatory features of arterial aging. Ang II increases within the aged arterial wall and activates matrix metalloproteinase type II (MMP2) (Wang et al., 2003; Jiang et al., 2008). Ang II appears to initiate growth-promoting signal transduction through ROS-sensitive tyrosine kinases (Frank & Eguchi, 2003; Touyz et al., 2003).

3. Gender differences on vascular aging

Although arteries from females are so exposed to mechanical and oxidative damage as arteries from males, they seem do not follow the same time course for vascular aging, or at least, they do not age in the same way. Experimental and clinical studies support the hypothesis that men are hemodynamically older than age-matched, premenopausal women (Messerli et al., 1987; Bairey Merz et al., 2006; Shaw et al., 2006). With aging, the progression of cardiovascular disease occurs at an earlier age and become more severe in males compared to age-matched premenopausal females (Taddei et al., 1996; Virdis et al., 2010).

Arterial stiffening and distensibility are established markers for vascular aging and have been found to progressively increase with aging in both men and women. Studies in rodents indicate that there are gender differences in aging vessels, with stiffness increasing more in male than in females (Ruiz-Feria et al., 2009; Chan et al., 2011). Also in nonhuman primates, aortic stiffness has shown to be increased more in old male monkeys than in old females (Qiu et al., 2007). However, gender-associated relationship with those markers in humans remains unclear and currently limited studies have addressed to the evaluation of age-related vascular changes in man and women separately. Even though, many studies have performed their analysis on aging correlation with arterial stiffness and distensibility in men and women separately, their statistical models generally mask the gender differences in the influence of these variables (Breithaupt-Grogler & Belz, 1999; Segers et al., 2007; Redheuil et al., 2010; Miyoshi et al., 2011). In most clinical studies using small population group, the data do not provide sufficient power to detect significant gender-related differences in the rate of age-dependent change in vascular wall structure. The field still misses a large multi-centric populational study to identify whether aging-related effects are modulated by gender.

When it comes to the endothelium, sexual dimorphism on endothelial dysfunction and the progression of cardiovascular disease has also been well documented in various animal models (Ouchi et al., 1987; Ashton & Balment, 1991; Dantas et al., 2004a). With aging, males exhibit signals of impairment on endothelium-dependent relaxation at earlier age than do females (Kauser & Rubanyi, 1995; Huang et al., 1997). Thus far, the mechanisms better established to explain the gender- and aging-associated differences involve: 1) increased NO production by females (Huang et al., 1997); and 2) increased oxidative stress in male blood vessels (Dantas et al., 2004a). In this area of age-associated effects, a translation of animal models to humans can be performed. Early clinical studies on gender- and aging-related effects on endothelium-dependent relaxation in forearm blood flow have identified a constant age-related decline in maximal vasodilation to acetylcholine per year (Taddei et al., 1996). In contrast, women were found to show a slight decrease per year in vasodilation to acetylcholine up to middle-age (around 50's). After that, the vascular decline in the responses to the endothelium-dependent vasodilator hasten, and even decline more quickly in comparison with men (Taddei et al., 1996).

Gender modulation of vascular tone is also observed in functional studies. Contractile responses are greater in the aorta of male than female rats (Stallone et al., 1991; Crews et al., 1999; Tostes et al., 2000). These differences may be related to the vasodilatory effects of estrogens (Crews et al., 1999; Kanashiro & Khalil, 2001) through a direct action on vascular smooth muscle (Jiang et al., 1992; Mugge et al., 1993; Gerhard & Ganz, 1995; Crews & Khalil, 1999). Expression of estrogen receptors in smooth muscle may vary depending on the gender and the gonadal status (Tamaya et al., 1993). The decreased vascular responses to constrictors may be related to 1) the higher relative abundance of estrogen receptors in females arteries (Collins et al., 1995), 2) estrogen-induced down-regulation of gene expression of vasoconstrictor receptors, such as Ang II (Nickenig et al., 2000), and 3) signaling mechanisms of vascular smooth muscle contraction downstream from receptor activation.

As intracellular free Ca^{2+} concentration ($[Ca^{2+}]i$) is important for the initiation of smooth muscle contraction (Horowitz et al., 1996), several studies have used isolated vascular preparations and smooth muscle cells from control and gonadectomized male and female animals to investigate the effect of estrogen on $[Ca^{2+}]i$ and the Ca^{2+}-mobilization mechanisms (i.e. Ca^{2+} release from the intracellular stores and Ca^{2+} entry from the extracellular space) (Zhang et al., 1994; Crews & Khalil, 1999; Crews et al., 1999; Murphy & Khalil, 1999; Murphy & Khalil, 2000; Novella et al., 2010).

Taken together those studies can suggest that, with aging, women are more protected against its deleterious consequences in the cardiovascular system than men. After menopause, however, this protection seems to be lost, since the incidence of cardiovascular disease increases considerably to levels similar (or higher) to those found in men. Because the onset of menopause is marked by the loss of endogenous estrogen production from the ovaries, estrogen is felt to confer the premenopausal protection.

4. Vascular aging in females: Effects of estrogen on vascular function and aging

In women, arterial aging includes an aggravating risk factor in comparison to men. The decrease in estrogen production by menopause is thought to contribute to increased cardiovascular risk. Although aging *per se* has detrimental effects in the vasculature of middle aged female, these effects seem to be potentiated by the lack of estrogen with menopause, and restored by estrogen replacement (Harman, 2004; Stice et al., 2009; Novella et al., 2010). For this reason it is particularly difficult to distinguish what would be the contribution of aging and the lack of estrogen in the control of vascular function in menopausal women.

Epidemiological observations and extensive basic laboratory research has shown that female sex hormones, and more specifically estrogen, has direct beneficial effects in the cardiovascular system (Staessen et al., 1989; Dantas et al., 1999; Tostes et al., 2003; Dantas et al., 2004b; Hinojosa-Laborde et al., 2004). Estrogen has been described to display a myriad of metabolic, hemodynamic, and vascular effects, which have been largely associated to cardiovascular protection in females. For instance, estrogen can promote cardiovascular protection by indirectly influence on the metabolism of lipoproteins or directly by acting on the modulation of molecular pathways in the vessel wall (Miller & Duckles, 2008). Receptors for estrogen have been identified biochemically and show a plentiful expression in both vascular smooth muscle and endothelium, reinforcing the idea that estrogen play a key role in the control of vascular function (Couse et al., 1997; Pau et al., 1998; Arnal et al., 2010).

When considering the major structural changes caused by aging, cross-sectional studies have shown that postmenopausal females taking hormone replacement therapy present lower arterial stiffness compared with their peers not taken estrogen (Moreau et al., 2003; Sumino et al., 2005; Sumino et al., 2006). Besides, radial artery distensibility fluctuates in accordance with estrogen levels during menstrual cycles (Giannattasio et al., 1999). Basic research using animal models for estrogen withdrawn and aging have proposed that estrogen play a modulatory role in the molecular mechanisms to prevent stiffening of arterial wall. As mentioned above, content of collagen and elastin into arterial wall is a key factor that contributes to arterial wall thickening and stiffening, and is mostly regulated by the activity of matrix metalloproteinases (MMP), a family of enzymes capable of degrading components of the extracellular matrix. During aging there is a marked decrease of MMP activity which results in increase of collagen accumulation and consequent stiffening. Data from studies in female rodents have found that estrogen replacement in ovariectomized animals increases MMP activity and restores structural properties of aged arteries similar to that of the young group (Zhang et al., 2000). Altogether these studies suggest that estrogen can exert a favorable modulatory effect on arterial stiffness with aging in females.

Endothelial dysfunction secondary to estrogen deprivation has been largely described and has been mostly associated with reductions in NO availability. Estrogen is known to increases NO bioavailability by mechanisms that involve either increase of NO generation directly or by decreasing O_2^- concentration, and thereby attenuating O_2^- mediated inactivation of NO. The mechanisms involved in estrogen-induced increases in NO availability include: 1) transcriptional stimulation of endothelial NO synthase (eNOS) gene expression (Huang et al., 1997; Sumi & Ignarro, 2003); 2) non-genomic activation of enzyme activity via a phosphatidylinositol-3-OH kinase (PI3-kinase)/phosphokinase B (PKB/AKT) mediated signaling pathway (Hisamoto et al., 2001); 3) increased $[Ca^{2+}]_i$ in endothelial cells (Rubio-Gayosso et al., 2000); 4) decreased production of eNOS endogenous inhibitor, ADMA (Monsalve et al., 2007), and 5) attenuated O_2^- concentrations (Wassmann et al., 2001; Dantas et al., 2002; Ospina et al., 2002).

In addition to NO, actions of estrogen in the vasculature also influence the metabolism of other endothelium-derived factors (EDF). Estrogen has been described to positively up-regulate the production of endothelium-derived relaxing factors (EDRF), such as PGI_2 (Sobrino et al., 2009; Sobrino et al., 2010) and the endothelium-derived hyperpolarizing factors (EDHF) (Golding & Kepler, 2001), both of which are important mediators of vascular relaxation in resistance-sized arteries. Concomitantly, a modulating role of estrogen on constrictor factors (EDCF) is observed. Studies have shown that the beneficial effects of estrogen on the endothelium can be partially explained by an inhibitory effect on the production or action of the COX-derived vasoconstrictor agents (PGH$_2$ and TXA$_2$) (Davidge & Zhang, 1998; Dantas et al., 1999; Novella et al., 2010) and endothelin- 1 (ET-1) (David et al., 2001).

Estrogen has been shown to be a modulator of contractile responses by directly interfering with Ca^{2+} into the vascular smooth muscle cells. Although some studies have shown that estrogen does not inhibit Ca^{2+} release from the intracellular stores (Crews & Khalil, 1999; Murphy & Khalil, 1999), others have described that, supraphysiological concentrations of estrogen inhibit Ca^{2+} influx from the extracellular space (Han et al., 1995; Crews & Khalil, 1999; Murphy & Khalil, 1999) by inhibiting Ca^{2+} entry through voltage-gated Ca^{2+} channels (Freay et al., 1997; Kitazawa et al., 1997; Crews & Khalil, 1999; Murphy & Khalil, 1999). The expression of the L-type Ca^{2+} channels in cardiac muscle is substantially increased in estrogen receptor-deficient mice (Johnson et al., 1997), suggesting that estrogen may regulate Ca^{2+} mobilization by a receptor-mediated system.

Although a genomic action of physiological concentrations of estrogen on the expression of the Ca^{2+} channels may underlie the reduced cell contraction and [Ca^{2+}]i observed in vascular smooth muscle cells of females, it is less likely to account for the acute inhibitory effects of 17β-estradiol on cell contraction and [Ca^{2+}]i *in vitro*. The acute nature of the vasorelaxant effects of exogenous estrogen may represent additional non-genomic effects of estrogen on the mechanisms of Ca^{2+} entry into vascular smooth muscle (Kitazawa et al., 1997; Crews & Khalil, 1999; Murphy & Khalil, 1999). Whether estrogen inhibits Ca^{2+} entry by a direct or indirect action on plasmalemmal Ca^{2+} channels remains unclear. Some studies have shown that estrogen blocks Ca^{2+} channels in smooth muscle cells (Zhang et al., 1994; Nakajima et al., 1995) and others have shown that estrogen activates large conductance Ca^{2+}-activated K$^+$ channels, which could lead to hyperpolarization and decreased Ca^{2+} entry through voltage-gated channels (White et al., 1995; Wellman et al., 1996). Estrogen may also decrease [Ca^{2+}]i by stimulating Ca^{2+} extrusion via the plasmalemmal Ca^{2+} pump (Prakash et al., 1999). However, this mechanism seems less likely because the rate of decay of [Ca^{2+}]i transients in smooth muscle incubated in Ca^{2+}-free solution are not affected by estrogen (Crews & Khalil, 1999; Murphy & Khalil, 1999).

Other systems critically involved in the control of vascular function are also known to undergo estrogen modulation. For example, estrogen has been described to exert direct modulation on the components of renin-angiotensin system (RAS), which is a key regulator of blood pressure and smooth muscle cell growth. Estrogen reduces production of the active hormone of the RAS, Ang II in part, by inhibiting angiotensin-converting enzyme (ACE) expression. ACE activity in the circulation and in tissues, including the kidney and aorta, is reduced upon chronic estrogen replacement in animal models of menopause as well as in postmenopausal women (Brosnihan et al., 1999; Seely et al., 2004). Furthermore, estrogen attenuates the expression and tissue response to type 1 (AT$_1$) angiotensin receptor in several cardiovascular tissues including the aorta, heart and kidney (Silva-Antonialli et al., 2000; Wu et al., 2003).

Because increased oxidative stress play a crucial role on aging-associated vascular damage, numerous studies have assessed the antioxidant potential of estrogens. Basic research in human cultured endothelial cells revealed an antioxidant effect of estradiol (Hermenegildo et al., 2002a). In addition, clinical experimental studies have shown that different estrogens are capable of reducing oxidation of LDL- cholesterol and consequently the development of atherosclerosis (Keaney, Jr. et al., 1994; Shwaery et al., 1998; Hermenegildo et al., 2001; Hermenegildo et al., 2002b). In addition to its antioxidant role, estradiol exerts a direct effect by restoring the ADMA levels rise induced by oxidized LDL in human cultured endothelial cells acting through estrogen receptor α. Estrogen also attenuates the deleterious effects induced by increased generation of ROS follow ischemia/reperfusion in distinct research models (Kim et al., 1996; Kim et al., 2006; Guo et al., 2010).

As a result of their phenolic molecular structure, several estrogens, such as 17β-estradiol, estrone or estriol, have been described to act as ROS scavengers by virtue of the hydrogen-donating capacity of their phenolic groups (Halliwell & Grootveld, 1987; Dubey & Jackson, 2001). However, in these studies the direct effect of estrogens as scavenger can only be observed at concentrations above 1 micromolar (Arnal et al., 1996; Kim et al., 1996). Considering that plasma concentrations of estrogen in physiological conditions are in the nanomolar range is likely that the direct action as a scavenger is not the main anti-oxidant mechanism by estrogen. In fact, studies have established that estrogen modulates ROS concentration a mechanism that involves interaction with its nuclear receptor to decrease oxidative proteins and/or increase antioxidant enzymes expression. Many studies have shown that changes in estrogen levels are associated with altered levels of anti-oxidant enzymes including glutathione peroxidase, catalase and superoxide dismutase (Capel et al., 1981; Robb & Stuart, 2011; Sivritas et al., 2011). Moreover, recent studies have shown a modulatory effect of estrogen on O_2^-, via modulation of NADH/NADPH oxidases and AT_1 receptor gene expression (major sources of O_2^- production) (Wassmann et al., 2001; Dantas et al., 2002).

Among all research on cellular aging process and its complication, there is a growing interest on mechanisms to delay or decrease telomere shortening by aging, and therefore, keeping cellular integrity and function (Allsopp et al., 1995). In this sense, few studies have explored the effects of estrogen on telomere shortening, and even fewer have addressed this issue in association with vascular aging. Mechanistic studies have found that estrogen treatment up-regulates transcription of hTERT, the catalytic subunit of human telomerase, in distinct cell lines, including endothelial cells (Farsetti et al., 2009). Intriguingly, activation of hTERT by NO signaling has also been reported (Vasa et al., 2000). Considering that estrogen augments NO production, one can suggest that estrogens doubly prevent vascular senescence: by directly interacting with its receptor and by increasing NO.

Although estrogen modulates several mechanisms that are closely associated with vascular aging, assuming that estrogen put a break on vascular aging in females would be rather speculative. There is no sufficient data available to correlate estrogen levels with a delay on progression of vascular aging and recent clinical trials have questioned the value of estrogen replacement therapy in protecting vascular function. The benefits of hormone replacement therapy on the life expectancy and vascular health of women have dramatically lost consensus since publication of the results of the Women's Health Initiative study (WHI) (Rossouw et al., 2002). The WHI trial did not find any cardiovascular benefit from estrogen in postmenopausal women and in fact, showed hormone replacement therapy was associated with increased risk to the cardiovascular system (Rossouw et al., 2002).

There is much controversy over the interpretation of WHI. Concerns raised include that the estrogens used in those trials are not naturally occurring and thus would not act identically to natural estrogens. Most importantly was the fact that the WHI, as well as the majority of clinical trial on hormone replacement therapy, studied a population of women that were estrogen deficient for, on average, 10 years before hormone replacement was initiated. Currently, it is not known if the vascular effects of estrogen are modified by aging in females. These observations, together with observational studies, have led scientists to create the so-called "timing hypothesis". This theory states that estrogen-mediated benefits to prevent cardiovascular disease only occur when treatment is initiated before the detrimental effects of aging are established on vascular wall (Harman, 2006). In this regard, few recent basic studies have shown that aging is associated with significant reductions in the direct estrogen-mediated mechanisms of vascular relaxation (Wynne et al., 2004; LeBlanc et al., 2009; Lekontseva et al., 2010). The lack of estrogen responses in those animals was not related to age-associated changes in the plasma levels of estrogen or activity of estrogen receptors, but rather by possible age-related changes in estrogen-mediated signaling pathways in the vasculature.

Moreover, recent clinical studies have revealed that different risk factors for cardiovascular disease in postmenopausal women were lower among women 50 to 59 years old at enrolment for estrogen replacement therapy (Manson et al., 2007; Sherwood et al., 2007). Nevertheless, the field lacks detailed research on the long-term effects by estrogen and how it modulates cardiovascular function during aging. It remains unclear to what extent the protective effects of estrogen replacement well described in young females can be extrapolated to older ones. The aging issue still needs to be addressed in both experimental and clinical studies, and together, these studies demonstrate that estrogen has complex biologic effects and may influence the risk of cardiovascular events and other outcomes through multiple pathways. Therefore aging of a giving organism should always be taken into account when the pharmacological and physiological responses by estrogens are determined.

5. Conclusion

We live in an aging society, with life expectancy far greater today than a century ago. The increasing incidence of older-age people in our society represents the culmination of centuries of medical, scientific, and social accomplishments. The challenge for modern medicine is how to increase the number of disease-free years in elderly people and improve quality of life in later years. However, a disproportionate number of people who reach old age suffer from cardiovascular diseases.

Clinical and basic studies have established that vascular aging in women does not follow the same chronology as in men. Men display a pattern of progressive vascular aging, while timing for vascular aging in women presents a clear hallmark, i.e. menopause. Several studies have shown that the incidence of cardiovascular diseases in premenopausal women is markedly low compared to age-matched men. After menopause, however, these figures increase to values that are close, or even higher, to those found in men. Cardiovascular disease is the primary cause of death among women after menopause. Despite this, there is still a concerning gap in the knowledge, understanding, and general awareness of mechanisms for cardiovascular aging in women.

It has become apparent that to improve diagnosis and treatment of vascular aging, the gender differences in cardiovascular control must be addressed. The impact of the menstrual cycle and hormonal replacement therapy on vascular function of females should also be taken into consideration. Different strategies have shown benefit in preventing, delaying or attenuating vascular aging. Nevertheless, it yet remains to be fully demonstrated whether vascular aging can be pharmacologically prevented. Further research efforts are needed to understand the causes and consequences of female vascular aging and propose new therapeutic strategies for the management of vascular senescence in women.

6. Acknowledgment

This work was supported by the Spanish Ministerio de Ciencia e Innovación, Instituto de Salud Carlos III - FEDER-ERDF (grants FIS PI10/00518, FIS PI080176 and Red HERACLES RD06/0009), Consellería de Sanidad, Generalitat Valenciana (grants AP 097/2011, AP 104/2011 and GE 027/2011), and Spanish Society of Cardiology (DN040480).

7. References

Adragao, T.; Pires, A.; Lucas, C.; Birne, R.; Magalhaes, L.; Goncalves, M., & Negrao, A.P. (2004). A simple vascular calcification score predicts cardiovascular risk in haemodialysis patients. *Nephrol Dial Transplant*, 19, 1480-1488.

Allsopp, R.C.; Chang, E.; Kashefi-Aazam, M.; Rogaev, E.I.; Piatyszek, M.A.; Shay, J.W., & Harley, C.B. (1995). Telomere shortening is associated with cell division in vitro and in vivo. *Exp Cell Res*, 220, 194-200.

Arnal, J.F.; Clamens, S.; Pechet, C.; Negre-Salvayre, A.; Allera, C.; Girolami, J.P.; Salvayre, R., & Bayard, F. (1996). Ethinylestradiol does not enhance the expression of nitric oxide synthase in bovine endothelial cells but increases the release of bioactive nitric oxide by inhibiting superoxide anion production. *Proc Natl Acad Sci U S A*, 93, 4108-4113.

Arnal, J.F.; Fontaine, C.; Billon-Gales, A.; Favre, J.; Laurell, H.; Lenfant, F., & Gourdy, P. (2010). Estrogen receptors and endothelium. *Arterioscler Thromb Vasc Biol*, 30, 1506-1512.

Ashton, N., & Balment, R.J. (1991). Sexual dimorphism in renal function and hormonal status of New Zealand genetically hypertensive rats. *Acta Endocrinol (Copenh)*, 124, 91-97.

Avolio, A.P.; Chen, S.G.; Wang, R.P.; Zhang, C.L.; Li, M.F., & O'Rourke, M.F. (1983). Effects of aging on changing arterial compliance and left ventricular load in a northern Chinese urban community. *Circulation*, 68, 50-58.

Avolio, A.P.; Deng, F.Q.; Li, W.Q.; Luo, Y.F.; Huang, Z.D.; Xing, L.F., & O'Rourke, M.F. (1985). Effects of aging on arterial distensibility in populations with high and low prevalence of hypertension: comparison between urban and rural communities in China. *Circulation*, 71, 202-210.

Bairey Merz, C.N.; Shaw, L.J.; Reis, S.E.; Bittner, V.; Kelsey, S.F.; Olson, M.; Johnson, B.D.; Pepine, C.J.; Mankad, S.; Sharaf, B.L.; Rogers, W.J.; Pohost, G.M.; Lerman, A.; Quyyumi, A.A., & Sopko, G. (2006). Insights from the NHLBI-Sponsored Women's Ischemia Syndrome Evaluation (WISE) Study: Part II: gender differences in presentation, diagnosis, and outcome with regard to gender-based pathophysiology of atherosclerosis and macrovascular and microvascular coronary disease. *J Am Coll Cardiol*, 47, S21-S29.

Barton, M. (2010). Obesity and aging: determinants of endothelial cell dysfunction and atherosclerosis. *Pflugers Arch*, 460, 825-837.

Bekaert, S.; De Meyer T.; Rietzschel, E.R.; De Buyzere, M.L.; De Bacquer D.; Langlois, M.; Segers, P.; Cooman, L.; Van Damme P.; Cassiman, P.; Van Criekinge W.; Verdonck, P.; De Backer, G.G.; Gillebert, T.C., & Van Oostveldt P. (2007). Telomere length and cardiovascular risk factors in a middle-aged population free of overt cardiovascular disease. *Aging Cell*, 6, 639-647.

Breithaupt-Grogler, K., & Belz, G.G. (1999). Epidemiology of the arterial stiffness. *Pathol Biol (Paris)*, 47, 604-613.

Briones, A.M.; Montoya, N.; Giraldo, J., & Vila, E. (2005). Ageing affects nitric oxide synthase, cyclooxygenase and oxidative stress enzymes expression differently in mesenteric resistance arteries. *Auton Autacoid Pharmacol*, 25, 155-162.

Brosnihan, K.B.; Senanayake, P.S.; Li, P., & Ferrario, C.M. (1999). Bi-directional actions of estrogen on the renin-angiotensin system. *Braz J Med Biol Res*, 32, 373-381.

Bucciarelli, P., & Mannucci, P.M. (2009). The hemostatic system through aging and menopause. *Climacteric*, 12 Suppl 1, 47-51.

Burnett, A.L. (2006). The role of nitric oxide in erectile dysfunction: implications for medical therapy. *J Clin Hypertens (Greenwich)*, 8, 53-62.

Burton, D.G. (2009). Cellular senescence, ageing and disease. *Age (Dordr)*, 31, 1-9.

Byon, C.H.; Javed, A.; Dai, Q.; Kappes, J.C.; Clemens, T.L.; Darley-Usmar, V.M.; McDonald, J.M., & Chen, Y. (2008). Oxidative stress induces vascular calcification through modulation of the osteogenic transcription factor Runx2 by AKT signaling. *J Biol Chem*, 283, 15319-15327.

Camici, G.G.; Shi, Y.; Cosentino, F.; Francia, P., & Luscher, T.F. (2011). Anti-aging medicine: molecular basis for endothelial cell-targeted strategies - a mini-review. *Gerontology*, 57, 101-108.

Capel, I.D.; Jenner, M.; Williams, D.C.; Donaldson, D., & Nath, A. (1981). The effect of prolonged oral contraceptive steroid use on erythrocyte glutathione peroxidase activity. *J Steroid Biochem*, 14, 729-732.

Chan, V.; Fenning, A.; Levick, S.P.; Loch, D.; Chunduri, P.; Iyer, A.; Teo, Y.L.; Hoey, A.; Wilson, K.; Burstow, D., & Brown, L. (2011). Cardiovascular changes during maturation and ageing in male and female spontaneously hypertensive rats. *J Cardiovasc Pharmacol*, 57, 469-478.

Chang, E., & Harley, C.B. (1995). Telomere length and replicative aging in human vascular tissues. *Proc Natl Acad Sci U S A*, 92, 11190-11194.

Clarke, M.C.; Figg, N.; Maguire, J.J.; Davenport, A.P.; Goddard, M.; Littlewood, T.D., & Bennett, M.R. (2006). Apoptosis of vascular smooth muscle cells induces features of plaque vulnerability in atherosclerosis. *Nat Med*, 12, 1075-1080.

Clempus, R.E., & Griendling, K.K. (2006). Reactive oxygen species signaling in vascular smooth muscle cells. *Cardiovasc Res*, 71, 216-225.

Collins, P.; Rosano, G.M.; Sarrel, P.M.; Ulrich, L.; Adamopoulos, S.; Beale, C.M.; McNeill, J.G., & Poole-Wilson, P.A. (1995). 17β-Estradiol attenuates acetylcholine-induced coronary arterial constriction in women but not men with coronary heart disease. *Circulation*, 92, 24-30.

Cooper, L.T.; Cooke, J.P., & Dzau, V.J. (1994). The vasculopathy of aging. *J Gerontol*, 49, B191-B196.

Couse, J.F.; Lindzey, J.; Grandien, K.; Gustafsson, J.A., & Korach, K.S. (1997). Tissue distribution and quantitative analysis of estrogen receptor-alpha (ERα) and estrogen receptor-beta (ERβ) messenger ribonucleic acid in the wild-type and ERα-knockout mouse. *Endocrinology*, 138, 4613-4621.

Crews, J.K., & Khalil, R.A. (1999). Antagonistic effects of 17β-estradiol, progesterone, and testosterone on Ca^{2+} entry mechanisms of coronary vasoconstriction. *Arterioscler Thromb Vasc Biol*, 19, 1034-1040.

Crews, J.K.; Murphy, J.G., & Khalil, R.A. (1999). Gender differences in Ca^{2+} entry mechanisms of vasoconstriction in Wistar-Kyoto and spontaneously hypertensive rats. *Hypertension*, 34, 931-936.

Dantas, A.P.; Franco, M.C.; Silva-Antonialli, M.M.; Tostes, R.C.; Fortes, Z.B.; Nigro, D., & Carvalho, M.H. (2004a). Gender differences in superoxide generation in microvessels of hypertensive rats: role of NAD(P)H-oxidase. *Cardiovasc Res*, 61, 22-29.

Dantas, A.P.; Franco, M.C.; Tostes, R.C.; Fortes, Z.B.; Costa, S.G.; Nigro, D., & Carvalho, M.H. (2004b). Relative contribution of estrogen withdrawal and gonadotropins increase secondary to ovariectomy on prostaglandin generation in mesenteric microvessels. *J Cardiovasc Pharmacol*, 43, 48-55.

Dantas, A.P.; Scivoletto, R.; Fortes, Z.B.; Nigro, D., & Carvalho, M.H. (1999). Influence of female sex hormones on endothelium-derived vasoconstrictor prostanoid generation in microvessels of spontaneously hypertensive rats. *Hypertension*, 34, 914-919.

Dantas, A.P.; Tostes, R.C.; Fortes, Z.B.; Costa, S.G.; Nigro, D., & Carvalho, M.H. (2002). In vivo evidence for antioxidant potential of estrogen in microvessels of female spontaneously hypertensive rats. *Hypertension*, 39, 405-411.

David, F.L.; Carvalho, M.H.; Cobra, A.L.; Nigro, D.; Fortes, Z.B.; Reboucas, N.A., & Tostes, R.C. (2001). Ovarian hormones modulate endothelin-1 vascular reactivity and mRNA expression in DOCA-salt hypertensive rats. *Hypertension*, 38, 692-696.

Davidge, S.T., & Zhang, Y. (1998). Estrogen replacement suppresses a prostaglandin H synthase-dependent vasoconstrictor in rat mesenteric arteries. *Circ Res*, 83, 388-395.

de Grey, A.D. (2006). Free radicals in aging: causal complexity and its biomedical implications. *Free Radic Res*, 40, 1244-1249.

De Meyer T.; Rietzschel, E.R.; De Buyzere, M.L.; Van Criekinge W., & Bekaert, S. (2011). Telomere length and cardiovascular aging: the means to the ends? *Ageing Res Rev*, 10, 297-303.

Dubey, R.K., & Jackson, E.K. (2001). Estrogen-induced cardiorenal protection: potential cellular, biochemical, and molecular mechanisms. *Am J Physiol Renal Physiol*, 280, F365-F388.

Dudzinski, D.M., & Michel, T. (2007). Life history of eNOS: partners and pathways. *Cardiovasc Res*, 75, 247-260.

Eaker, E.; Chesebro, J.H.; Sacks, F.M.; Wenger, N.K.; Whisnant, J.P., & Winston, M. (1994). Special report: cardiovascular disease in women. Special writing group. *Heart Dis Stroke*, 3, 114-119.

Eaker, E.D.; Chesebro, J.H.; Sacks, F.M.; Wenger, N.K.; Whisnant, J.P., & Winston, M. (1993). Cardiovascular disease in women. *Circulation*, 88, 1999-2009.

Elliott, R.J., & McGrath, L.T. (1994). Calcification of the human thoracic aorta during aging. *Calcif Tissue Int*, 54, 268-273.

Erdely, A.; Greenfeld, Z.; Wagner, L., & Baylis, C. (2003). Sexual dimorphism in the aging kidney: Effects on injury and nitric oxide system. *Kidney Int*, 63, 1021-1026.

Erusalimsky, J.D. (2009). Vascular endothelial senescence: from mechanisms to pathophysiology. *J Appl Physiol*, 106, 326-332.

Eskurza, I.; Myerburgh, L.A.; Kahn, Z.D., & Seals, D.R. (2005). Tetrahydrobiopterin augments endothelium-dependent dilatation in sedentary but not in habitually exercising older adults. *J Physiol*, 568, 1057-1065.

Farsetti, A.; Grasselli, A.; Bacchetti, S.; Gaetano, C., & Capogrossi, M.C. (2009). The telomerase tale in vascular aging: regulation by estrogens and nitric oxide signaling. *J Appl Physiol*, 106, 333-337.

Frank, G.D., & Eguchi, S. (2003). Activation of tyrosine kinases by reactive oxygen species in vascular smooth muscle cells: significance and involvement of EGF receptor transactivation by angiotensin II. *Antioxid Redox Signal*, 5, 771-780.

Freay, A.D.; Curtis, S.W.; Korach, K.S., & Rubanyi, G.M. (1997). Mechanism of vascular smooth muscle relaxation by estrogen in depolarized rat and mouse aorta. Role of nuclear estrogen receptor and Ca^{2+} uptake. *Circ Res*, 81, 242-248.

Gerhard, M., & Ganz, P. (1995). How do we explain the clinical benefits of estrogen? From bedside to bench. *Circulation*, 92, 5-8.

Giannattasio, C.; Failla, M.; Grappiolo, A.; Stella, M.L.; Del, B.A.; Colombo, M., & Mancia, G. (1999). Fluctuations of radial artery distensibility throughout the menstrual cycle. *Arterioscler Thromb Vasc Biol*, 19, 1925-1929.

Godfrey, M.; Nejezchleb, P.A.; Schaefer, G.B.; Minion, D.J.; Wang, Y., & Baxter, B.T. (1993). Elastin and fibrillin mRNA and protein levels in the ontogeny of normal human aorta. *Connect Tissue Res*, 29, 61-69.

Golding, E.M., & Kepler, T.E. (2001). Role of estrogen in modulating EDHF-mediated dilations in the female rat middle cerebral artery. *Am J Physiol Heart Circ Physiol*, 280, H2417-H2423.

Griendling, K.K., & Alexander, R.W. (1997). Oxidative stress and cardiovascular disease. *Circulation*, 96, 3264-3265.

Guo, J.; Krause, D.N.; Horne, J.; Weiss, J.H.; Li, X., & Duckles, S.P. (2010). Estrogen-receptor-mediated protection of cerebral endothelial cell viability and mitochondrial function after ischemic insult in vitro. *J Cereb Blood Flow Metab*, 30, 545-554.

Halliwell, B., & Grootveld, M. (1987). The measurement of free radical reactions in humans. Some thoughts for future experimentation. *FEBS Lett*, 213, 9-14.

Han, S.Z.; Karaki, H.; Ouchi, Y.; Akishita, M., & Orimo, H. (1995). 17β-Estradiol inhibits Ca^{2+} influx and Ca^{2+} release induced by thromboxane A_2 in porcine coronary artery. *Circulation*, 91, 2619-2626.

Harman, S.M. (2004). What do hormones have to do with aging? What does aging have to do with hormones? *Ann N Y Acad Sci*, 1019, 299-308.

Harman, S.M. (2006). Estrogen replacement in menopausal women: recent and current prospective studies, the WHI and the KEEPS. *Gend Med*, 3, 254-269.

Harrison, D.G. (1997). Endothelial function and oxidant stress. *Clin Cardiol*, 20, II-7.

Hayashi, T.; Yano, K.; Matsui-Hirai, H.; Yokoo, H.; Hattori, Y., & Iguchi, A. (2008). Nitric oxide and endothelial cellular senescence. *Pharmacol Ther*, 120, 333-339.

Hermenegildo, C.; Garcia-Martinez, M.C.; Tarin, J.J., & Cano, A. (2002a). Estradiol reduces $F_{2\alpha}$-isoprostane production in cultured human-endothelial cells. *Am J Physiol Heart Circ Physiol*, 283, H2644-H2649.

Hermenegildo, C.; Garcia-Martinez, M.C.; Tarin, J.J.; Llacer, A., & Cano, A. (2001). The effect of oral hormone replacement therapy on lipoprotein profile, resistance of LDL to oxidation and LDL particle size. *Maturitas*, 38, 287-295.

Hermenegildo, C.; Garcia-Martinez, M.C.; Valldecabres, C.; Tarin, J.J., & Cano, A. (2002b). Transdermal estradiol reduces plasma myeloperoxidase levels without affecting the LDL resistance to oxidation or the LDL particle size. *Menopause*, 9, 102-109.

Herrera, M.D.; Mingorance, C.; Rodriguez-Rodriguez, R., & Alvarez de Sotomayor M. (2010). Endothelial dysfunction and aging: an update. *Ageing Res Rev*, 9, 142-152.

Hill, J.M.; Zalos, G.; Halcox, J.P.; Schenke, W.H.; Waclawiw, M.A.; Quyyumi, A.A., & Finkel, T. (2003). Circulating endothelial progenitor cells, vascular function, and cardiovascular risk. *N Engl J Med*, 348, 593-600.

Hinojosa-Laborde, C.; Craig, T.; Zheng, W.; Ji, H.; Haywood, J.R., & Sandberg, K. (2004). Ovariectomy augments hypertension in aging female Dahl salt-sensitive rats. *Hypertension*, 44, 405-409.

Hisamoto, K.; Ohmichi, M.; Kurachi, H.; Hayakawa, J.; Kanda, Y.; Nishio, Y.; Adachi, K.; Tasaka, K.; Miyoshi, E.; Fujiwara, N.; Taniguchi, N., & Murata, Y. (2001). Estrogen induces the Akt-dependent activation of endothelial nitric-oxide synthase in vascular endothelial cells. *J Biol Chem*, 276, 3459-3467.

Horowitz, A.; Menice, C.B.; Laporte, R., & Morgan, K.G. (1996). Mechanisms of smooth muscle contraction. *Physiol Rev*, 76, 967-1003.

Huang, A.; Sun, D.; Kaley, G., & Koller, A. (1997). Estrogen maintains nitric oxide synthesis in arterioles of female hypertensive rats. *Hypertension*, 29, 1351-1356.

Jacobson, A.; Yan, C.; Gao, Q.; Rincon-Skinner, T.; Rivera, A.; Edwards, J.; Huang, A.; Kaley, G., & Sun, D. (2007). Aging enhances pressure-induced arterial superoxide formation. *Am J Physiol Heart Circ Physiol*, 293, H1344-H1350.

Jiang, C.; Sarrel, P.M.; Poole-Wilson, P.A., & Collins, P. (1992). Acute effect of 17β-estradiol on rabbit coronary artery contractile responses to endothelin-1. *Am J Physiol*, 263, H271-H275.

Jiang, L.; Wang, M.; Zhang, J.; Monticone, R.E.; Telljohann, R.; Spinetti, G.; Pintus, G., & Lakatta, E.G. (2008). Increased aortic calpain-1 activity mediates age-associated angiotensin II signaling of vascular smooth muscle cells. *PLoS One*, 3, e2231.

Johnson, B.D.; Zheng, W.; Korach, K.S.; Scheuer, T.; Catterall, W.A., & Rubanyi, G.M. (1997). Increased expression of the cardiac L-type calcium channel in estrogen receptor-deficient mice. *J Gen Physiol*, 110, 135-140.

Kanashiro, C.A., & Khalil, R.A. (2001). Gender-related distinctions in protein kinase C activity in rat vascular smooth muscle. *Am J Physiol Cell Physiol*, 280, C34-C45.

Kauser, K., & Rubanyi, G.M. (1995). Gender difference in endothelial dysfunction in the aorta of spontaneously hypertensive rats. *Hypertension*, 25, 517-523.

Keaney, J.F., Jr.; Shwaery, G.T.; Xu, A.; Nicolosi, R.J.; Loscalzo, J.; Foxall, T.L., & Vita, J.A. (1994). 17β-estradiol preserves endothelial vasodilator function and limits low-density lipoprotein oxidation in hypercholesterolemic swine. *Circulation*, 89, 2251-2259.

Kielstein, J.T.; Bode-Boger, S.M.; Frolich, J.C.; Ritz, E.; Haller, H., & Fliser, D. (2003). Asymmetric dimethylarginine, blood pressure, and renal perfusion in elderly subjects. *Circulation*, 107, 1891-1895.

Kim, J.; Kil, I.S.; Seok, Y.M.; Yang, E.S.; Kim, D.K.; Lim, D.G.; Park, J.W.; Bonventre, J.V., & Park, K.M. (2006). Orchiectomy attenuates post-ischemic oxidative stress and ischemia/reperfusion injury in mice. A role for manganese superoxide dismutase. *J Biol Chem*, 281, 20349-20356.

Kim, J.H.; Bugaj, L.J.; Oh, Y.J.; Bivalacqua, T.J.; Ryoo, S.; Soucy, K.G.; Santhanam, L.; Webb, A.; Camara, A.; Sikka, G.; Nyhan, D.; Shoukas, A.A.; Ilies, M.; Christianson, D.W.; Champion, H.C., & Berkowitz, D.E. (2009). Arginase inhibition restores NOS coupling and reverses endothelial dysfunction and vascular stiffness in old rats. *J Appl Physiol*, 107, 1249-1257.

Kim, Y.D.; Chen, B.; Beauregard, J.; Kouretas, P.; Thomas, G.; Farhat, M.Y.; Myers, A.K., & Lees, D.E. (1996). 17 beta-Estradiol prevents dysfunction of canine coronary endothelium and myocardium and reperfusion arrhythmias after brief ischemia/reperfusion. *Circulation*, 94, 2901-2908.

Kitazawa, T.; Hamada, E.; Kitazawa, K., & Gaznabi, A.K. (1997). Non-genomic mechanism of 17β-oestradiol-induced inhibition of contraction in mammalian vascular smooth muscle. *J Physiol*, 499 (Pt 2), 497-511.

Kliche, K.; Jeggle, P.; Pavenstadt, H., & Oberleithner, H. (2011). Role of cellular mechanics in the function and life span of vascular endothelium. *Pflugers Arch*, 462, 209-217.

Konova, E.; Baydanoff, S.; Atanasova, M., & Velkova, A. (2004). Age-related changes in the glycation of human aortic elastin. *Exp Gerontol*, 39, 249-254.

Lakatta, E.G. (2003). Arterial and cardiac aging: major shareholders in cardiovascular disease enterprises: Part III: cellular and molecular clues to heart and arterial aging. *Circulation*, 107, 490-497.

Lakatta, E.G., & Levy, D. (2003). Arterial and cardiac aging: major shareholders in cardiovascular disease enterprises: Part I: aging arteries: a "set up" for vascular disease. *Circulation*, 107, 139-146.

Lamas, S.; Lowenstein, C.J., & Michel, T. (2007). Nitric oxide signaling comes of age: 20 years and thriving. *Cardiovasc Res*, 75, 207-209.

LeBlanc, A.J.; Reyes, R.; Kang, L.S.; Dailey, R.A.; Stallone, J.N.; Moningka, N.C., & Muller-Delp, J.M. (2009). Estrogen replacement restores flow-induced vasodilation in coronary arterioles of aged and ovariectomized rats. *Am J Physiol Regul Integr Comp Physiol*, 297, R1713-R1723.

Lekontseva, O.N.; Rueda-Clausen, C.F.; Morton, J.S., & Davidge, S.T. (2010). Ovariectomy in aged versus young rats augments matrix metalloproteinase-mediated vasoconstriction in mesenteric arteries. *Menopause*, 17, 516-523.

Lerner, D.J., & Kannel, W.B. (1986). Patterns of coronary heart disease morbidity and mortality in the sexes: a 26-year follow-up of the Framingham population. *Am Heart J*, 111, 383-390.

Lu, D., & Kassab, G.S. (2011). Role of shear stress and stretch in vascular mechanobiology. *J R Soc Interface*, 8, 1379-1385.

Luo, G.; Ducy, P.; McKee, M.D.; Pinero, G.J.; Loyer, E.; Behringer, R.R., & Karsenty, G. (1997). Spontaneous calcification of arteries and cartilage in mice lacking matrix GLA protein. *Nature*, 386, 78-81.

Lyle, A.N., & Griendling, K.K. (2006). Modulation of vascular smooth muscle signaling by reactive oxygen species. *Physiology (Bethesda)*, 21, 269-280.

Manson, J.E.; Allison, M.A.; Rossouw, J.E.; Carr, J.J.; Langer, R.D.; Hsia, J.; Kuller, L.H.; Cochrane, B.B.; Hunt, J.R.; Ludlam, S.E.; Pettinger, M.B.; Gass, M.; Margolis, K.L.; Nathan, L.; Ockene, J.K.; Prentice, R.L.; Robbins, J., & Stefanick, M.L. (2007). Estrogen therapy and coronary-artery calcification. *N Engl J Med*, 356, 2591-2602.

Messerli, F.H.; Garavaglia, G.E.; Schmieder, R.E.; Sundgaard-Riise, K.; Nunez, B.D., & Amodeo, C. (1987). Disparate cardiovascular findings in men and women with essential hypertension. *Ann Intern Med*, 107, 158-161.

Michel, T., & Vanhoutte, P.M. (2010). Cellular signaling and NO production. *Pflugers Arch*, 459, 807-816.

Miller, S.J.; Watson, W.C.; Kerr, K.A.; Labarrere, C.A.; Chen, N.X.; Deeg, M.A., & Unthank, J.L. (2007). Development of progressive aortic vasculopathy in a rat model of aging. *Am J Physiol Heart Circ Physiol*, 293, H2634-H2643.

Miller, V.M., & Duckles, S.P. (2008). Vascular actions of estrogens: functional implications. *Pharmacol Rev*, 60, 210-241.

Miyoshi, H.; Mizuguchi, Y.; Oishi, Y.; Iuchi, A.; Nagase, N.; Ara, N., & Oki, T. (2011). Early detection of abnormal left atrial-left ventricular-arterial coupling in preclinical patients with cardiovascular risk factors: evaluation by two-dimensional speckle-tracking echocardiography. *Eur J Echocardiogr*, 12, 431-439.

Monsalve, E.; Oviedo, P.J.; Garcia-Perez, M.A.; Tarin, J.J.; Cano, A., & Hermenegildo, C. (2007). Estradiol counteracts oxidized LDL-induced asymmetric dimethylarginine production by cultured human endothelial cells. *Cardiovasc Res*, 73, 66-72.

Moon, S.K.; Thompson, L.J.; Madamanchi, N.; Ballinger, S.; Papaconstantinou, J.; Horaist, C.; Runge, M.S., & Patterson, C. (2001). Aging, oxidative responses, and proliferative capacity in cultured mouse aortic smooth muscle cells. *Am J Physiol Heart Circ Physiol*, 280, H2779-H2788.

Moreau, K.L.; Donato, A.J.; Seals, D.R.; DeSouza, C.A., & Tanaka, H. (2003). Regular exercise, hormone replacement therapy and the age-related decline in carotid arterial compliance in healthy women. *Cardiovasc Res*, 57, 861-868.

Mugge, A.; Riedel, M.; Barton, M.; Kuhn, M., & Lichtlen, P.R. (1993). Endothelium independent relaxation of human coronary arteries by 17β-oestradiol in vitro. *Cardiovasc Res*, 27, 1939-1942.

Murphy, J.G., & Khalil, R.A. (1999). Decreased $[Ca^{2+}]_i$ during inhibition of coronary smooth muscle contraction by 17β-estradiol, progesterone, and testosterone. *J Pharmacol Exp Ther*, 291, 44-52.

Murphy, J.G., & Khalil, R.A. (2000). Gender-specific reduction in contractility and $[Ca^{2+}]_i$ in vascular smooth muscle cells of female rat. *Am J Physiol Cell Physiol*, 278, C834-C844.

Nakajima, T.; Kitazawa, T.; Hamada, E.; Hazama, H.; Omata, M., & Kurachi, Y. (1995). 17β-Estradiol inhibits the voltage-dependent L-type Ca^{2+} currents in aortic smooth muscle cells. *Eur J Pharmacol*, 294, 625-635.

Nakano-Kurimoto, R.; Ikeda, K.; Uraoka, M.; Nakagawa, Y.; Yutaka, K.; Koide, M.; Takahashi, T.; Matoba, S.; Yamada, H.; Okigaki, M., & Matsubara, H. (2009). Replicative senescence of vascular smooth muscle cells enhances the calcification through initiating the osteoblastic transition. *Am J Physiol Heart Circ Physiol*, 297, H1673-H1684.

Newby, A.C. (2006). Matrix metalloproteinases regulate migration, proliferation, and death of vascular smooth muscle cells by degrading matrix and non-matrix substrates. *Cardiovasc Res*, 69, 614-624.

Nickenig, G.; Strehlow, K.; Wassmann, S.; Baumer, A.T.; Albory, K.; Sauer, H., & Bohm, M. (2000). Differential effects of estrogen and progesterone on AT(1) receptor gene expression in vascular smooth muscle cells. *Circulation*, 102, 1828-1833.

Novella, S.; Dantas, A.P.; Segarra, G.; Novensa, L.; Bueno, C.; Heras, M.; Hermenegildo, C., & Medina, P. (2010). Gathering of aging and estrogen withdrawal in vascular dysfunction of senescent accelerated mice. *Exp Gerontol*, 45, 868-874.

O'Rourke, M.F., & Hashimoto, J. (2007). Mechanical factors in arterial aging: a clinical perspective. *J Am Coll Cardiol*, 50, 1-13.

O'Rourke, M.F., & Nichols, W.W. (2005). Aortic diameter, aortic stiffness, and wave reflection increase with age and isolated systolic hypertension. *Hypertension*, 45, 652-658.

Ospina, J.A.; Krause, D.N., & Duckles, S.P. (2002). 17β-estradiol increases rat cerebrovascular prostacyclin synthesis by elevating cyclooxygenase-1 and prostacyclin synthase. *Stroke*, 33, 600-605.

Ouchi, Y.; Share, L.; Crofton, J.T.; Iitake, K., & Brooks, D.P. (1987). Sex difference in the development of deoxycorticosterone-salt hypertension in the rat. *Hypertension*, 9, 172-177.

Pau, C.Y.; Pau, K.Y., & Spies, H.G. (1998). Putative estrogen receptor beta and alpha mRNA expression in male and female rhesus macaques. *Mol Cell Endocrinol*, 146, 59-68.

Pereira, T.M.; Nogueira, B.V.; Lima, L.C.; Porto, M.L.; Arruda, J.A.; Vasquez, E.C., & Meyrelles, S.S. (2010). Cardiac and vascular changes in elderly atherosclerotic mice: the influence of gender. *Lipids Health Dis*, 9, 87.

Prakash, Y.S.; Togaibayeva, A.A.; Kannan, M.S.; Miller, V.M.; Fitzpatrick, L.A., & Sieck, G.C. (1999). Estrogen increases Ca^{2+} efflux from female porcine coronary arterial smooth muscle. *Am J Physiol*, 276, H926-H934.

Price, J.M.; Hellermann, A.; Hellermann, G., & Sutton, E.T. (2004). Aging enhances vascular dysfunction induced by the Alzheimer's peptide beta-amyloid. *Neurol Res*, 26, 305-311.

Qiu, H.; Depre, C.; Ghosh, K.; Resuello, R.G.; Natividad, F.F.; Rossi, F.; Peppas, A.; Shen, Y.T.; Vatner, D.E., & Vatner, S.F. (2007). Mechanism of gender-specific differences in aortic stiffness with aging in nonhuman primates. *Circulation*, 116, 669-676.

Redheuil, A.; Yu, W.C.; Wu, C.O.; Mousseaux, E.; de Cesare A.; Yan, R.; Kachenoura, N.; Bluemke, D., & Lima, J.A. (2010). Reduced ascending aortic strain and distensibility: earliest manifestations of vascular aging in humans. *Hypertension*, 55, 319-326.

Reid, J.D., & Andersen, M.E. (1993). Medial calcification (whitlockite) in the aorta. *Atherosclerosis*, 101, 213-224.

Robb, E.L., & Stuart, J.A. (2011). Resveratrol interacts with estrogen receptor-beta to inhibit cell replicative growth and enhance stress resistance by upregulating mitochondrial superoxide dismutase. *Free Radic Biol Med*, 50, 821-831.

Rodriguez-Manas, L.; El-Assar, M.; Vallejo, S.; Lopez-Doriga, P.; Solis, J.; Petidier, R.; Montes, M.; Nevado, J.; Castro, M.; Gomez-Guerrero, C.; Peiro, C., & Sanchez-Ferrer, C.F. (2009). Endothelial dysfunction in aged humans is related with oxidative stress and vascular inflammation. *Aging Cell*, 8, 226-238.

Rosamond, W.; Flegal, K.; Furie, K.; Go, A.; Greenlund, K.; Haase, N.; Hailpern, S.M.; Ho, M.; Howard, V.; Kissela, B.; Kittner, S.; Lloyd-Jones, D.; McDermott, M.; Meigs, J.; Moy, C.; Nichol, G.; O'Donnell, C.; Roger, V.; Sorlie, P.; Steinberger, J.; Thom, T.; Wilson, M., & Hong, Y. (2008). Heart disease and stroke statistics--2008 update: a report from the American Heart Association Statistics Committee and Stroke Statistics Subcommittee. *Circulation*, 117, e25-146.

Ross, R. (1993). The pathogenesis of atherosclerosis: a perspective for the 1990s. *Nature*, 362, 801-809.

Rossouw, J.E.; Anderson, G.L.; Prentice, R.L.; LaCroix, A.Z.; Kooperberg, C.; Stefanick, M.L.; Jackson, R.D.; Beresford, S.A.; Howard, B.V.; Johnson, K.C.; Kotchen, J.M., & Ockene, J. (2002). Risks and benefits of estrogen plus progestin in healthy postmenopausal women: principal results From the Women's Health Initiative randomized controlled trial. *JAMA*, 288, 321-333.

Rubio-Gayosso, I.; Sierra-Ramirez, A.; Garcia-Vazquez, A.; Martinez-Martinez, A.; Munoz-Garcia, O.; Morato, T., & Ceballos-Reyes, G. (2000). 17β-estradiol increases intracellular calcium concentration through a short-term and nongenomic mechanism in rat vascular endothelium in culture. *J Cardiovasc Pharmacol*, 36, 196-202.

Ruiz-Feria, C.A.; Yang, Y.; Thomason, D.B.; White, J.; Su, G., & Nishimura, H. (2009). Pulse wave velocity and age- and gender-dependent aortic wall hardening in fowl. *Comp Biochem Physiol A Mol Integr Physiol*, 154, 429-436.

Santhanam, L.; Christianson, D.W.; Nyhan, D., & Berkowitz, D.E. (2008). Arginase and vascular aging. *J Appl Physiol*, 105, 1632-1642.

Schmidt, R.J.; Beierwaltes, W.H., & Baylis, C. (2001). Effects of aging and alterations in dietary sodium intake on total nitric oxide production. *Am J Kidney Dis*, 37, 900-908.

Schwartz, S.M. (1997). Smooth muscle migration in atherosclerosis and restenosis. *J Clin Invest*, 100, S87-S89.

Seals, D.R.; Jablonski, K.L., & Donato, A.J. (2011). Aging and vascular endothelial function in humans. *Clin Sci (Lond)*, 120, 357-375.

Seely, E.W.; Brosnihan, K.B.; Jeunemaitre, X.; Okamura, K.; Williams, G.H.; Hollenberg, N.K., & Herrington, D.M. (2004). Effects of conjugated oestrogen and droloxifene on the renin-angiotensin system, blood pressure and renal blood flow in postmenopausal women. *Clin Endocrinol (Oxf)*, 60, 315-321.

Segers, P.; Rietzschel, E.R.; De Buyzere, M.L.; Vermeersch, S.J.; De Bacquer D.; Van Bortel, L.M.; De Backer G.; Gillebert, T.C., & Verdonck, P.R. (2007). Noninvasive (input) impedance, pulse wave velocity, and wave reflection in healthy middle-aged men and women. *Hypertension*, 49, 1248-1255.

Semba, R.D.; Najjar, S.S.; Sun, K.; Lakatta, E.G., & Ferrucci, L. (2009). Serum carboxymethyl-lysine, an advanced glycation end product, is associated with increased aortic pulse wave velocity in adults. *Am J Hypertens*, 22, 74-79.

Shaw, L.J.; Bairey Merz, C.N.; Pepine, C.J.; Reis, S.E.; Bittner, V.; Kelsey, S.F.; Olson, M.; Johnson, B.D.; Mankad, S.; Sharaf, B.L.; Rogers, W.J.; Wessel, T.R.; Arant, C.B.; Pohost, G.M.; Lerman, A.; Quyyumi, A.A., & Sopko, G. (2006). Insights from the NHLBI-Sponsored Women's Ischemia Syndrome Evaluation (WISE) Study: Part I: gender differences in traditional and novel risk factors, symptom evaluation, and gender-optimized diagnostic strategies. *J Am Coll Cardiol*, 47, S4-S20.

Sherwood, A.; Bower, J.K.; McFetridge-Durdle, J.; Blumenthal, J.A.; Newby, L.K., & Hinderliter, A.L. (2007). Age moderates the short-term effects of transdermal 17β-estradiol on endothelium-dependent vascular function in postmenopausal women. *Arterioscler Thromb Vasc Biol*, 27, 1782-1787.

Shwaery, G.T.; Vita, J.A., & Keaney, J.F., Jr. (1998). Antioxidant protection of LDL by physiologic concentrations of estrogens is specific for 17β-estradiol. *Atherosclerosis*, 138, 255-262.

Silva-Antonialli, M.M.; Fortes, Z.B.; Carvalho, M.H.; Scivoletto, R., & Nigro, D. (2000). Sexual dimorphism in the response of thoracic aorta from SHRs to losartan. *Gen Pharmacol*, 34, 329-335.

Sivritas, D.; Becher, M.U.; Ebrahimian, T.; Arfa, O.; Rapp, S.; Bohner, A.; Mueller, C.F.; Umemura, T.; Wassmann, S.; Nickenig, G., & Wassmann, K. (2011). Antiproliferative effect of estrogen in vascular smooth muscle cells is mediated by Kruppel-like factor-4 and manganese superoxide dismutase. *Basic Res Cardiol*, 106, 563-575.

Sobrino, A.; Mata, M.; Laguna-Fernandez, A.; Novella, S.; Oviedo, P.J.; Garcia-Perez, M.A.; Tarin, J.J.; Cano, A., & Hermenegildo, C. (2009). Estradiol stimulates vasodilatory and metabolic pathways in cultured human endothelial cells. *PLoS One*, 4, e8242.

Sobrino, A.; Oviedo, P.J.; Novella, S.; Laguna-Fernandez, A.; Bueno, C.; Garcia-Perez, M.A.; Tarin, J.J.; Cano, A., & Hermenegildo, C. (2010). Estradiol selectively stimulates endothelial prostacyclin production through estrogen receptor-α. *J Mol Endocrinol*, 44, 237-246.

Spina, M.; Garbisa, S.; Hinnie, J.; Hunter, J.C., & Serafini-Fracassini, A. (1983). Age-related changes in composition and mechanical properties of the tunica media of the upper thoracic human aorta. *Arteriosclerosis*, 3, 64-76.

Squadrito, G.L., & Pryor, W.A. (1998). Oxidative chemistry of nitric oxide: the roles of superoxide, peroxynitrite, and carbon dioxide. *Free Radic Biol Med*, 25, 392-403.

Staessen, J.; Bulpitt, C.J.; Fagard, R.; Lijnen, P., & Amery, A. (1989). The influence of menopause on blood pressure. *J Hum Hypertens*, 3, 427-433.

Stallone, J.N.; Crofton, J.T., & Share, L. (1991). Sexual dimorphism in vasopressin-induced contraction of rat aorta. *Am J Physiol*, 260, H453-H458.

Stice, J.P.; Eiserich, J.P., & Knowlton, A.A. (2009). Role of aging versus the loss of estrogens in the reduction in vascular function in female rats. *Endocrinology*, 150, 212-219.

Su, Y.; Kondrikov, D., & Block, E.R. (2005). Cytoskeletal regulation of nitric oxide synthase. *Cell Biochem Biophys*, 43, 439-449.

Su, Y.; Kondrikov, D., & Block, E.R. (2007). Beta-actin: a regulator of NOS-3. *Sci STKE*, 2007, e52.

Sumi, D., & Ignarro, L.J. (2003). Estrogen-related receptor alpha 1 up-regulates endothelial nitric oxide synthase expression. *Proc Natl Acad Sci U S A*, 100, 14451-14456.

Sumino, H.; Ichikawa, S.; Kasama, S.; Kumakura, H.; Takayama, Y.; Sakamaki, T., & Kurabayashi, M. (2005). Effect of transdermal hormone replacement therapy on carotid artery wall thickness and levels of vascular inflammatory markers in postmenopausal women. *Hypertens Res*, 28, 579-584.

Sumino, H.; Ichikawa, S.; Kasama, S.; Takahashi, T.; Kumakura, H.; Takayama, Y.; Kanda, T., & Kurabayashi, M. (2006). Different effects of oral conjugated estrogen and transdermal estradiol on arterial stiffness and vascular inflammatory markers in postmenopausal women. *Atherosclerosis*, 189, 436-442.

Taddei, S.; Virdis, A.; Ghiadoni, L.; Mattei, P.; Sudano, I.; Bernini, G.; Pinto, S., & Salvetti, A. (1996). Menopause is associated with endothelial dysfunction in women. *Hypertension*, 28, 576-582.

Takenouchi, Y.; Kobayashi, T.; Matsumoto, T., & Kamata, K. (2009). Gender differences in age-related endothelial function in the murine aorta. *Atherosclerosis*, 206, 397-404.

Tamaya, T.; Wada, K.; Nakagawa, M.; Misao, R.; Itoh, T.; Imai, A., & Mori, H. (1993). Sexual dimorphism of binding sites of testosterone and dihydrotestosterone in rabbit model. *Comp Biochem Physiol Comp Physiol*, 105, 745-749.

Thompson, G.R., & Partridge, J. (2004). Coronary calcification score: the coronary-risk impact factor. *Lancet*, 363, 557-559.

Thorin, E., & Thorin-Trescases, N. (2009). Vascular endothelial ageing, heartbeat after heartbeat. *Cardiovasc Res*, 84, 24-32.

Toda, T.; Tsuda, N.; Nishimori, I.; Leszczynski, D.E., & Kummerow, F.A. (1980). Morphometrical analysis of the aging process in human arteries and aorta. *Acta Anat (Basel)*, 106, 35-44.

Tostes, R.C.; David, F.L.; Carvalho, M.H.; Nigro, D.; Scivoletto, R., & Fortes, Z.B. (2000). Gender differences in vascular reactivity to endothelin-1 in deoxycorticosterone-salt hypertensive rats. *J Cardiovasc Pharmacol*, 36, S99-101.

Tostes, R.C.; Nigro, D.; Fortes, Z.B., & Carvalho, M.H. (2003). Effects of estrogen on the vascular system. *Braz J Med Biol Res*, 36, 1143-1158.

Touyz, R.M. (2003). Reactive oxygen species in vascular biology: role in arterial hypertension. *Expert Rev Cardiovasc Ther*, 1, 91-106.

Touyz, R.M.; Cruzado, M.; Tabet, F.; Yao, G.; Salomon, S., & Schiffrin, E.L. (2003). Redox-dependent MAP kinase signaling by Ang II in vascular smooth muscle cells: role of receptor tyrosine kinase transactivation. *Can J Physiol Pharmacol*, 81, 159-167.

Tracy, R.P. (2006). The five cardinal signs of inflammation: Calor, Dolor, Rubor, Tumor ... and Penuria (Apologies to Aulus Cornelius Celsus, De medicina, c. A.D. 25). *J Gerontol A Biol Sci Med Sci*, 61, 1051-1052.

Van Craenenbroeck, E.M., & Conraads, V.M. (2010). Endothelial progenitor cells in vascular health: focus on lifestyle. *Microvasc Res*, 79, 184-192.

Vasa, M.; Breitschopf, K.; Zeiher, A.M., & Dimmeler, S. (2000). Nitric oxide activates telomerase and delays endothelial cell senescence. *Circ Res*, 87, 540-542.

Virdis, A.; Ghiadoni, L.; Giannarelli, C., & Taddei, S. (2010). Endothelial dysfunction and vascular disease in later life. *Maturitas*, 67, 20-24.

Wallace, D.C. (2005). A mitochondrial paradigm of metabolic and degenerative diseases, aging, and cancer: a dawn for evolutionary medicine. *Annu Rev Genet*, 39, 359-407.

Wang, M.; Takagi, G.; Asai, K.; Resuello, R.G.; Natividad, F.F.; Vatner, D.E.; Vatner, S.F., & Lakatta, E.G. (2003). Aging increases aortic MMP-2 activity and angiotensin II in nonhuman primates. *Hypertension*, 41, 1308-1316.

Wassmann, S.; Baumer, A.T.; Strehlow, K.; van Eickels M.; Grohe, C.; Ahlbory, K.; Rosen, R.; Bohm, M., & Nickenig, G. (2001). Endothelial dysfunction and oxidative stress during estrogen deficiency in spontaneously hypertensive rats. *Circulation*, 103, 435-441.

Wassmann, S.; Wassmann, K., & Nickenig, G. (2004). Modulation of oxidant and antioxidant enzyme expression and function in vascular cells. *Hypertension*, 44, 381-386.

Watanabe, M.; Sawai, T.; Nagura, H., & Suyama, K. (1996). Age-related alteration of cross-linking amino acids of elastin in human aorta. *Tohoku J Exp Med*, 180, 115-130.

Wellman, G.C.; Bonev, A.D.; Nelson, M.T., & Brayden, J.E. (1996). Gender differences in coronary artery diameter involve estrogen, nitric oxide, and Ca^{2+}-dependent K^+ channels. *Circ Res*, 79, 1024-1030.

White, R.E.; Darkow, D.J., & Lang, J.L. (1995). Estrogen relaxes coronary arteries by opening BKCa channels through a cGMP-dependent mechanism. *Circ Res*, 77, 936-942.

Wong, A.J.; Pollard, T.D., & Herman, I.M. (1983). Actin filament stress fibers in vascular endothelial cells in vivo. *Science*, 219, 867-869.

Wu, Z.; Maric, C.; Roesch, D.M.; Zheng, W.; Verbalis, J.G., & Sandberg, K. (2003). Estrogen regulates adrenal angiotensin AT_1 receptors by modulating AT_1 receptor translation. *Endocrinology*, 144, 3251-3261.

Wynne, F.L.; Payne, J.A.; Cain, A.E.; Reckelhoff, J.F., & Khalil, R.A. (2004). Age-related reduction in estrogen receptor-mediated mechanisms of vascular relaxation in female spontaneously hypertensive rats. *Hypertension*, 43, 405-412.

Xiong, Y.; Yuan, L.W.; Deng, H.W.; Li, Y.J., & Chen, B.M. (2001). Elevated serum endogenous inhibitor of nitric oxide synthase and endothelial dysfunction in aged rats. *Clin Exp Pharmacol Physiol*, 28, 842-847.

Yoon, H.J.; Cho, S.W.; Ahn, B.W., & Yang, S.Y. (2010). Alterations in the activity and expression of endothelial NO synthase in aged human endothelial cells. *Mech Ageing Dev*, 131, 119-123.

Yoshida, Y.I.; Eda, S., & Masada, M. (2000). Alterations of tetrahydrobiopterin biosynthesis and pteridine levels in mouse tissues during growth and aging. *Brain Dev*, 22 Suppl 1, S45-S49.

Zhang, F.; Ram, J.L.; Standley, P.R., & Sowers, J.R. (1994). 17β-Estradiol attenuates voltage-dependent Ca^{2+} currents in A7r5 vascular smooth muscle cell line. *Am J Physiol*, 266, C975-C980.

Zhang, Y.; Stewart, K.G., & Davidge, S.T. (2000). Estrogen replacement reduces age-associated remodeling in rat mesenteric arteries. *Hypertension*, 36, 970-974.

4

Molecular Biomarkers of Aging

Sergio Davinelli[1], Sonya Vasto[2], Calogero Caruso[2],
Davide Zella[3] and Giovanni Scapagnini[1]
[1]*Department of Health Sciences, University of Molise, Campobasso,*
[2]*Department of Pathobiology and Biomedical Methodologies,*
Immunosenescence Unit, University of Palermo, Palermo,
[3]*Department of Biochemistry and Molecular Biology,*
Institute of Human Virology-School of Medicine,
University of Maryland, Baltimore, MD,
[1,2]*Italy*
[3]*USA*

1. Introduction

In the Western World, the public perception of advanced aging involves the inability to survive due to chronic diseases and the combined loss of mobility, sensory functions, and cognition with an exponential growth of health costs. Therefore, biomarkers of human aging are urgently needed to assess the health state of elderly and the possible therapeutic interventions. Aging is considered a process that changes the performances of most physiological systems and increases susceptibility to diseases and death. The aging phenotype is a complex interaction of stochastic, environmental, genetic and epigenetic variables. However, these variables do not create the aging phenotype but favour the lost of molecular fidelity and therefore as the random accumulation of damages in the human organism's cells, tissues, or whole organism during life increases, the probability of disease and death also augments in proportion (Candore et al., 2008). What a biomarker for aging should be or predict is quite broadly defined. A biomarker should not only (i) reflect some basic property of aging, but also (ii) be reproducible in cross-species comparison, (iii) change independently of the passage of chronological time (so that the biomarker indicates biological rather than chronological age), (iv) be obtainable by non invasive means, and (v) be measurable during a short interval of life span. A biomarker should reflect the underlying aging process rather than disease (Warner et al., 2004). In addition, a set of biomarkers should be based on mechanisms described by major theories of aging. A sustained number of biomarkers are currently under investigation, such as inflammatory markers, markers of oxidative stress or markers of telomere shortening but the definition of biomarker is strictly related to the understanding of the mechanisms of aging and we might not be able to define an ideal biomarker yet. Moreover, the biomarkers of aging discussed in literature, are associated not only to age but also to diseases Accordingly, it is crucial to monitor basic mechanisms that underlies the aging process. Noteworthy, a recent study reported that biomarkers of cardiovascular diseases (CVD) and diabetes are useful predictors of healthy aging (Crimmins et al., 2008).

Another problem, which is probably even more challenging, is to understand if a biomarker validated for rodents could be applied equally to humans.

Notably, it should be highlighted that mammalian cells have developed highly refined inducible systems against a variety of stressful conditions; upon stimulation, each one of these systems can be engaged concertedly to alleviate and hinder the manifestation of a distinctive age-related disorder. In this context, increasing scientific evidence supports a pivotal role for the heat shock proteins in the protection against oxidative stress and inflammation. Heat shock response is a fundamental cellular survival pathway, involving both transcriptional and post-trascriptional regulation. The impairment of this regulatory mechanism might directly contribute to the defective cellular stress response to oxidative stress and deregulation of inflammatory processes, which characterizes senescence.

In the present chapter, we will focus on the importance of biomarkers involved in inflammatory responses, oxidative stress but also markers based on immunosenescence. Additionally, we will describe the major experimental methods that are available in biogerontology for the interpretation of the aging phenotypes. In summary, we will present an overview on the current knowledge of the complex molecular and biological events leading to cellular senescence and how we can measure this progression to possibly improve our quality of life.

2. Aging and the immune system

Aging is accompanied by a general dysregulation in immune system function, commonly referred to as immunosenescence. This progressive deterioration affects both innate and adaptive immunity, although accumulating evidence indicates that the adaptive arm of the immune system may exhibit more profound changes. Most of our current understanding of immune senescence stems from clinical and rodent studies. Studies have suggested that aging is associated with increase permeability of mucosal barriers, decreased phagocytic activity of macrophages and dendritic cells (DCs), reduced natural killer (NK) cell cytotoxicity, and dysregulated production of soluble mediators such as cytokines and chemokines (Weiskopf et al., 2009). These alterations could lead to increased pathogen invasion and poor activation of the adaptive immune response mediated by T and B-lymphocytes. The age-related changes which occur in the adaptive and innate immune response are summarized in Table 1.

Aging, is also associated with quantitative and qualitative changes within the naive CD4+T-cell compartment (Aspinall & Andrew, 2000; Fulop et al., 2006; Kilpatrick et al., 2008). Decreased numbers of recent thymic emigrants (RTE), shortened telomeres, hyporesponsiveness to stimulation, decreased proliferative capacity, reduced IL-2 production, alterations in signal transduction and changes in cell surface phenotype (Whisler et al, 1996; Fulop et al., 2006; Kilpatrick et al., 2008) have all been reported. These changes likely contribute to the poor response to vaccines and increased susceptibility to infectious diseases and neoplasms reported for older adults (Webster, 2000; Effros, 2000; Herndler-Brandstetter et al., 2006).

Aging causes a shift in the ratio of naive to memory T-cells, with associated changes in the cytokine profile that favor increases in pro-inflammatory interleukin-1β (IL-1β), interleukin-6 (IL-6), interferon γ (IFNγ), tumor necrosis factor (TNFα), and transforming growth factor

(TGFβ) (Sansoni et al., 2008). The production of IL-6, but not IL-1β or TNF-α, by peripheral blood mononuclear cells increases in the elderly (Roubenoff et al., 1998), and IL-1β production increases in peripheral blood mononuclear cells in older animals (Chung et al., 2006). In contrast, IL-1β levels are higher and IL-6 levels lower in the livers of old rats than young rats (Rikans et al., 1999).

As the hematopoietic system ages, the immune function deteriorates, the lymphoid potential diminishes, and the incidence of myeloid leukemia increases (Rossi et al., 2005). Aging leads to increased stem cell dysfunction, and as a result leukemia can develop in failed attempts by the bone marrow to return to a homeostatic condition after stress or injury. Stem cells leave the hibernation state and undergo self-renewal and expansion to prevent premature hematopoietic stem cell (HSC) exhaustion under conditions of hematopoietic stress (Walkey et al., 2005). HSCs in older mice produce a decreased number of progenitors per cell, decreased self-renewal and increased apoptosis with stress (Janzen et al., 2006).

The remaining stem cells divided more rapidly as if to compensate for those that were lost. Stimulating old stem cells to grow more rapidly, perhaps by stress such as infrared (IR), puts stem cells at greater risk of becoming cancer cells because of acquired DNA damage.

Metabolically active senescent cells, identified by the biomarkers of cellular aging, such as the γ-H2AX foci and perhaps the senescence-associated β-galactosidase (SA-β-gal) enzyme, accumulate in aging primates (Herbig et al., 2006). Cellular senescence can be induced in one of two ways. Firstly, reactive oxygen species (ROS) may contribute to the plentiful single-strand breaks (SSBs) and double-strand breaks (DSBs) present in senescent cells (Sedelnikova et al., 2004); this is a form of telomere-independent stress-induced senescence. Alternatively, telomere-dependent uncapping of telomere DNA causes replicative senescence. An increase in oxidative stress is a more probable cause of HSC senescence than telomere erosion (Beauséjour et al., 2007). High doses of IR lead to apoptosis of HSCs, while lower doses cause HSCs to senesce and lose the ability to clone themselves (Wang et al., 2006). Furthermore, irradiated normal human fibroblasts and tumor cell lines can also lose their clonogenic potential and undergo to accelerated senescence (Mirzayans et al., 2005). The inhibition of tumorigenesis by cellular senescence is oncogene-induced and linked to increased expression of tumor suppressor genes cyclin-dependent kinase inhibitor 2A (p16INK4a or CDKN2A) and tumor protein 53 (TP53) via the DNA damage response (Bartkova et al., 2006). Recent research points to the p16INK4a protein being an important aging biomarker as its concentrations in peripheral blood exponentially increase with chronological age, reducing stem cell self-renewal (Liu et al., 2009). The few articles published to date linking radiation's health effects and p16INK4a can be paradoxical with regard to aging. A Chinese study showed the cumulative radiation dose of radon gas among uranium miners to be positively associated with the aberrant promoter methylation and inactivation of the p16INK4a and O6-methylguanine-DNA methyltransferase (MGMT) genes in sputum, perhaps indicating the early DNA damage and a greater susceptibility to lung cancer (Su et al., 2006).

The number and proliferation potential of stem cell populations, including those of the intestinal crypt and muscle, decrease with age, leading to a progressive deterioration of tissue and organ maintenance and function (Schultz et al., 1982; Martin et al., 1998). Macromolecular damage in general and DNA damage in particular, accumulate in HSCs with age (Rossi et al., 2007). The reduced ability to repair DNA DSBs leads to a progressive loss of HSCs and bone marrow cellularity during aging (Nijnik et al., 2007). In addition, high radiation dose (>12.5 Gy) from 45Ca, a bone-targeting beta-ray emitter (Barranco et al., 1969), resulted in marked reduction in marrow cellularity, similar to the one observed in normal aging indicating a possible contribution of the DNA–repair mechanisms to the aging process.

	IMMUNE BIOMARKERS	AGE-RELATED INCREASE	AGE-RELATED DECREASE
INNATE IMMUNE SYSTEM	Cytokines and Chemokines	Serum levels of IL6, IL1β and TNFα	
	NK cells	Total number of cells	Proliferative response
	Dendritic cells		Capacity to stimulate antigen specific T-cells
	Neutrophils		Bactericidal activity; Oxidative burst
	Macrophages		Phagocytic capacity
ADAPTIVE IMMUNE SYSTEM	T-lymphocytes	Release of proinflammatory cytokines	Number of naive T-cells; Diversity of the T-cell repertoire
	B-lymphocytes	Autoreactive serum antibodies	Number of naive B-cells; Antibody affinity; Generation of B-cell precursors

Table 1. Age-related changes in the innate and adaptive immune system.

3. Oxidative stress and inflammation as causes of aging

To date, there are several theories which attempt to explain the process of aging, such as telomere theory, caloric restriction, and evolutionary theory. The oxidative stress hypothesis/free radical theory of aging, updated by Harman in 2006 (Harman, 2006) offers a possible biological explanation of the entire aging process. In a biological context, a condition of oxidative stress occurs when there is an imbalance between oxidant molecules and antioxidant defensive molecules. Such critical balance is disrupted when antioxidants are depleted or if the formation of ROS increases beyond the ability of the antioxidative systems. Additionally, the free radical theory proposed that the production of intracellular ROS is the major determinant of life span.

However, the most critical problem is to find a correlation between oxidative biomarkers amounts and human health. Nevertheless, according to the free radical theory of aging, oxidative stress increases with increasing age resulting in oxidative DNA damage, protein oxidation and lipid peroxidation.

One of major risk markers of oxidative damage of nucleic acids is the 8-hydroxy-29-deoxyguanosine (8-OHdG). So far, 8-OHdG is the most studied oxidative DNA lesion and it is formed when ROS act on deoxyguanine in DNA (Ravanat et al., 2000). 8-OHdG can alter gene expression, inhibits methylation and its mutagenic potential leads to GC → AT conversion. The formation of 8-OHdG in leukocyte DNA and the excretion of 8-OHdG into urine have been frequently measured to assess oxidative stress in humans. However, even though interesting results have been obtained with 8-OHdG, several studies have associated aging with a progressive loss of antioxydant defence.

Recently, several findings have emphasized the importance of lipid peroxidation in relation to the role of caloric restriction and the extension of longevity (Sanz et al., 2006). Lipid peroxidation products have also been shown to be mutagenic and carcinogenic and has been implicated as the underlying mechanisms in numerous disorders including aging. Notably, lipid oxidation not only causes membrane disruption but also produces aldehydic species, such as malondialdehyde (MDA), able to perpetrate further damage by binding to and modifying proteins. Although producing contradictory results, the measure of lipid peroxidation is an example of biomarkers of oxidative stress. The measurement of MDA is very easy to perform, fast and not expensive. MDA is often utilized to evaluate human aging and in numerous studies MDA was significantly higher in healthy elderly, confirming the presence of increased lipoperoxidation in old age.

Another important product generated by lipid peroxidation is 4-hydroxy-2-nonenal (HNE) that reacts with nucleic acids, proteins, and phospholipids inducing many cytotoxic, mutagenic, and genotoxic effects (Uchida, 2003). Low-density lipoproteins (LDL) seems to be another good marker because oxidised LDL appears to be involved in the development of various pathological conditions aging related. Measurements of LDL could be obtained *in vivo* by measuring oxidised LDL particles in blood using immunological methods with appropriate specificity.

In addition, phosphatidylcholine hydroperoxides (PCOOH) measured in blood or tissue is also an acceptable marker of lipid peroxidation.

Recently, isoprostanes (IsoPs), compounds that are produced *in vivo* by free radical-induced peroxidation of arachidonic acid, have been also proposed to assess the oxidative stress status but we have only few experimental evidence and convincing outcomes have not emerged yet (Montuschi et al., 2007). Particularly, the analysis of F2-isoprostanes has revealed a role for free radicals and oxidant injury in a wide variety of human diseases. However, it must be taking into account that the measurement of F2-isoprostanes represents a snapshot of oxidant stress at a discrete point in time. Indeed, F2-isoprostanes are cleared rapidly from the circulation. However, such molecules that are stable isomers of prostaglandin F2, seems to be the best reliable marker and it has been proposed as most affidable index of systemic or " whole body" oxidative stress over time.

Closely related to oxidative stress is the protein oxidation. The main molecular characteristic of aging is the progressive accumulation of damages in macromolecules and age related damage in proteins have been reported in cells, tissues and organs (Rattan, 2006). The measurement of the protein oxidation is a clinically important factor for the prediction of the aging process and age-related diseases. The most widely studied oxidative stress-induced modification to proteins is the formation of carbonyl derivatives. Carbonyl formation can occur through a variety of mechanisms including direct oxidation of certain amino-acid side chains and oxidation-induced peptide cleavage. Furthermore, advanced oxidation protein products considered as biomarkers to estimate the degree of oxidative modifications of proteins and carbonyl groups may be introduced into proteins by reactions with aldehydes, reactive carbonyl derivatives or through their oxidation products with lysine residues of proteins. Although all organs and all proteins can potentially be modified by oxidative stress, certain tissues and specific protein may be especially sensitive. For istance, recent studies characterized oxidatively modified proteins in the brain and identified specific proteins that are oxidatively modified in Alzheimer' s disease (Butterfield & Sultana, 2007).

Aging is accompanied by chronic low-grade inflammation status and inflammatory mediators may be usefull to monitor aging processes. Molecular activation of pro-inflammatory genes by altered redox signaling pathways will eventually lead to inflamed tissues and organs. Accordingly, molecular inflammation is an important biological component of aging. In this perspective, nuclear factor of kappa light polypeptide gene enhancer in B-cells (NF-kβ) is a transcription factor that plays a pivotal role in modulating cellular signaling of oxidative stress-induced molecular inflammation. For example, stimulus-mediated phosphorylation and the subsequent proteolytic degradation of nuclear factor of kappa light polypeptide gene enhancer in B-cells inhibitor (Ikβ) allows the release and nuclear translocation of NF-κB, where transactivates several target genes such as forkhead box (FOXO), IL-lβ, IL-6, TNFa, adhesion molecules, cyclooxygenase-2 (COX-2) and nitric oxide synthases inducible (iNOS), all key players in inflammation.

Aging is associated with activation of both the innate and the adaptive immune system. As mentioned above, during aging increased blood levels of proinflammatory cytokines such as IL-6 and TNFa can be observed. In healthy elderly populations, high circulating levels of TNFa and IL-6 predict mortality, in a manner independent from comorbidity (Bruunsgaard & Pedersen, 2003).

Additionally, an inflammatory response appears to be the prevalent triggering mechanism driving tissue damage associated with different age-related diseases and the definition of "inflamm-aging" has been coined to explain the underlining inflammatory changes common to most age-associated diseases (Licastro et al. 2005).

Finally, reduced glutathione (GSH) is a major intracellular non-protein -SH compound and is accepted as the most important intracellular hydrophilic antioxidant (Melov, 2002). Glutathione system is the most important endogenous defense system against oxidative stress in body. Under oxidative conditions GSH is reversibly oxidized to glutathione disulfide (GSSG). A recent study on age-related changes in GSH in rat brain suggests a significant age-related reduction in the GSH level in all regions of the brain, associated with an increase in GSH oxidation to GSSG and decrease in the GSH/GSSG ratio (Zhu et al., 2006).

4. Methods for analysis of biological aging

The aging research requires multi- and transdisciplinary approaches and new high-throughput technologies are continually in development, increasing exponentially the amount of biological informations in aging research and elucidating complex unknown mechanisms. Although there have been extraordinary advances in study related to gene expression, proteomic and functional data, one challenge in aging research is to bring together this large variety of data that are still fragmented. Here, we provide a brief description of the main technological approaches for biomarkers discovery and for analyzing the molecular and cellular changes involved in aging cells. We also describe the major databases, computational tools and bioinformatics methods that are available in biogerontology for data interpretation of the aging phenotypes.

Analysis of gene-expression data has led to remarkable progress in many biomedical disciplines, including gerontology. Numerous methods have been developed for this analysis, but the emergence of high-throughput expression profiling and sequencing, such as microarray technology (Blalock et al., 2003) or more recently next-generation sequencing RNA-Seq have become diffusely used leading to breakthroughs in the investigations of aging (Twine et al., 2011). The application of microarray technology to gerontological studies has improved our understanding of mechanisms of aging (Golden & Melov, 2007) allowing to elucidate molecular differences associated with aging and to scan the entire genome for genes that change expression with age. The core principle of a microarray experiment is the hybridization of RNA/DNA strands of at least two different conditions, such as normal and disease or different ages, with a microarray chip. Briefly, the data collection is followed by bioinformatics analysis that require background noise subtraction, normalization and identification of statistically significant changes with dedicated software packages that calculate through multiple statistical measurements the significance for each gene. Ultimately, an application of transcriptomic microarray reported the first assessment of age-related alterations in gene expression in a large population-based cohort suggesting that modification of messenger RNA (mRNA) processing may comprise an important feature of human aging (Harries et al., 2011). In addition, an example of differential expression analysis in aging is the comparison of Ames dwarf mice $Prop1^{df/df}$ versus $Prop1^{+/+}$ and Little mice $Ghrhr^{lit/lit}$ versus $Ghrhr^{+/lit}$ (Amador-Noguez et al., 2004). In both cases, the mutants show delayed aging with significantly increased lifespan and the authors found 1125 and 1152 differentially expressed genes in these mutants, respectively, using analysis of variance (ANOVA). There is a growing number of age-related molecular repositories and one database of gene expression profiles during aging is the Gene Aging Nexus, which features a compilation of aging microarray data and microarray datasets across different platforms and species (http://gan.usc.edu/) (Pan et al., 2007).

The genomic convergence approach is a new powerful method alternative to genome-wide association studies that combines trascriptional profiling, expression of quantitative trait mapping and gene association. Briefly, microarray technology are used to identify genes that show age-related changes in expression. In the next step single nucleotide polymorphisms (SNPs) are tested for association with the expression of age-regulated genes and finally the expression of quantitative trait loci (eQTLs) are tested for association with a phenotype of aging (Wheeler et al., 2009).

Currently, basic methods to understand biological processes and to identify possible candidate biomarkers for a specific pathology are shifted toward "omics" approaches, where all classes of biological compounds can be analyzed by respective "omic" techniques. To date, proteomic investigations have special relevance to aging-related research since altered protein interactions may have a key role in aging-related diseases. Different proteomic technology platforms were applied to define the proteomes and conventional techniques as two-dimensional gel electrophoresis (2DE), surface-enhanced laser desorption/ionization (SELDI), liquid chromatography-mass spectrometry (LC-MS), capillary electrophoresis (CE) coupled to mass spectrometry (CE-MS) and protein arrays are also associated to aging related studies. The proteomic analysis require adequate tools for data analysis and there are several bioinformatic approaches in proteomics. Many algorithms have now been designed to handle the increasing amount of data that are available thanks to proteomic analysis and numerous computational approaches and software tools have been developed to automatically assign candidate peptide sequences to fragment ion spectra, for example, SEQUEST, MASCOT or ProteinProspector. In addition, quantitative proteomics based on stable isotope labeling, such as isotope-coded affinity tags (ICAT) or stable isotope labeling by amino acids in cell culture (SILAC) represents a promising approach for aging studies providing important information to interpret protein biomarkers of age-related disease (Zhang et al., 2005). The investigation of changes in metabolite fluxes or the analysis of all metabolites in high-throughput fashion, called metabolomics (or metabonomics), is an attractive and expanding field in aging research. One goal of the metabolomics is to assess the impact of metabolite concentrations on aging phenotyope. Several studies have emerged with metabolomics approach traditionally using nuclear magnetic resonance (NMR) and recently MS techniques. An example of the application of MS-based metabolomics in aging research is given by Lawton et al. which analyzed the plasma of 269 individuals and discovered that age significantly altered the concentrations of over 100 metabolites (Lawton et al., 2008).

In addition, there are available for metabolomics researchers interested in aging databases for metabolite identification, such as METLIN that contains information on metabolites, as well as MS data (http://metlin.scripps.edu/) or The Human Metabolome Database (http://www.hmdb.ca/) with information on small molecule human metabolites.

Taking into account that the human aging phenotype is a highly polygenic trait which involves changes in genes involved in multiple processes and results from a combination of different factors, systems biology approach is particularly powerful in studies of aging. To date it seems to be the only method able to define and connect the large volumes of experimental data generated by "omics" fields. The final aim of the systems biology of aging is to generate an integrative approach which elucidates the molecular mechanisms of aging and to characterize this phenotype at systemic/organism level. In addition to quantify and integrate data produced by high-throughput technologies, the systems biology approach combines data-driven modelling and hypothesis-driven experimental studies in order to link aging phenotypes and its causes. One area in which sistems biology can be applied to aging research is the generation of a conceptual whole cell model that considers the dynamic behaviors of cellular metabolism. The whole cell representation is structured into subcellular entities not only connected by protein-protein interactions but also by process related to metabolism, oxidative stress or trascriptional regulation. The goal of aging cell modelling is to build a conceptual framework through the simulation of dynamic system and to make predictions about the aging phenotype. The systems biology community has developed tools and modeling platforms to facilitate the representation of metabolic and signaling pathways among biological processes and allowing the understanding of complex phenomena such as aging. Systems Biology Markup Language (SBML) is the main language for coding biological models and currently there are two softwares that support construction of models in SMBL, CellDesigner and JDesigner (Oda et al., 2005). Recently, by using SBML, McAuley et al. generated an *in silico* brain aging model which may help to predict aging-related brain changes in older people (McAuley et al., 2009). Most of genes and proteins exert their functions within a complex network of interactions and another applications of systems biolgy is the assemblage of interactomes. The building of interaction networks allow to define changes in interaction of proteins implicated in aging process that are involved in maintaining the integrity of the human genome. In the protein-protein interaction (PPI) networks, each protein is a node and each interaction an edge and the first attempt towards constructing a "human longevity network" via analysis of human PPIs was made by Budovsky et al. in 2007 (Budovsky et al., 2007). According to the BioGrid database, the authors constructed a "core longevity network" that comprises 153 longevity-associated proteins and 33 non-longevity-associated proteins that have connections with at least five longevity-associated proteins or more. Therefore, network-based approaches are notably valuable for deciphering complex biological systems providing insights about aging, longevity and age-related disease. In addition, there are several collections of online resources available for biogerontologists that can also serve for the visualization of protein-protein interactions. Databases focused on genes related to aging and/or longevity include GeneAge and AnAge featured by the Human Aging Genomic Resources (http://genomics.senescence.info/) that is a collection of databases and tools designed for understanding the genetics of human aging. GeneAge is a curated reference database of different searchable data sets of genes associated with the human aging phenotype. One possible approach of GeneAge is the visualizzation of protein-protein interactions with one or more genes as query but additional ways can be used to build genes and protein interaction networks in conjunction with data stored in interaction databases such as IntAct (Kerrien et al., 2007). There is an expanding number of age-related repositories, such as NetAge (http://netage-project.org/) which provides information on microRNA-regulated protein-protein interaction networks that are involved in aging and related processes. Furthermore, one gene expression database is AGEMAP (Zahn et al., 2007), which allows to analize multiple genes and mechanisms affect aging decribing changes in expression levels in different mouse tissues. Finally, one database on human aging which will be available to the public is MARK-AGE (http://www.mark-age.eu/), a large-scale integrated project supported by the European Community. The aim of this project is to conduct a population study in order to identify a set o biomarkers of aging.

The coordinate assessment of genotypes, trascriptional and proteomic profiles in association with system and computational biology strategies will be able to reach a comprehensive model for the study of human aging and longevity but also for healthy aging.

5. Conclusion

Biomarkers of aging are an hot topic and have the ability to improve our life. Various parameters are directly affected and altered during aging and several indicators are being used to evaluate the aging process. However, not all can be used as biomarker of aging because many of them are influenced by different factors such as diet, enviroment or type of tissue. In addition, some biomarkers are dependent by the methods used to measure them. Accordingly, there is not yet a "pure" biomarkers of aging and the markers discussed are related not only to age but also to disease.

Furthermore, it is a subject of debate whether the determination of biological age markers is really addressing aging itself, or if it indicates stress induced acceleration of the age process by exogenic factors.

Aging is a multi-dimensional process and these biomarkers could be used to monitor and identify the development of age-associated disease providing new anti-aging strategies.

6. Acknowledgment

The authors would like to thank all their collegues for helpful comments regarding this manuscript and the respective departments for the support.

7. Abbreviations

The abbreviations used are: CVD, cardiovascular diseases; DCs, dendritic cells; NK, natural killer; RTE, thymic emigrants; IL-1β, Interleukin-1β; IL-6, Interleukin-6; IFNγ, interferon γ; TNFα, tumor necrosis factor α; HSC, hematopoietic stem cell; SA-β-gal, Senescence-Associated β-galactosidase; ROS, reactive oxygen species; SSBs, single-strand breaks; DSBs, double-strand breaks; IR, infrared; *p16INK4a* or CDKN2A, cyclin-dependent kinase inhibitor 2A; TP53, tumor protein 53;MGMT,O6-methylguanine-DNAmethyltransferase;8-OHdG,8-hydroxy-29 deoxyguanosine; MDA, malondialdehyde; HNE, 4-hydroxy-2-nonenal; LDL, low-density lipoproteins; PCOOH, phosphatidylcholine hydroperoxides; IsoPs, isoprostanes; NF-kβ, nuclear factor of kappa light polypeptide gene enhancer in B-cells; Ikβ, nuclear factor of kappa light polypeptide gene enhancer in B-cells inhibitor; FOXO, forkhead box; COX-2, cyclooxygenase-2; GSH, reduced glutathione; GSSG, glutathione disulfide; mRNA, messenger RNA; SNPs, single nucleotide polymorphisms; eQTLs, expression of quantitative trait loci; 2DE, two-dimensional gel electrophoresis; SELDI, surface-enhanced laser desorption/ionization; MS, mass spectrometry; LC, liquid chromatography; CE, capillary electrophoresis; NMR, nuclear magnetic resonance; SBML, Systems Biology Markup Language; PPI, protein-protein interaction.

8. References

Amador-Noguez D, Yagi, K, Venable S., Darlington G. (2004). Gene expression profile of long-lived Ames dwarf mice and Little mice. *Aging Cell* 3(6):423-41.

Aspinall R. & Andrew D. (2000). Thymic involution in aging. *J Clin Immunol* 20(4): 250–256.

Barranco SC., Beers RF. Jr., Merz T. (1969). Marrow cell injury following Ca45 uptake in bone: changes in marrow and peripheral blood cellularity. *Am J Roetgenol Radium Ther Nucl Med* 106(4):794-801.

Bartkova J., Rezaei N., Liontos M., Karakaidos P., Kletsas D., Issaeva N., Vassiliou LV., Kolettas E., Niforou K., Zoumpourlis VC., Takaoka M., Nakagawa H., Tort F., Fugger K., Johansson F., Sehested M., Andersen CL., Dyrskjot L., Ørntoft T., Lukas J., Kittas C., Helleday T., Halazonetis TD., Bartek J., Gorgoulis VG. (2006). Oncogene-induced senescence is part of the tumorigenesis barrier imposed by DNA damage checkpoints. *Nature* 444(7119):633-637.

Beauséjour C. (2007). Bone marrow-derived cells: the influence of aging and cellular senescence. *Handb Exp Pharmacol* 180:67-88.

Blalock EM., Chen KC., Sharrow K, Herman JP., Porter NM., Foster TC., Landfield PW. (2003). Gene microarrays in hippocampal aging: statistical profiling identifies novel processes correlated with cognitive impairment. *J Neurosci* 23(9):3807-19.

Bruunsgaard H. & Pedersen BK. (2003). Age-related inflammatory cytokines and disease. *Immunol Allergy Clin North Am* 23(1):15-39.

Budovsky A, Abramovich A, Cohen R, Chalifa-Caspi V, Fraifeld V. (2007). Longevity network: construction and implications. *Mech Ageing Dev* 128(1):117-24.

Butterfield DA. & Sultana R. (2007). Redox proteomics identification of oxidatively modified brain proteins in Alzheimer's disease and mild cognitive impairment: insights into the progression of this dementing disorder. *J Alzheimers Dis* 12(1):61–72.

Candore G., Balistreri C.R., Listì F., Grimaldi M.P., Vasto S., Colonna-Romano G., Franceschi C., Lio D., Caselli G., Caruso G. (2006). Immunogenetics, gender, and longevity. *Ann N Y Acad Sci.* 1089:516-37.

Chung HY., Cesari M., Anton S., Marzetti E., Giovannini S., Seo AY., Carter C., Yu BP., Leeuwenburgh C. (2009). Molecular inflammation: underpinnings of aging and agerelated diseases. *Ageing Res Rev* 8(1):18-30.

Crimmins E., Vasunilashorn S., Kim J.K., Alley D. (2008). Biomarkers related to aging in human populations. *Adv Clin Chem* 46:161-216.

Effros RB. (2000). Long-term immunological memory against viruses. *Mech Aging Dev* 121 (1-3):161–171.

Fulop T., Larbi A., Douziech N., Levesque I., Varin A., Herbein G. (2006). Cytokine receptor signalling and aging. *Mech Ageing Dev* 127(6):526-37.

Golden TR., Melov S. (2007). Gene expression changes associated with aging in C. elegans. *WormBook* 12:1-12.

Harman D. (2006). Free radical theory of aging: An update. *Ann N Y Acad Sci* 1067: 1–12.

Harries LW., Hernandez D., Henley W., Wood AR., Holly AC., Bradley-Smith RM., Yaghootkar H., Dutta A., Murray A., Frayling TM., Guralnik JM., Bandinelli S., Singleton A., Ferrucci L., Melzer D. (2011). Human aging is characterized by focused changes in gene expression and deregulation of alternative splicing. *Aging Cell* doi: 10.1111/j.1474-9726.2011.00726.x.

Herbig U., Ferreira M., Condel L., Carey D., Sedivy JM. (2006). Cellular senescence in aging primates. *Science* 311(5765):1257.

Herndler-Brandstetter D., Cioca DP., Grubeck-Loebenstein B. (2006). Immunizations in the elderly: do they live up to their promise? *Wien med Wochenschr* 156(5-6): 130–141.

http://gan.usc.edu/
http://metlin.scripps.edu/
http://www.hmdb.ca/
http://genomics.senescence.info/
http://netage-project.org/
http://www.mark-age.eu/

Janzen V., Forkert R., Fleming HE., Saito Y., Waring MT., Dombkowski DM., Cheng T., DePinho RA., Sharpless NE., Scadden DT. (2006). Stem-cell ageing modified by the cyclin-dependent kinase inhibitor p16INK4a. *Nature* 443(7110):421-426.

Kerrien S., Alam-Faruque Y., Aranda B., Bancarz I., Bridge A., Derow C., Dimmer E., Feuermann M., Friedrichsen A., Huntley R., Kohler C., Khadake J., Leroy C., Liban A., Lieftink C., Montecchi-Palazzi L., Orchard S., Risse J., Robbe K., Roechert B., Thorneycroft D., Zhang Y., Apweiler R., Hermjakob H. (2007). IntAct open source resource for molecular interaction data. *Nucleic Acids Res* 35:D561–D565.

Kilpatrick RD., Rickabaugh T., Hultin LE., Hultin P., Hausner MA., Detels R., Phair J., Jamieson BD. (2008). Homeostasis of the naive CD4+ T cell compartment during aging. *J Immunol* 180(3):1499–1507.

Lawton KA., Berger A., Mitchell M., Milgram KE., Evans AM., Guo L., Hanson RW., Kalhan SC., Ryals JA., Milburn MV. (2008). Analysis of the adult human plasma metabolome. *Pharmacogenomics* 9(4):383-97.

Licastro F., Candore G., Lio D., Porcellini E., Colonna-Romano G., Franceschi C., Caruso C. (2005). Innate immunity and inflammation in Aging: a key for understanding age-related diseases. *Immun Aging* 18;2:8.

Liu Y., Sanoff HK., Cho H., Burd CE., Torrice C., Ibrahim JG., Thomas NE., Sharpless NE. (2009). Expression of *p16INK4a* in peripheral blood T-cells is a biomarker of human aging. *Aging Cell* 8(4):439 448.

Martin K., Kirkwood TB., Potten CS. (1998). Age changes in stem cells of murine small intestinal crypts. *Exp Cell Res* 241(2):316-323.

McAuley MT., Kenny RA., Kirkwood TB., Wilkinson DJ., Jones JJ., Miller VM. (2009). A mathematical model of aging-related and cortisol induced hippocampal dysfunction. *BMC Neurosci* 10:26.

Melov S. (2002). Animal models of oxidative stress, aging and therapeutic antioxidant interventions. *Int J Biochem Cell Biol* 34(11):1395–1400.

Mirzayans R., Scott A., Cameron M., Murray D. (2005). Induction of accelerated senescence by gamma radiation in human solid tumor-derived cell lines expressing wild-type *TP53*. *Radiat Res* 163(1):53-62.

Montuschi P., Barnes P., Roberts LJ 2nd. (2007). Insights into oxidative stress: the isoprostanes. *Curr Med Chem* 14(6):703-17.

Nijnik A., Woodbine L., Marchetti C., Dawson S., Lambe T., Liu C., Rodrigues NP., Crockford TL., Cabuy E., Vindigni A., Enver T., Bell JI., Slijepcevic P., Goodnow CC., Jeggo PA., Cornall RJ. (2007). DNA repair is limiting for haematopoietic stem cells during ageing. *Nature* 447(7145):686-690.

Oda K., Matsuoka Y., Funahashi A., Kitano H. (2005). A comprehensive pathway map of epidermal growth factor receptor signaling. *Mol Syst Biol* 1:2005.0010.

Pan F., Chiu CH., Pulapura S., Mehan MR., Nunez-Iglesias J., Zhang K., Kamath K., Waterman MS., Finch CE., Zhou XJ. (2007). Gene aging nexus: a web database and data mining platform for microarray data on aging. *Nucleic Acids Res* 35: D756-D9.

Rattan SI. (2006). Theories of biological aging: Genes, proteins and free radicals. *Free Radic Res* 40(12):10–12.

Ravanat JL., Di Mascio P., Martinez GR., Medeiros MH., Cadet J. (2000). Singlet oxygen induces oxidation of cellular DNA. *J Biol Chem* 275(51): 40601–40604.

Rikans LE., DeCicco LA., Hornbrook KR., Yamano T. (1999). Effect of age and carbon tetrachloride on cytokine concentrations in rat liver. *Mech Ageing Dev* 108(2):173-182.

Rossi DJ., Bryder D., Seita J., Nussenzweig A., Hoeijmakers J., Weissman IL. (2007) Deficiencies in DNA damage repair limit the function of haematopoietic stem cells with age. *Nature* 447(7145):725 729.

Rossi DJ., Bryder D., Zahn JM., Ahlenius H., Sonu R., Wagers AJ., Weissman IL. (2005). Cell intrinsic alterations underlie hematopoietic stem cell aging. *Proc Natl Acad Sci U S A* 102(26):9194-9199.

Roubenoff R., Harris TB., Abad LW., Wilson PW., Dallal GE., Dinarello CA. (1998). Monocyte cytokine production in an elderly population: effect of age and inflammation. *J Gerontol A Biol Sci Med Sci* 53(1):M20-6.

Sansoni P., Vescovini R., Fagnoni F., Biasini C., Zanni F., Zanlari L., Telera A., Lucchini G., Passeri G., Monti D., Franceschi C., Passeri M. (2008). The immune system in extreme longevity. *Exp Gerontol* 43(2):61-65.

Sanz A., Pamplona R., Barja G. (2006). Is the mitochondrial free radical theory of aging intact? *Antioxid Redox Signal* 8(3-4):582–599

Schultz E. & Lipton BH. (1982). Skeletal muscle satellite cells: changes in proliferative potential as function of age. *Mech Ageing Dev* 20(4):377-383.

Sedelnikova OA., Horikawa I., Zimonjic DB., Popescu NC., Bonner WM., Barrett JC. (2004). Senescing human cells and ageing mice accumulate DNA lesions with unrepairable double-strand breaks. *Nat Cell Biol* 6(2):168-170.

Su S., Jin Y., Zhang W., Yang L., Shen Y., Cao Y., Tong J. (2006). Aberrant promoter methylation of *p16INK4a* and *O6-methylguanine-DNA methyltransferase* genes in workers at a Chinese uranium mine. *J Occup Health* 48(4):261 266.

Twine NA., Janitz K., Wilkins MR., Janitz M. (2011). Whole transcriptome sequencing reveals gene expression and splicing differences in brain regions affected by Alzheimer's disease. *PLoS One* 6(1):e16266.

Uchida K. (2003). 4-Hydroxy-2-nonenal: a product and mediator of oxidative stress. *Prog Lipid Res* 42(4):318–343.

Walkley CR., Fero ML., Chien WM., Purton LE., McArthur GA. (2005). Negative cell-cycle regulators cooperatively control self-renewal and differentiation of haematopoietic stem cells. *Nat Cell Biol* 7(2):172-178.

Wang Y., Schulte BA., LaRue AC., Ogawa M., Zhou D. (2006). Total body irradiation selectively induces murine hematopoietic stem cell senescence. *Blood* 107(1):358-366.

Warner HR. (2004) Current status of efforts to measure and modulate the biological rate of aging. *J Gerontol A Biol Sci Med Sci* 59(7):692-6.

Webster RG. (2000). Immunity to influenza in the elderly. *Vaccine* 18(16): 1686–1689.

Weiskopf D., Weinberger B., Grubeck-Loebenstein B. (2009). The aging of the immune system. *Transpl Int* 22(11):1041–1050.

Wheeler HE., Metter EJ., Tanaka T., Absher D., Higgins J., Zahn JM., Wilhelmy J., Davis RW., Singleton A., Myers RM., Ferrucci L., Kim SK. (2009). Sequential use of transcriptional profiling, expression quantitative trait mapping, and gene association implicates MMP20 in human kidney aging. *PLoS Genet* 5(10):e1000685.

Whisler RL., Newhouse YG., Bagenstose SE. (1996). Age-related reductions in the activation of mitogen-activated protein kinases p44mapk/ERK1 and p42mapk/ ERK2 in human T cells stimulated via ligation of the T cell receptor complex. *Cell Immunol* 168(2): 201–210.

Zahn JM., Poosala S., Owen AB., Ingram DK., Lustig A., Carter A., Weeraratna AT., Taub DD., Gorospe M., Mazan-Mamczarz K., Lakatta EG., Boheler KR., Xu X., Mattson

MP., Falco G., Ko MS., Schlessinger D., Firman J., Kummerfeld SK., Wood WH 3rd., Zonderman AB., Kim SK., Becker KG. (2007). AGEMAP: a gene expression database for aging in mice. *PLoS Genet* 3(11):e201.

Zhang J., Goodlett DR., Peskind ER., Quinn JF., Zhou Y., Wang Q., Pan C., Yi E., Eng J., Aebersold RH., Montine TJ. (2005). Quantitative proteomic analysis of age-related changes in human cerebrospinal fluid. *Neurobiol Aging* 26(2):207-27.

Zhu Y., Carvey PM., Ling Z. (2006). Age-related changes in glutathione and glutathione-related enzymes in rat brain. *Brain Res* 1090(11):35–44.

5

New Targets for the Identification of an Anti-Inflammatory Anti-Senescence Activity

Patrizia d'Alessio[1], Annelise Bennaceur-Griscelli[2],
Rita Ostan[3] and Claudio Franceschi[3]

[1]University Paris Sud-11 ESTEAM Stem Cell Core Facility and AISA Therapeutics, Evry,
[2]University Paris Sud-11 ESTEAM Stem Cell Core Facility and Inserm U935, Villejuif,
[3]Department of Experimental Pathology, University of Bologna, Bologna,
[1,2]France
[3]Italy

> "L'esprit étant précisément une force qui peut
> tirer d'elle même plus qu'elle ne contient,
> rendre plus qu'elle ne reçoit,
> donner plus qu'elle n'a."
>
> Henri Bergson (1911)
> La conscience et la vie

1. Introduction

Aging is necessarily associated to cell senescence, but is it its sole motor or phenomenon? Replicative cell senescence can be accelerated by stress, inflammation and uneven life conditions. We talk about stress-induced premature senescence when cell metabolism is exposed to a systemic profile of cortisol and catecholamines and inflammation is maintained by a high profile of cytokines. These latter are detrimental for cells and stem-cells, accelerating the senescence of both.

Aging is increasingly an issue in developed countries as life expectancy increases and birth rate decreases. These demographic trends have led to a strong increase of age-related pathologies, and an understanding of immune senescence promises to limit the development and progression of these diseases. Thus, "immunosenescence" is the term coined for the age-associated decrease in immune competence that renders individuals more susceptible to disease and increases morbidity and mortality due to infectious diseases in the elderly compared with the young (Franceschi et al., 2000). The main observed result at old age is a decrease in adaptive immunity and an increase of low-grade chronic inflammatory status, which has been referred to as "inflammaging" (De Martinis et al., 2005).

Inflammaging is pivotal in many ways and determines whether an individual will become either a healthy centenarian or will have to face sickness and depression for much of his/her life. It is well known that major age related pathologies such as cardiovascular disease, metabolic syndrome and frailty are associated with a progressive low grade inflammation process. While these diseases are thoroughly investigated, the role of the inflammatory process in other important age-related diseases involving the CNS such as depression and dementia, is relatively neglected. The importance of this topic is reinforced by the hypothesis that inflammation likely plays a major role also in CNS diseases such as chronic mood disorders and schizophrenia (Franceschi et al., 2001).

Ever-increasing longevity has produced the ambition for a personalized lifestyle and consequently the introduction of new standards of health. In this context, the aging population is comparable to a new subject, to which science and society will have to bring up new responses and solutions. Mainly based on prevention and maintenance, several indicators have been identified as preventing degenerative issues, aimed at healthy aging. Environmental quality and the role of nutrition are important elements of this strategy.

The importance of microbiome (Biagi et al., 2010) in the maintenance of host health has been recognized for years. Recent studies suggest an association between inflammation status and the presence of chronic disease in the elderly. They also indicate that an altered host-gut microbiota might contribute to maintaining a low systemic inflammatory status in the elderly. Nevertheless, the aging process and longevity in particular also depend on genetic and metabolic stability, as well as resistance to stress. However, if stress persists, it may lead to chronic disease, thus accelerating aging. Moderate stress, independent of conventional risk factors, can also induce a potent alteration in health as stressful life conditions induce a systemic pro-inflammatory status, consequently shortening life quality and lifespan.

In the present work, we would like to focus on the relationship between cell senescence at a daily scale and long-term consequence such as the loss of performance of organs, tissues and cells, and aging as it appears with osteoporosis, memory loss and immune-senescence (Ostan et al., 2008).

Inflammation is a critical defense mechanism, that, uncontrolled, contributes to chronic conditions with inflammatory pathogenesis. Markers of inflammation indicate vascular endothelial activation and dysfunction (d'Alessio, 2004; AISA Patent Family n°1).

Chronic inflammation appears to be determinant, in that it also affects functional aspects of stem cells, which are crucial to the maintenance of long term homeostasis of organs and tissues during lifetime. In fact, as reports from mice models and more recently from humans including centenarians (Bagnara et al., 2000) have confirmed, stem cells also undergo aging, as do somatic differentiated cells (Chambers et al., 2007; Ergen & Goodell, 2009). Most of the results have shown that the stem cell compartment becomes compromised, not by a quantitative loss, but by the progressive loss of function. These age-related changes have been suggested to be due to factors intrinsic to the stem cell (SC) including epigenetic changes and expression of transcription factors.

Aging can also be due to extrinsic (Rossi, 2008) environmental factors including the microenvironment of the SC local context (Bagnara et al., 2000). Very recent evidence has shown that there is a dialogue between the niche and the stem cell compartment leading to the maintenance of cellular homeostasis (Rossi et al., 2005). One of the key factors of the degeneration of the stem cell within its niche is certainly the inflammatory challenge (Chambers et al., 2007). Moreover, the dysfunction of stem cells is probably the result of epigenetic events. We consider the usefulness of studying strategies that could partially reverse it. An indirect evidence linking stress and lifespan comes from the studies of Linda Buck showing that human anti-depressant drug mianserin and serotonin receptor antagonists ser-3 and ser-4 were able to increase lifespan in *C. elegans* (Petrascheck et al., 2007). Other natural substances (such as curcumin) that have been characterized for their anti-inflammatory activity could be of the same value. AISA terpens have shown their incidence on replicative senescence in cells, as well as anti-inflammatory and anti-stress effects in pre-clinical and clinical studies (Bisson et al., 2008).

2. Current theories

2.1 "Inflammaging" and "SIPS"

The concept based on observations in immunology, linking inflammation with aging was proposed by Claudio Franceschi (Franceschi et al., 2000; Franceschi et al., 2007), whereas the concept based on proteomic analysis, linking stress to the appearance of premature senescence (SIPS, abbreviation of Stress Induced Premature Senescence) was proposed by Olivier Toussaint. These two researchers were showing that *aging stands in a biologically relevant link to inflammation*. According to the stochastic theories of aging, damage that accumulates with time in the cellular components is responsible for cellular aging. Some sort of premature senescence would appear when the damage level is artificially increased due to the presence of stressing agents at sub-cytotoxic level. Several models have shown that after sub-cytotoxic long-term stresses, human diploid fibroblasts (HDFs) display biomarkers of replicative senescence (RS), which led to the concept of SIPS (Dierick et al., 2002) as compared to telomere-dependent RS, changes accounting for "molecular scars" of sub-cytotoxic stresses.

2.2 Judith Campisi and her double-edged sword theory of cellular senescence

Pr. Campisi is internationally known for the work she has performed on cellular aging, genome stability and tumor suppressor genes during the last 20 years. After 11 years at the Lawrence Berkeley National Laboratory, she now works closely with several laboratories at the Buck Institute to understand and manipulate the cell phenotypes of characteristic of aging, cancer and age-related degeneration (Campisi, 2011). Trying to understand the cellular and molecular biology of aging, she has studied the importance of the cellular senescence, cell death and the effects of DNA damage regarding premature aging and cancer. Campisi's recent work indicates an interesting new insight on cell aging (senescence/death) and both cancer (hyper-proliferation) and degenerative diseases (Bazarov et al., 2010; de Keizer et al., 2010).

2.3 ROS, mitochondria decline and DNA damage

2.3.1 Harmann's theory on free radical damage

Harmann's theory on free radical damage formulated in 1956 has generated several important insights and further raised the importance of DNA damage and DNA repair for aging and longevity respectively (reparosome dependent mechanisms).

2.3.2 Miroslav Radman's vaccination

Radman's work accounts for several mechanisms of DNA repair that *E. coli* have extrapolated, ex. gr. the exceptional resistance of *D. radiodurans* to DNA damaging agents (radiations and chemicals) and to desiccation (Zahradka et al., 2006). He described the global mechanism of DNA fragment reassembly as a two-stage process, which involves mutually dependent DNA replication and recombination events (Babic et al., 2008), and defined key steps in this most efficient and precise DNA repair process, assigned gene and protein function to critical repair steps, and showed the kinetics of the key steps.

2.4 Tom Kirkwood's systems-biology approach

Starting from the damage theory published in *Cell* in 2005, the extent of investment of organism's genome in survival stands at the heart of the 'disposable soma' theory, formulated in early 1977 in *Nature*. What is possible instead, according to this scientist, is to slow the rate at which damage accumulates, given the malleability of the aging process (Kirkwood, 2008). *Here we are conceptually very close to a putative concept of 'reversibility' that has nothing to do with repair and that we claim to be able to demonstrate.* As Kirkwood says 'the devil is in the details', i.e. where should the critical point be situated when random damage becomes damage oriented to the development of frailty? Indeed the concept of 'robustness' as opposed to 'vulnerability' is at the heart of systems biology (Kitano, 2007). In spite of numerous affecting defects in cells, tissues and molecules, none of them contributes to characterize the senescent phenotype. Moreover the systematic stochasticity of all events with resulting variability and increasing multiplicity, impairs the identification of a unifying element for a true explanation of the increasing lifespan phenomenon in humans, described as "healthy aging".

2.5 Linda B. Buck's (2004 Nobel Prize) bio-products targeting mechanisms of olfaction and lifespan in *C. elegans*

Determinants of aging and lifespan by Buck's laboratory on the short-lived nematode *Caenorhabditis elegans* (*C. elegans*) have identified a number of genes that can influence the lifespan of this organism (Petrascheck et al., 2007). Looking for the identification of chemicals that would increase *C. elegans* lifespan, studies on the endogenous targets of Buck's chemicals provided additional insights into the underlying mechanisms of aging. By conducting a high-throughput screening, she identified 100 compounds that increase *C. elegans* lifespan when given only during adulthood. The animal's lifespan can be increased about 30 percent by mianserin, a drug used as an antidepressant in humans. This effect requires a specific serotonin receptor, SER-3 or SER-4, a receptor for another neurotransmitter, octopamine. *The drug increases lifespan via mechanisms linked to dietary restriction.* Curiously, the drug does not appear to reduce food intake. One possible explanation for these findings is that the inhibitory effect of mianserin on SER-3 and SER-4 mimics a reduction in food intake and thereby triggers anti-aging mechanisms associated with dietary restriction. This approach is particularly interesting because of the concomitant effects on senescence and mood – via the management of the inflammatory reaction – such as observed for bio-products identified by us.

2.6 Sirtuins anti-inflammatory action

Sirtuins (Dali-Youcef et al., 2007) (LP Guarente laboratory) exhibit protection of DNA from metabolic damage and are therefore thought to affect regulatory systems of longevity (Donmez

& Guarente, 2010). The idea that sirtuins can affect inflammation comes from the evidence that they are able to inactivate NF-κB, which is not according to us, a sufficient element to claim for an anti-inflammatory and thus potentially anti-aging activity. These results, shown into a variety of lower organisms, have been transferred into a transgenic mice model by M. Serrano (Spanish National Cancer Center, Madrid) who confirmed that Sir-1 improves healthy aging concluding about a consequent anti-aging activity. We think that sirtuins may protect from metabolic syndrome (Herranz et al., 2009) and decrease spontaneous cancer.

2.7 Interaction of biological and social factors

There is increasing recognition that intra-uterine life can influence the health of the newborn well into his/hers adult years, although the mechanisms through which these occur are, as yet, unclear. These evidences form the basis of the Barker Hypothesis, formulated on human aging (de Kretser, 2010), which links under-nutrition in utero, leading to low birth weight, with an increased risk of hypertension, coronary artery disease, stroke, diabetes and the metabolic syndrome in adulthood. All these syndromes may be the result of impaired nephrogenesis and a greater susceptibility to renal disease, impaired development of the endothelium and increased sensitivity to glucocorticoid hormones (Froy & Miskin, 2010; Kolokotrones et al., 2010). Given that the *in utero* 'environmental status' affects the organ function many years later, there is a strong possibility that the mechanism will involve imprinting of genes. But also, concerning the social impact on aging, the psychological stress in adult life has shown its risk for development of psychiatric diseases, based on the recognition of molecular makers of aging (von Zglinicki et al., 2001). In women aged 20–50 years, those with the highest levels of psychological stress had the shortest telomeres and the lowest telomerase activity in peripheral blood leukocytes, and showed the highest levels of oxidative stress with consequent impairment of SC repair capacity and generation of metabolic syndrome diseases (Mathieu et al., 2010; Ingram & Mussolino, 2010).

2.8 Stem cell based approach to the study of intrinsic senescence

2.8.1 Telomere shortening and its implication for SC niches' aging

Telomeres are specialized structures that adorn the ends of human chromosomes, essential for the integrity of chromosomes. These nucleoprotein caps are maintained by the enzyme telomerase. The importance of adequate telomerase activity and maintenance of telomere length for both replicative potential in culture and aging in organisms was initially inferred from studies of primary human fibroblasts. In culture, division of fibroblasts results in progressive telomere attrition, culminating in a state of proliferative arrest — or cellular senescence — after a finite number of cell divisions, a barrier known as the Hayflick limit. Moreover, enforced expression of TERT, the catalytic subunit of telomerase, in cultured human fibroblasts stabilized telomere length and endowed the cells with unlimited replicative potential without engendering malignant properties. The remarkable capacity of experimentally induced telomerase activity to circumvent senescence and allow indefinite growth has been documented in many other human cell types. Telomere dynamics bear relevance to the processes of aging, and human population studies have correlated decreased telomere length in peripheral blood leukocytes with higher mortality rates in individuals who are more than 60 years old; a recent large cohort study did report a positive link between telomere length and years of healthy life (Sahin & DePinho, 2010); another recent study on centenarians and their offspring found a positive link between telomere length and longevity; in particular, those with longer telomeres had an overall improved health profile (with decreased age-associated disease and better cognitive function and lipid profiles) with respect to controls (Atzmon et al., 2010).

2.8.2 SC's aging is relevant to cell senescence

De Pinho's lab at Boston Harvard Medical School has established that we age, in part, because our self-renewing stem cells grow old as a result of heritable intrinsic events, such as DNA damage, as well as extrinsic forces, such as changes in their supporting niches. Mechanisms that suppress the development of cancer, such as senescence and apoptosis, which rely on telomere shortening and the activities of p53 and p16 (INK4a), may also induce an unwanted consequence: a decline in the replicative function of certain stem-cell types with advancing age. This decreased regenerative capacity appears to contribute to some aspects of mammalian aging, with new findings pointing to a *'stem-cell hypothesis' for human age-associated conditions such as frailty, atherosclerosis and type 2 diabetes*. This approach is of particular interest for us, because of the possibility to look at the role of reprogrammed SC in the identification of cell/tissue signature of characterized by increased repair activity.

2.8.3 Epigenetic deregulation responsible for SC decline

Goodell M, Texas and Scadden D, have characterized inflammatory stress responsible for niche component degeneration (Chambers et al., 2007). In the continuity of this work, organ specific targets have been looked at. Muscle and adipose tissue are concerned with age-related muscle dysfunction (Degens, 2007). Stem cell reprogramming, as well as the adipocyte role in promoting accelerated cell senescence and aging on the base of its capacity to stock pro-inflammatory cytokines, have to be taken into account (Naveiras et al., 2009).

3. Our approach: Cell senescence as multi-factorial process

Cell senescence is a pleiotropic process, initially determined by genetic and environmental conditions. In the experimental work presented here, we focus on replicative senescence phenomena in a nearly non dividing human cell type, the endothelial cell lining the vascular wall. Among its numerous functions, endothelium is also implicated in the regulation of several steps of the inflammatory process. Two scenarios are prone to accelerate senescence in inflammation.

1. One is the tissue repair sequence that follows the primary neutralization of the microbial or traumatic agent. In this case intense neo-angiogenesis of low quality as well as tissue replacement take place.
2. On the other hand, it is possible that the incoming *noxa* will not be neutralized (here the stress model is extremely useful to illustrate endogenous cell senescence) and the pro-inflammatory stimulation persists.

Thereby, a high concentration of pro-inflammatory cytokines occurring during the inflammatory response, will promote the appearance of premature senescence of endothelial cells, independently of the age of the subject. When chronic inflammatory disease develops, it contributes to the deterioration of endothelial cell function, further increasing their premature senescence. This ancient if not universal mechanism is conserved in stem cell, niches undergoing senescence by the same (pro-inflammatory) mechanisms. In fact, the cytokine TNF-α (Tumor Necrosis Factor-α) is mainly responsible for the major cell modifications of the senescent cell that we have identified in three categories:

1. loss of contact inhibition;
2. overexpression of cell adhesion molecules (d'Alessio, 2004);
3. modification of cell morphology sustained by the development of stress fibers into a large mono-directional fiber system (AISA Patent Family n°1).

Based on this "senescent" phenotype, we have launched a bio-guided research aiming at the reversibility of these characteristics. We have characterized a family of molecules able to inhibit the expression of inflammatory markers and, in particular, of adhesion molecule expression in endothelial cells following TNF-α stimulation. We have named these molecules "AISA" (Anti Inflammatory Senescence Actives), because not only a change in cell shape occurred but also the consequences of replicative senescence were reversed (*in vitro* studies). Moreover, *in vivo* studies with one of these compounds ("AISA 5203-L") showed an exceptional capacity to restore both the colon's enterocytes and dermis from pro-inflammatory agonists and toxic substances (AISA Patent Family n°3). Most probably the protective effect is due to the capacity of AISA 5203-L to inhibit circulating Tumor Necrosis Factor- α (TNF- α), Interlukin-6 (IL-6) and Interlukin-1 (IL-1), most relevant to the inflammatory reaction in relationship to the aging process by *inflammaging*. AISA 5203-L was used as treatment (oral administration) in an rat model for non-pathologic stress (defined by anxiety situations such as isolation or separation). It showed a compelling anti-stress activity, as measured by a FOB (Functional Observation Battery), whereby analgesic effects were associated to enhanced motility and less irritability.

4. Relevance of chronic inflammation

As humans grow older, systemic inflammation can inflict devastating degenerative effects throughout the body. Chronic inflammation is an underlying cause of many apparently unrelated, age-related diseases. This fact is often overlooked, yet persuasive scientific evidence exists that correcting a chronic inflammatory disorder will enable many of the infirmities of aging to be prevented or reversed.

When we envisage a link between aging and recurrent or chronic inflammation, we refer to pathological consequences of inflammation in well-documented medical literature. Regrettably, the origins as well as the consequences of systemic inflammation continue to be ignored. By following specific prevention protocols (such as weight loss), the inflammatory stimulation could be significantly reduced. An important role in preventing the onset of a chronic inflammatory condition has been attributed either to the practice of a physical activity or to the prescription of a personalized diet, or both. In the frame of the EU Capacities study RISTOMED (www.ristomed.eu), AISA Therapeutics treatment associated as dietary supplementation to a controlled diet in a cohort of elderly otherwise healthy individuals (65-85 years) was validated as anti-inflammatory medical food.

5. Low grade inflammation and cell degeneration

The immune function also is affected in aging. As lymphocyte function decreases, macrophages take over concomitantly with an enhanced secretion of inflammatory cytokines, such as TNF-α and IL-6. This mechanism is of vital importance for tissue defense from microorganisms (and anti-infectious defense is crucial in the elderly), but it also contributes to the progression of many degenerative diseases. Rheumatoid arthritis is a classic autoimmune disorder in which exceeding levels of cytokines such as IL-6, IL-1β and/or IL-8 are known to cause or contribute to the inflammatory syndrome.

Chronic inflammation is also involved in diseases associated to the metabolic syndrome resulting in atherosclerosis, heart valve dysfunction, obesity, diabetes, congestive heart failure, and digestive system diseases. Cancer and Alzheimer's disease have both been shown to benefit from a systemic inflammation for their progression. In aged people with multiple degenerative diseases, the inflammatory marker C-reactive protein is often elevated, indicating the presence of an underlying inflammatory condition. Moreover, when a cytokine blood profile is conducted on people in a weakened condition, an excess level of one or more of the inflammatory cytokines, e.g., TNF- α , IL-6, IL-1 β , as well as IL-8, are usually found.

6. Systemic markers of cell senescence

In 2000 the New England Journal of Medicine published several studies showing that the blood indicators of inflammation are strong predictive factors for determining susceptibility to undergo a heart attack. Many international studies subsequently validated this first communication (Ridker et al., 1997; Harris et al., 1999; Walston et al., 2002; Ziccardi et al., 2002; Clément et al., 2004).

Again, C-reactive protein represents a critical inflammatory marker. This marker indicates an increased risk for destabilized atherosclerotic plaque (here we are beyond senescence) and abnormal arterial clotting, which can lead to an acute heart attack. One of these studies (Ridker et al., 1997) showed that people with high levels of C-reactive protein were almost three times as likely to die from a heart attack. This also implicates that elevated C-reactive protein, IL-6 and other inflammatory cytokines indicate significantly greater risks of contracting or dying from specific diseases (heart attack, stroke, Alzheimer's disease).

Moreover, C-reactive protein and IL-6 could also predict the risk of all-cause mortality as addressed by a study conducted on a sample of 1,293 healthy elderly people (Harris et al., 1999) followed for a period of 4.6 years. Higher IL-6 levels were associated with a twofold greater risk of death. Higher C-reactive protein was also associated with a greater risk of death, but to a lesser extent than elevated IL-6. Subjects with both high C-reactive protein and IL-6 were 2.6 times more likely to die during follow up than those with low levels of both of these measurements of inflammation.

Thus it would seem that C-reactive protein and IL-6 may be useful for identification of high-risk subgroups for anti-inflammatory interventions. Indeed, in 2003, the American Heart Association and Centers for Disease Control & Prevention (CDC) jointly endorsed the C-reactive protein test and screen for coronary-artery inflammation to identify patients at risk for heart attack. Interestingly, together with other relevant markers, C-reactive protein has been importantly diminished by AISA treatment.

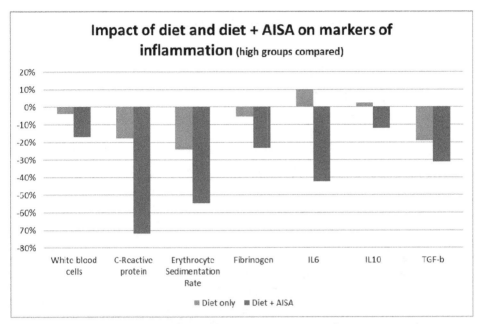

Fig. 1. In the aforementioned RISTOMED study, relevant pro-inflammatory markers characterizing the specific degree of the inflammatory reaction in the studied population, as well as the capacity by AISA compounds to lower them, have been measured. In particular, CRP (aspecific marker) and IL-6 (specific marker, relevant in arthritis) were significantly lowered. (For each inflammatory marker the T1 data were taken as baseline and the difference T56-T1 was expressed as a % of T1).

7. Frailty and inflammatory profiles

Results addressing the role of inflammation during aging were further developed by a new study on almost 5,000 elderly people (Walston et al., 2002) that have compared frail seniors to their healthier counterparts for the presence of increased inflammation markers. Associated to the elevated blood inflammatory markers, these frail seniors also tended to show an enhanced clotting activity, muscle weakness, fatigue and disability when compared to the not frail elderly people. For the moment, we are not able to document to what extent these clinical outcomes are the origin or the consequence of inflammatory status, but once we recognize that they are interdependent, we can address them by prevention and treatment.

Collectively, these studies should motivate public health policies as well as conscious individuals to monitor their inflammatory status. If C-reactive protein is elevated, then the Inflammatory Cytokine Test Panel would be also highly recommended. Secondly, all those who suffer from any type of chronic disease may also consider to access to the Inflammatory Cytokine Test Panel in order to identify the specific inflammatory mediator that is causing or contributing to their health problem.

Pro-Inflammatory Citokines

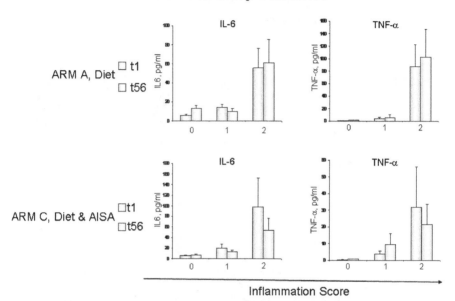

Fig. 2. The histograms show the levels of IL-6 and TNF-α in two arms of Ristomed project: the diet (Arm A) and of the diet with AISA (Arm C). Subjects were clustered on the basis of the inflammation score. Considering "low" and "medium" inflamed subjects (IS=0 and IS=1, the left and centre pairs in each chart), diet (arm A) and diet with AISA (Arm C) seems to have no effect on IL-6 and TNF-α levels. Considering "high inflamed" subjects (IS=2, the right hand pair in each chart) Arm A seems to slightly increase IL-6 and TNF-α whereas arm C (terpens) lowers them.

8. Anti-inflammation to decrease cell senescence

For a long time the identification and production of anti-inflammatory drugs has concentrated on symptom remission, ignoring vascular mechanisms of inflammation and their consequences in the long term, such as increased cell replicative senescence, promoting degenerative disease. Now we know that the presence of pro-inflammatory markers in blood may be in part responsible for degenerative diseases, characteristic of aging and reposing on accelerated cell senescence. Through their clinical relevance, inflammatory markers can witness, via endothelial dysfunction, their implication in the occurring of disease (Edelman, 1993; Vanier, 2005; Farhadi et al., 2003; Garcìa-Cardena & Gimbrone, 2006; Ingber, 2003; Marconi et al., 2003).

In the past, monoterpens, sesquiterpenes, diterpens and triterpens (Zhang et al., 2006) have been characterized by several authors (Vigushin et al., 1998; Crowell et al., 1992; Hardcastle et al., 1999) as potential anti-cancer drugs on the basis of *in vitro* and *in vivo* studies, but their role as anti-inflammatory drugs has remained elusive.

In 2002, following our bio-guided selection, performed by means of our *in vitro* cell biology screening platform, we were able to identify 4 molecules out of 2000 as able to reverse inflammatory markers in senescent endothelial cells. Focusing on functional criteria, we were aiming at the identification of non-toxic molecules able to inhibit *in vitro* the hallmarks of the inflammatory response, such as the expression of adhesion molecules (ICAM-1, VCAM-1, selectins), as well as the concomitant actin polymerization in endothelial cells. The four monoterpens were contained in plants extracts (kindly provided by the University of Hanoi) originating from regional medicinal plants.

9. Pre-clinical studies

After an acute toxicity study and several dose-response studies aiming at the appreciation of the therapeutic window, pre-clinical studies were performed on a female rat TNBS induced colitis model and a murine SHK TPA model. These studies showed that the inhibition of adhesion molecules were comparable in the *in vitro* / *in vivo* experiments (Yamada et al., 1992; Medeiros et al., 2007; AISA Patent Family n°3).

Our studies also allowed us to establish that the therapeutic window corresponded to the *in vivo* pharmacological active dose of 10 mg/kg given either *per os* or applied topically. Moreover, our *in vivo* data showed that plasma concentration of TNF-α is greatly reduced by the administration of AISA 5203-L and the score of post-lesional tissue regeneration was comparable with that of ibuprofen. Unlike ibuprofen, AISA 5203-L also importantly contributes to mood matching. Finally, on the quite differentiated capacity to elicit adhesion molecule expression following TNF stimulation by different steroid and non steroid anti-inflammatory drugs (Zhang et al., 2006), AISA molecules do persistently inhibit their expression.

10. AISA pertinence for combating cell senescence

Monoterpenes are a class of isoprenoid molecules derived from the anabolism of acetate by the mevalonic acid branch biosynthetic pathways of plants. *d*-Limonene for example, a major component of orange peel oil, is formed by the cyclization of the 10-carbon isoprene intermediate geranylpyrophosphate. Interest in *d*-Limonene came from the ability of the compound to inhibit carcinogenesis in the murine benzo(*a*)pyrene-induced skin tumor model and inhibition of dibenzopyrene-induced s.c. sarcomas. The mechanisms by which *d*-Limonene and other cyclic monoterpenes inhibit tumor growth have not been firmly established. Geranylpyrophosphate, the isoprene intermediate from which these compounds are derived, is required for synthesis of cholesterol, coenzyme Q (ubiquinone), and substrates used in the isoprenylation of several cellular proteins. Crowell *et al.* found that *d*-Limonene and other monoterpenes inhibited isoprenylation of M_r 21,000–26,000 proteins, including p21[ras] and other members of the ras family of GTP-binding proteins involved in signal transduction and growth regulation. The post-translational isoprenylation of these and other proteins is an essential covalent modification required for protein localization and function. For example, farnesylation is required for plasma membrane association and signaling function of p21[ras]. Other intracellular proteins require isoprenylation by addition of a farnesyl (15-carbon) or geranylgeranyl (20-carbon) group to the COOH terminus for localization to a cellular compartment or for interaction with other proteins. The four molecules, identified by the AISA Therapeutics cell biology platform for their specific anti-inflammatory activity, following an *in vitro* screening on endothelial targets associating cyto – protective and adhesion inhibiting activities, turned out to be monoterpens: more, geraniol, geranyl acetate, d-Limonene and iso-menthone are intimately linked by a metabolic loop.

Although data available emphasized the anti-cancer activities of geraniol and *d*-Limonene, we were tempted to find out about the *in vitro / in vivo* consistency of our data in models adapted to the study of acute and chronic inflammation. In confirmation to our *in vitro* results, the capacity of geraniol (AISA 5202-G) to inhibit the adhesion of leukocytes following TNF-α stimulation had already been established. As for *d*-Limonene, in consideration of the efficacy of its metabolite, perillyl alcohol (POH), already tested in clinical trials in patients with refractory solid malignancies (Miller et al., 2011), it seems plausible that it plays the role of a precursor. In conclusion, the complex sequence of events of the inflammatory response including endothelial adhesion molecule expression for the vascular recruitment of leukocytes to the site of injury, concomitant with actin polymerization challenges the signaling pathway of the rho GTPase family (Xu et al., 2009; Burridge & Wennerberg, 2004; Millan & Ridley, 2005; Dillon & Goda, 2005). The activation of these proteins requires a post-translational iso-prenylation. *We think that the same mechanisms of action of the anti-cancer effects reported for geraniol and d-Limonene could equally be at the origin of their anti-inflammatory properties, here reported.* This shared mechanism between cancer and inflammation again suggests the existence of a mechanism connecting stem cell biology and cancer proliferation.

11. Why stress is relevant for cell senescence

Important effects on mood in presence of stress situations had been documented by us in a rodent model thus motivating our choice to explore more in detail this unexpected effect (AISA Patent Family n°2; MacPhail, 1987; Shibeshi et al., 2007; Esler et al., 2008; Querè et al., 2009; May et al., 2009; Chandola et al., 2008).

As established by our Functional Observation Battery (FOB), *d*-Limonene was able to substantially contribute to pain tolerance and mood stabilization. However, the most intriguing result, was the fact that the stressed animal (by a so-called non-pathological stress stimulating anxiety, comparable to maternal deprivation), instead of developing a freezing attitude, following oral administration of *d*-Limonene, developed a "ludic" activity, starting to play with the wheel next to it. This is particularly interesting when compared to other mood or anxiety treating molecules, displaying substantially a hypnotic effect. Moreover the Ristomed study results obtained for quality of life assessment, SF-36v2™ Health Survey, *Summary* (PCS) *Mental Component Summary* (MCS), General Health Questionnarie-12 (GHQ-12) and mood by the State-Trait Anxiety Inventory-X (STAI-X) and Center for Epidemiologic Studies Depression Scale (CES-D) by use of questionnaires were interestingly confirming our findings on mood modulation, especially in females.

Arm		N	FEMALES - CES-D	mean	SD	differences in pairs		
						mean	SD	Sig. (2-code)
A	Diet	17	CES-D (T1)	8,94	7,267	3,06	5,080	0,025
		17	CES-D (T56)	5,88	5,183			
B	Diet & Probiotic	16	CES-D (T1)	8,19	5,671	2,06	5,579	0,160
		16	CES-D (T56)	6,13	5,620			
C	Diet & Terpene	16	CES-D (T1)	14,94	11,457	6,13	9,069	0,016
		16	CES-D (T56)	8,81	7,035			
D	Diet & Argan Oil	18	CES-D (T1)	8,06	8,335	2,06	3,208	0,015
		18	CES-D (T56)	6,00	7,507			

Arm		N	CES-D	mean	SD	differences in pairs		
						mean	SD	Sig. (2-code)
A	Diet	31	CES-D (T1)	6,71	6,659	2,55	4,280	0,002
		31	CES-D (T56)	4,16	4,670			
B	Diet & Probiotic	31	CES-D (T1)	6,74	5,416	2,13	4,924	0,022
		31	CES-D (T56)	4,61	5,149			
C	Diet & Terpene	29	CES-D (T1)	10,41	10,304	4,03	7,351	0,006
		29	CES-D (T56)	6,38	6,264			
D	Diet & Argan Oil	32	CES-D (T1)	8,28	7,809	1,66	5,033	1,66
		32	CES-D (T56)	6,63	7,487			

Fig. 3. CES-D evaluation showed (the analysis stratified by arm of study), that the significant differences between T1-T56 were in the arm "A" (p = 0,002), "B" (p = 0,022) and "C" (P = 0,006) but not in arm "D"; analyzing separately male and female, the significant statistical difference was confirmed for males only in the arm "A" (p = 0,38) and for females in the arm "A" (p = 0,025), "C" (p = 0,016) and "D" (P = 0,015) but not in arm "B", but by far the highest score difference was observed in arm "C" (mean difference score T1-T56 = 6,13 ± 9,069) that show the greatest improvement of how the subject feels and behaves in the preceding week.

12. Discussion

In summary, we claim that links between inflammation, senescence and stress so far addressed in a fragmentary way should be considered by an integrated approach to better elucidate the senescence process. Therefore the identification of molecules able to prove anti- inflammatory effective on replicative senescence having a subtle but tangible effect on mood became a way to put this link in evidence. In this regard, at the end of this chapter, I would evoke the historical and almost anecdotal properties of such molecules in food and recipes throughout the ages.

12.1 How inflammation takes advantage from ongoing cell senescence

Inflammatory diseases are numerous and systemic inflammation is a silent companion of stress and age. On the other hand, psychological stress in response to pain appears as an important customer of inflammation. Pharmacological strategies trying to inhibit inflammatory symptoms and related clinical episodes have gone far, and when properly prescribed can be considered successful despite recent side effects reported for several of them. But disease is, independently from its etio-pathology, a stressing agent by itself, able to anticipate inflammation by not yet totally unraveled mechanisms. Unfortunately, a sustained anti-inflammatory treatment is inevitably associated with adverse effects, thus opening a field of research and development for new, less or non-toxic and better tolerated anti-inflammatory strategies. In particular, we could provide evidence that the expression of vascular adhesion molecules is challenged by the most frequently-used anti-inflammatory steroid and non-steroid drugs when compared to the effect of a triterpen contained in an edible plant used by Chinese populations since centuries to prevent rheumatoid arthritis (Zhang et al., 2006).

12.2 How inflammation and stress define the senescent phenotype

Recently, much attention has been given to stress as promoter of disease and syndromes implied in health decline and we have addressed this issue in a review and a research article (d'Alessio, 2004; Bisson et al., 2008).

Indeed, compounds found in natural substances, mostly plants, have acquired a new status as valid pharmacological candidates for the development of new drugs preventing, maintaining and curing on the basis of body integrity and substantially addressing wellness more than health. We think that if the aging process depends on genetic stability, metabolic control, and resistance to stress, longevity in particular seems related to the latter. If responses to stress anticipate adaptation to an unacceptable disparity between real or imagined personal experience and expectation, they include adaptive stress, anxiety, and depression. However, if stress persists, it may lead to chronic diseases, ranging from inflammation and cancer to degenerative diseases. If in the past only extreme stress was acknowledged to induce immune and vascular alterations, such as infection or hypertension, now it is known that also moderate stress independent of conventional risk factors can induce a potent alteration of health conditions and consequently shorten life quality and lifespan. If inflammation is a critical defense mechanism, that, uncontrolled, contributes to chronic conditions with inflammatory pathogenesis, stressful life conditions turn out to induce a diffuse (systemic) pro-inflammatory status. Moreover, if sub-clinical chronic inflammation is an important pathogenic factor in the development of metabolic syndrome, a cluster of common pathologies, including cardiovascular disease, will include markers associated with endothelial activation and dysfunction.

13. Perspectives

In fact, the comprehension of the mechanisms underlying inflammation and neuro-inflammation in aging and age-related disease is of particular importance as far as public health is concerned, since these diseases are characterized by a high rate of prevalence in the western countries and have a great impact in terms of social and economic costs. A better understanding of the mechanisms that cause such diseases will help to design new therapeutic approaches, particularly useful in the early phases of the diseases.

In particular, the correlation of data from the analysis of inflammatory mechanisms and new treatments based on iPSC (induced Pluripotent Stem Cells), will provide potentially useful markers to researchers and clinicians for possible new targets for treatments based upon lowering of pro-inflammatory status in age-related diseases.

Tissue stem cells' fate and age-related phenomena are quite related. The anatomical and physiological changes associated with advancing age emerge with variable onset, pace and severity in individuals, and affect organs and tissue types both with highly mitotic and quiescent profiles. In the whole organism, the hallmarks of aging include loss of muscle mass (sarcopenia), decreased musculoskeletal mobility, reduction in bone mass (osteoporosis), thinning and reduced elasticity of skin (wrinkling). The aging haematopoietic system exhibits progressive altered immune profiles. During lifetime, our bodies possess a remarkable ability for extensive and sustained tissue renewal. This continuous self-renewal capacity is maintained by reservoirs of somatic tissue stem cells (Sharpless & DePinho, 2007). These tissue stem cells have garnered increasing attention in aging and regenerative research given accumulating evidence that age-associated physiological decline, particularly in highly proliferative organs, parallels blunted proliferative responses and misdirected differentiation of resident tissue stem cells.

By a multidisciplinary approach using reprogrammed stem cell lines new bioassays for the identification of specific cell signatures, both genetic and epi-genetic, could be designed. For example, crossing cell lines from dementia-free healthy centenarians with Down syndrome's would allow to identify intrinsic or extrinsic maintenance mechanisms. Advanced post-genomic techniques may also be aimed to the definition of a signature of response of cells to different challenges (e.g: inflammation effectors, pathogens). In particular :

We need to validate of the hypothesis of "inflammaging" : characterization of the contribution of inflammation of the acceleration of senescence within biobank cell collections (dementia free centenarians vs. Down syndrome) and its consequences for phenotype stability.

We need to add new evidences linking inflammation and aging ("extrinsic aging") which does encompass the stem cell compartment as well, contributing to new insight on the "intrinsic" cell senescence mechanisms as they may depend on whole organisms compliance.

In addition, we need to validate the relevance of bio-products as cyto-protectants on engineered differentiated cells from patients specific iPSC : can they enhance their maintenance contributing to the inhibition of the "intrinsic" aging mechanism and / or the accelerated replicative senescence due to inflammatory challenge ("extrinsic" aging) ?

Finally, validate the possibility that *ex vivo* cell collections from dementia-free centenarians, as well as Down's syndrome could be characterized at the biologic, genetic and epigenetic level, characterizing their inflammatory phenotype (by system biology approach).

Coming towards an end, the reader should be alerted to a few conclusive remarks (Galeno, 1973; Issuree et al., 2009; Atzei, 2004).

Today our study contributes to enhance evidence for the relevance of a specific class of molecules contained in substances which may have been used either in the domesticated fruit and vegetable environment as food, such as the oleocantal ibuprofen–like molecule contained in olive oil (Beauchamp et al., 2005), or as ritual substances, such as the incense and myrrh (Nomicos, 2007) containing anti-inflammatory and mood modulating terpens. We presume that for centuries these raw materials were integrated in sacred recipes devoted to the maintenance of health and the prevention of aging, because of their content in biological active molecules, displaying their curing properties, either as anti-inflammatory remedies or inducing mood modulation allowing an enhanced perception of life.

But the question we will not be able to avoid is, if cell senescence can be moderated by plant molecules, we eat and drink (caffeine), drugs as potent as man ever has known, raising the problem of a new pharmaco-vigilance and a global change in our assumption on what is healthy and what is not, regardless of the elasticity of our lifespan.

14. References

AISA Patent Family n°1, "Composition for treating or preventing cell degeneration using at least one molecule capable of maintaining adhesion molecule expression reversibility and inhibit vascular endothelium actin fiber polymerization" (FR 2 869 230, WO 2005/105074, EP1748771, US-2009-0012162-A1).

AISA Patent Family n°2 « Use of a monoterpen to treat or prevent stress » (EP1990047, WO 2008/138905).

AISA Patent Family n°3 « Use of a monoterpene to increase tissue repair » (EP2042167, WO 2009/040420).

Atzei, A.D. (2004) « Le piante nella tradizione popolare della Sardegna »   (Plants of the popular tradition in Sardinia) *Carlo Delfino ed.*, Sassari, Italy.

Atzmon, G., Cho, M., Cawthon, R.M., Budagov, T., Katz, M., Yang, X., Siegel, G., Bergman, A., Huffman, D.M., Schechter, C.B., Wright, W.E., Shay, J.W., Barzilai, N., Govindaraju, D.R., and Suh, Y. (2010) Evolution in health and medicine Sackler colloquium: Genetic variation in human telomerase is associated with telomere length in Ashkenazi centenarians. *Proc Natl Acad Sci U S A.* 107 Suppl 1, pp. 1710-7.

Babic, A., Linder, A.B., Vulic, M., Stewart, E.J., & Radman, M. (2008) Direct visualization of horizontal gene transfer. *Science*, 319(5869), pp. 1533-6.

Bagnara, G.P., Bonsi, L., Strippoli, P., Bonifazi, F., Tonelli, R., D'Addato, S., Paganelli, R., Scala, E., Fagiolo, U., Monti, D., Cossarizza, A., and, Bonafè, M., & Franceschi, C. (2000) Hemopoiesis in healthy old people and centenarians: well-maintained responsiveness of CD34+ cells to hemopoietic growth factors and remodeling of cytokine network. *J Gerontol A Biol Sci Med Sci.* , 55, pp. B61-6.

Bazarov, A., Van Sluis, M., Hines, W.C., Bassett, E., Beliveau, A., Campeau, E., Mukhopadhyay, R., Lee, W.J., Melodyev, S., Zaslavsky, Y., Lee, L., Rodier, F., Chicas, A., Lowe, S.W., Benhattar, J., Ren, B., Campisi, J., & Yaswen, P. (2010) p16 (INK4a)-mediated suppression of telomerase in normal and malignant human breast cells. *Aging Cell*, 9, pp. 736-46.

Beauchamp, G.K., Keast, R.S., Morel, D., Lin, J., Pika, J., Han, Q., Lee, C.H., Smith, A.B., & Breslin, P.A. (2005) Phytochemistry: ibuprofen-like activity in extra-virgin olive oil. *Nature*, 437, pp. 45-6.

Biagi, E., Nylund, L., Candela, M., Ostan, R., Bucci, L., Pini, E., Nikkïla, J., Monti, D., Satokari, R., Franceschi, C., Brigidi, P., & De Vos, W. (2010) Through Ageing, and Beyond: Gut Microbiota and Inflammatory Status in Seniors and Centenarians. *PLoS ONE* 5, e10667 *PLoS One*, 17;5(5): e10667. Erratum in: PLoS One. 2010;5(6).

Bisson, J.F., Menut, C., & d'Alessio, P. (2008) New pharmaceutical interventions in aging. *Rejuvenation Research*, 11, 399-407.

Burridge, K., & Wennerberg, K. (2004) Rho and Rac take center stage. *Cell*, 116, pp. 167–179.

Campisi, J. (2011) Cellular senescence: putting the paradoxes in perspective. *Current opinion in genetics & development*, 21(1), pp. 107-12Epub ahead of print.

Chambers, S.M., Shaw, C.A., Gatza, C., Fisk, C.J., Donehower, L.A., & Goodell, M.A. (2007) Aging Hematopoietic stem Cells decline in Function and Exhibit Epigenetic Dysregulation. *PLOS Biology*, 5, pp. 1750-62.

Chandola, T., Britton, A., Brunner, E., Hemingway, H., Malik, M., Kumari, M., Badrick, E., Kivimaki, M., & Marmot, M. (2008) Work stress and coronary heart disease: what are the mechanisms? *European Heart Journal*, 29, pp. 640-8.

Clément, K., Viguerie, N., Poitou, C., Carette, C., Pelloux, V., Curat, C.A., Sicard, A., Rome, S., Benis, A., Zucker, J.D., Vidal, H., Laville, M., Barsh, G.S., Basdevant, A., Stich, V., Cancello, R., & Langin, D. (2004) Weight loss regulates inflammation-related genes in white adipose tissue of obese subjects. *FASEB J*,. 18, pp. 1657-69.

Crowell, P.L., Lin, S., Vedejs, E., & Gould, M.N. (1992) Identification of metabolites of the anti-tumor agent *d*-Limonene capable of inhibiting protein isoprenylation and cell growth. *Cancer Chemother Pharmacol*, 31, pp. 205–212.

d'Alessio, P. (2004) Aging and the endothelium. *Journal of Experimental Gerontology*, 39, pp.165-171.

Dali-Youcef, N., Lagouge, M., Froelich, S., Koehl, C., Schoojans, K., & Auwerx, J. (2007) Sirtuins: The "magnificient seven", function, metabolism and longevity. *Annals of Medicine*, 39, pp. 335-45.

de Keizer, P.L.J., Laberge, R.M., & Campisi, J. (2010) p53: Pro-aging or pro-longevity? " *Aging (Albany NY)*, 2, pp. 377-9.

de Kretser, D.M. (2010) Determinants of male health: the interaction of biological and social factors. *Asian Journal of Andrology*, 12, pp. 291-7.

De Martinis, M., Franceschi, C., Monti, D., & Ginaldi, L. (2005) Inflamm-ageing and lifelong antigenic load as major determinants of ageing rate and longevity. *FEBS Lett.* , 579, pp. 2035-9.

Degens, H. (2007) Age-related skeletal muscle dysfunction: causes and mechanisms. *J. Musculoskelet Neuronal Interact*, 7, pp. 246-52.

Dierick, J.F., Eliaers, F., Remacle, J., Raes, M., Fey, S.J., Larsen, P.M., & Toussaint, O. (2002) Stress-induced premature senescence and replicative senescence are different phenotypes, proteomic evidence. *Biochemical Pharmacology*, 64, pp. 1011-7.

Dillon, C., & Goda, Y. (2005) The actin cytoskeleton: integrating form and function at the synapse. *Annu Rev Neurosci.*, 28, pp. 25-55.

Donmez, G., & Guarente, L. (2010) Aging and disease: connections to sirtuins. *Aging Cell*, 9, pp. 285-90.

Edelman, G. (1993) The golden age for adhesion. Cell Adhes Commun., 1, pp. 1-7.

Ergen, A.V.V, & Goodell, M.A. (2009) Mechanisms of hematopoietic stem cell aging. *Experimental Gerontology*, 45, pp. 286-90.

Esler, M., Eikelis, N., Schlaich, M., Lambert, G., Alvarenga, M., Dawood, T., Kaye, D., Barton, D., Pier, C., Guo, L., Brenchley, C., Jennings, G., & Lambert, E. (2008) Chronic mental stress is a cause of essential hypertension: presence of biological markers of stress. *Clinical and Experimental Pharmacology and Physiology*, 35, pp. 498-502.

Farhadi, A., Banan, A., & Keshavarzian, A. (2003) Role of cytoskeletal structure in modulation of intestinal permeability. In : *Archives Iranian Medicine*, 6, pp. 49-53.

Franceschi, C., Bonafè, M., Valensin, S., Olivieri, F., De Luca, M., Ottaviani, E., & De Benedictis, G. (2000) An evolutionary perspective on immunosenescence. *Ann N Y Acad Sci*, 908, pp. 244-54.

Franceschi, C., Valensin, S., Lescai, F., Olivieri, F., Licastro, F., Grimaldi, L.M., Monti, D., De Benedictis, G., & Bonafe, M. (2001) Neuroinflammation and the genetics of Alzheimer's disease: the search for a pro-inflammatory phenotype. *Aging (Milano)*, 13, pp. 163-70.

Franceschi, C., Capri, .M., Monti, D., Giunta, S., Olivieri,i F., Sevini, F., Panourgia, M.P., Invidia, L., Celani, L., Scurti, M., Cevenini, E., Castellani, G.C, & Salvioli, S. (2007) Inflammaging and anti-inflammaging: a systemic perspective on aging and longevity emerged from studies in humans. *Mech Ageing Dev.*, 128, pp. 92-105.

Froy, O., & Miskin, R. (2010) Effect of fedding feeding regiments on circadian rhythms: Implications for aging and longevity. *Aging*, 2, pp. 7-27.

Galeno (Galen) *La dieta dimagrante* (The thinning diet), pp. 45-47 Paravia ed., 1973, Torino, Italy.

García-Cardeña, G., & Gimbrone, M.A. Jr. (2006) Biomechanical modulation of endothelial phenotype: implications for health and disease. *Handb Exp Pharmacol*, 2, pp. 79-95.

Hardcastle, I.R., Rowlands, M.G., Barber, A.M., Grimshaw, R.M., Mohan, M.K., Nutley, B.P., & Jarman, M. (1999) Inhibition of protein prenylation by metabolites of limonene. *Biochem Pharmacol*, 57, pp. 801- 809.

Harris, T.B., Ferrucci, L., Tracy, R.P., Corti, M.C., Wacholder, S., Ettinger, W.H. Jr, Heimovitz, H., Cohen, H.J., & Wallace, R. (1999) Associations of elevated interleukin-6 and C-reactive protein levels with mortality in the elderly. *Am J Med*, 106(5), pp. 506-12.

Herranz, D., Munos-Martin, M., Canamero, M., Mulero, F., Martinez-Pastor, B., Fernandez-Capetillo, O., & Serrano, M. (2009) Sirt1 improves healthy ageing and protects from metabolic syndrome-associated cancer. *Nature Communications*, 1, pp. 1-8.

Ingber, D.E. (2003) Mechanobiology and diseases of mechanotransduction. *Ann Med*,. 35, pp. 564-77.

Ingram, D.D., & Mussolino, M.E. (2010) Weight loss from maximum body weight and mortality: the Third National Health and Nutrition Examination Survey Linked Mortality File. *International Journal of Obesity*, 34, pp. 1044-50.

Issuree, P.D., Pushparaj, P.N., Pervaiz, S., & Melendez, A.J. (2009) Resveratrol attenuates C5a-induced inflammatory responses *in vitro* and *in vivo* by inhibiting phospholipase D and sphingosine kinase activities. *FASEB J*, 23, pp. 2412-2424.

Kirkwood, T.B.L. (2008) Healthy old age. *Nature*, 455, pp. 739-40.

Kitano, H. (2007) Towards a theory of biological robustness. *Mol Syst Biol*, 3:137.

Kolokotrones, T., Savage, V., Deeds, E.J., & Fontana, W. (2010) Curvature in metabolic scaling. *Nature*, 464, pp. 753-6.

MacPhail, R.C., (1987) Observational batteries and motor activity. *Zentralbl Bakteriol Mikrobiol Hyg B.*, 185, pp. 21-7.

Marconi, A., Darquenne, S., Boulmerka, A., Mosnier, M., & d'Alessio, P. (2003) Naftidrofuryl-driven regulation of endothelial ICAM-1 involves Nitric Oxide. *Free Rad Biol Med*, 34, pp. 616-625.

Mathieu, P., Lemieux, I.I., & Després, J.P. (2010) Obesity, Inflammation, and Cardiovascular Risk. *Clinical Pharmacology & Therapeutics*, 87, pp. 407-16.

May, L., van den Biggelaar, A.H., van Bodegom, D., Meij, H.J., de Craen, A.J., Amankwa, J., Frölich, M., Kuningas, M., & Westendorp, R.G. (2009) Adverse environmental conditions influence age-related innate immune responsiveness. *Immun Ageing*, 30, pp. 6-7.

Medeiros, R., Otuki, M.F., Avellar, M.C., & Calixto, J.B. (2007) Mechanisms underlying the inhibitory actions of the pentacyclic triterpene alpha-amyrin in the mouse skin inflammation induced by phorbol ester 12-O- tetradecanoylphorbol-13-acetate. *Eur J Pharmacol*, 559, pp. 227-35.

Millan, J., & Ridley, A.J. (2005) Rho GTPases and leucocyte-induced endothelial remodelling. *Biochem J*, 385, pp. 329-337.

Miller, J.A., Thompson, P.A., Hakim, I.A., Sherry Chow, H.H., & Thomson, C.A. (2011) d-Limonene: a bioactive food component from citrus and evidence for a potential role in breast cancer prevention and treatment. *Oncol Rev*, 5, pp. :31-42.

Naveiras, O., Nardi, V., Wenzel, P.L., Hauschka, P.V., Fahey, F., & Daley, G.Q. (2009) Bone-Marrow adipocytes as negative regulators of the haematopoietic microenvironment. *Nature*, 460, pp. 259-63.

Nomicos, E.Y. (2007) Myrrh: medical marvel or myth of the Magi? *Holist Nurs Pract*,. 21, pp. 308-23.

Ostan, R. , Bucci, L., Capri, M., Salvioli, S., Scurti, M., Pini, E., Monti, D., & Franceschi, C. (2008) Immunosenescence and immunogenetics of human longevity. et al. (2008) Immunosenescence and immunogenetics of human longevity. *Neuroimmunomodulation*, 15, pp. 224-40.

Petrascheck, M., Ye, X., & Buck, L.B. (2007) An antidepressant that extends lifespan in adult *Caenorhabditis elegans. Nature* 450, pp. 553-7.

Queré, N., Noël, E., Lieutaud, A. & d'Alessio, P., (2009) Fasciatherapy combined with Pulsology induces changes in blood turbulence potentially beneficial for the Endothelium. *J Bodyw Mov Ther.*, 3, pp. 239-45.

Ridker, P.M., Cushman, M., Stampfer, M.J., Tracy, R.P., & Hennekens, C.H. (1997) Inflammation, aspirin, and the risk of cardiovascular disease in apparently healthy men. *N Engl J Med*, 336(14) pp. 973-9. Erratum in: *N Engl J Med* 1997; 337(5): 356.

Rossi, D.J., D Bryder, D., Zahn, J.M., Ahlenius, H., Sonu, R., Wagers, A.J., & Weissman, I.L. (2005) Intrinsic alterations underlie hematopoietic stem cell aging. *Proc Natl Acad Sci U S A*, 102, pp. 9194-99.

Rossi, D.J., Jamieson, C.H., & Weissman, I.L. (2008) Stems cells and the pathways to aging and cancer. *Cell*, 132, pp. 681-96.

Sahin, E., & DePinho, R.A. (2010) Linking functional decline of telomeres, mitochondria and stem cells during ageing. *Nature*, 464, pp. 520-8.

Sharpless, N.E., & DePinho, R.A. (2007) How stem cells age and why this makes us grow old. *Nature*, 8, pp. 703-13.

Shibeshi, W.A., Young-Xu, Y., & Blatt, C.M. (2007) Anxiety worsens prognosis in patients with coronary artery disease. *Journal of American College of Cardiology*, 49, pp. 2021-2027.

Toussaint, O., Royer, V., Salmon, M., & Remacle, J. (2002) Stress-induced premature senescence and tissue ageing. *Biological Pharmacology*, 64, pp. 1007-9.

Vanier, B. (2005) Intercellular adhesion molecule-1 (ICAM-1) in ulcerative colitis. *Digestive Diseases Sciences*, 54, 313-327.

Vigushin, D.M., Poon, G.K., Boddy, A., English, J., Halbert, G.W., Pagonis, C., Jarman, M., & Coombes, R.C. (1998) Phase I and pharmacokinetic study of d-limonene in patients with advanced cancer. Cancer Research Campaign Phase I/II Clinical Trials Committee. *Cancer Chemother Pharmacol*, 42, pp. 111–117.

von Zglinicki, T., Bürkle, A., & Kirkwood, T.B.L. (2001) Stress, DNA damage and ageing - an integrative approach. *Experimental Gerontology*, 36, pp. 1049-62.

Walston, J., McBurnie, M.A., Newman, A., Tracy, R.P., Kop, W.J., Hirsch, C.H., Gottdiener, J., & Fried, L.P. (2002) Cardiovascular Health Study. Frailty and activation of the inflammation and coagulation systems with and without clinical comorbidities: results from the Cardiovascular Health Study. *Arch Intern Med*, 162(20), pp. 2333-41.

Xu, Y., Li, J., Ferguson, G.D., Mercurio, F., Khambatta, G., Morrison, L., Lopez-Girona, A., Corral, L.G., Webb, D.R., Bennett, B.L., & Xie, W. (2009) Immunomodulatory drugs reorganize cytoskeleton by modulating Rho GTPases. *Blood*, 114, pp. 338-45.

Yamada, Y., Marshall, S., Specian, R.D., & Grisham, M.B. (1992) A comparative analysis of two models of colitis in rats. *Gastroenterology*, 102, pp. 1524-34.

Zahradka, K., Slade, D., Bailone, A., Sommer, S., Averbeck, D., Petranovic, M., Linder, A.B., & Radman, M. (2006) Reassembly of shattered chromosomes in *Deinococcus radiodurans*. *Nature*, 443(7111), pp. 569-73.

Zhang, D.H., Marconi, A., Xu, L.M., Yang, C.X., Sun, G.W., Feng, X.L., Ling, C.Q., Qin, W.Z., Uzan, G., & d'Alessio, P. (2006) Tripterine inhibits the expression of adhesion molecules in activated endothelial cells. *Journ Leuko Biol*, 80, pp. 309-319.

Ziccardi, P., Nappo, F., Giugliano, G., Esposito, K., Marfella, R., Cioffi, M., D'Andrea, F., Molinari, A.M., & Giugliano, D. (2002) Reduction of inflammatory cytokine concentrations after weight loss over one year. *Circulation*, 105, pp. 804-9.

6

The Emerging Role of Centromere/Kinetochore Proteins in Cellular Senescence

Kayoko Maehara
Department of Maternal-Fetal Biology
National Research Institute for Child Health and Development, Tokyo,
Japan

1. Introduction

Cellular senescence is an irreversible growth arrest triggered by several types of stress, including DNA damage, oxidative stress, telomere shortening, and oncogene activation (Ben-Porath & Weinberg, 2005; Collado et al., 2007; Deng et al., 2008; Hayflick & Moorhead, 1961; Serrano et al., 1997). Although how senescence is initiated remains to be determined, it has been shown to be triggered by certain defects in chromosome integrity, such as telomere shortening (Ben-Porath & Weinberg, 2005; Deng et al., 2008). In contrast to telomere shortening, the roles of which in senescence have been studied extensively, alterations in the centromere/kinetochore structure involved in senescence program remain to be elucidated. This chapter presents a discussion of the emerging roles of centromere/kinetochore proteins, particularly Centromere protein A (CENP-A, the centromere-specific variant of histone H3), in senescence.

2. Crucial roles of centromere/kinetochore proteins in mitosis

The genome of a cell is duplicated and segregated into two daughter cells during cell division (Fig. 1). Accurate chromosome segregation during cell division is essential for genome integrity and this process is mainly achieved by the structural/functional integrity of the microtubule spindle apparatus (kinetochore-microtubule interactions) and spindle assembly checkpoint (SAC) signaling (Cleveland et al., 2003; Musacchio, & Salmon, 2007; Tanaka, 2010). Spindle microtubules emanating from spindle pole bodies (centrosomes) attach to chromosomes via specialized structures called kinetochores where more than 100 proteins assemble at the centromeric region of each chromosome during mitosis. This interaction is monitored by the SAC signaling pathway to ensure high-fidelity chromosome segregation (Musacchio, & Salmon, 2007). Chromosome missegregation arising from defects in the structural integrity of the microtubule spindle apparatus and the SAC signaling pathway leads to aneuploidy, i.e., chromosome gain or loss (Compton, 2011). Aneuploidy is thought to be a major cause of congenital disorders. High rates of aneuploidy have been observed in various cancers and aneuploidy is speculated to be involved in tumorigenesis.

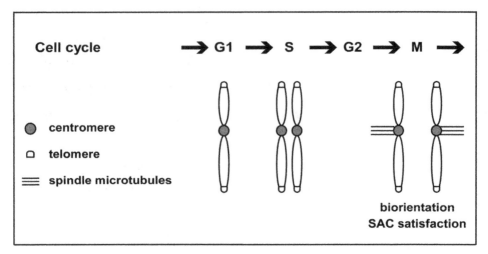

Fig. 1. Chromosome cycle and cell cycle (Adapted from Maehara, 2011)

In higher eukaryotes, the DNA sequence does not generally determine the functional centromeres except in the budding yeast *Saccharomyces cerevisiae*, in which the centromere, a 125-bp DNA element, is specified by its sequence. Centromeres in other organisms lack sequence specificity, but many of the proteins localizing at centromeres are well conserved across species (Fig. 2).

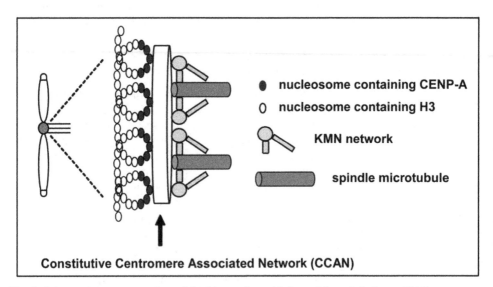

Fig. 2. Schematic representation of the kinetochore (Adapted from Maehara, 2011)

Among the numerous kinetochore-associated proteins identified to date, CENP-A represents an excellent candidate as an epigenetic marker of functional centromeres for several reasons. First, CENP-A is an evolutionarily conserved centromere-specific histone H3 variant (Blower & Karpen, 2001; Buchwitz et al, 1999; Earnshaw & Rothfield, 1985; Meluh et al., 1998; Palmer et al., 1987; Stoler et al., 1995; Takahashi et al., 2000). Canonical nucleosomes in chromosome arms consist of 146 bp of DNA wrapped around a histone octamer comprised of two subunits of each of H2A, H2B, H3, and H4. H3 is replaced with the H3 variant CENP-A at the centromeres. Second, many centromere-associated proteins are recruited to the centromere in a CENP-A-dependent manner (Foltz et al., 2006; Izuta et al., 2006; Obuse et al., 2004a, Okada et al., 2006). Third, neocentromeres, which are established as functional centromeres at ectopic chromosomal loci devoid of alpha satellite repeats, have been shown to contain CENP-A (Marshall et al., 2008). Thus, CENP-A seems to be an identifier of the functional centromere. Studies in a variety of organisms have indicated that CENP-A plays a crucial role in organizing kinetochore chromatin for precise chromosome segregation. Another conserved centromere protein, CENP-B, binds to a specific centromeric DNA sequence, the 17-bp "CENP-B box" in type I alpha satellite repeats in mammals (alphoid DNA in humans) (Earnshaw et al., 1987; Masumoto et al., 1989). CENP-B is essential for heterochromatin formation of pericentromeres and is thought to be important for the proper organization of kinetochore chromatin (Nakagawa et al., 2002; Nakano et al., 2008; Okada et al., 2007), although CENP-B is not essential for viability in higher eukaryotes (Hudson et al., 1998; Kapoor et al., 1998; Perez-Castro et al., 1998). In addition to the above proteins, several studies using proteomic approaches have identified 15 proteins known as the Constitutive Centromere Associated Network (CCAN) (Foltz et al., 2006; Izuta et al., 2006; Okada et al., 2006). Several of these proteins have DNA binding activity or associate directly with CENP-A. The KMN network (KNL1, Mis12 complex, and Ndc80 complex) is also important as it forms the interface for kinetochore-microtubule attachment (Cheeseman et al., 2006; Obuse et al., 2004b; Ruchaud et al., 2007). SAC is a surveillance mechanism that is capable of delaying anaphase if not all chromosomes have established biorientation within the spindle (Musacchio & Salmon, 2007). It should be noted that many other centromere/kinetochore-associated proteins not mentioned in this chapter also have crucial roles in mitosis. Thus, multiple biological processes including kinetochore, microtubule functions, the mitotic spindle apparatus, and SAC signaling pathway ensure high-fidelity chromosome segregation during cell division.

3. The emerging roles of centromere/kinetochore proteins in senescence

The importance of kinetochore in regulating proper chromosome segregation has been well established. Next, I highlight some recent work on the roles of centromere/kinetochore proteins in senescence in mammals.

3.1 SAC proteins are involved in the senescence program

The SAC signaling pathway monitors the attachment of spindle microtubules to kinetochores. Core components of SAC include Mad1, Mad2, Bub1, Bub3, and BubR1, and many other proteins are also involved in this checkpoint. The onset of anaphase is triggered by activation of the anaphase-promoting complex/cyclosome (APC/C), which degrades cyclin B and securin. SAC generates an inhibitory signal to block APC/C in the presence of unaligned chromosomes and stalls for time to establish biorientation. Mouse models have been generated with manipulation of the genes encoding SAC proteins. Complete loss of SAC proteins, including Mad2, Bub1, BubR1, Bub3, and Rae1, caused early embryonic lethality (Baker et al., 2004; Baker et al., 2006; Dobles et al., 2000; Jeganathan et al., 2007; Wang et al., 2004). These gene knockout studies revealed the essential nature of mammalian mitotic checkpoint proteins for viability. Intriguingly BubR1-insufficient ($Bub\ 1b^{H/H}$) mice, in which the levels of BubR1 are about 10% those in normal animals, develop progressive aneuploidy along with a variety of progeroid features, including short lifespan, cachectic dwarfism, lordokyphosis, cataracts, loss of subcutaneous fat, and impaired wound healing (Baker et al., 2004). Consistent with the features of premature aging of $Bub\ 1b^{H/H}$ mice, mouse embryonic fibroblasts derived from $Bub\ 1b^{H/H}$ mice show rapid senescence. Both premature aging and cellular senescence observed in $Bub\ 1b^{H/H}$ mice are attenuated by inactivation of p16, a tumor suppressor and an effector of senescence (Baker et al., 2008). In humans, biallelic mutations in $BUB1B$ encoding BubR1 cause mosaic variegated aneuploidy (MVA) (Hanks et al., 2004). MVA is a rare recessive condition characterized by constitutional mosaic aneuploidy, growth retardation, microcephaly, and predisposition to cancers such as rhabdomyosarcoma, Wilms tumor, and leukemia. Although aneuploidy and cataracts are common features detected in both $Bub\ 1b^{H/H}$ mice and individuals with MVA, MVA patients do not have typical features of premature aging. The difference in phenotype between $Bub\ 1b^{H/H}$ mice and individuals with MVA may be explained by the degree of BubR1 defects. A recent study indicated that mutations in $CEP57$ also cause MVA (Snape et al., 2011). CEP57 is a centrosomal protein and is involved in nucleating and stabilizing microtubules. This suggests that $BUB1B$ mutations underlie only a proportion of MVA, and other genes involved in regulating chromosome segregation may cause the disease. Bub3/Rae1-haploinsufficient mice have been reported to display an array of early aging-associated phenotypes (Baker et al., 2006) and Bub1 suppression in human fibroblasts activates a p53-dependent premature senescence response (Gjoerup et al., 2007). These studies involving the manipulation of SAC genes demonstrated that low levels of several SAC proteins play crucial roles in regulating commitment to the senescent state, although it remains to be determined how individual components of this checkpoint control cell viability and cell fate.

3.2 The roles of constitutively centromere-localized proteins in senescence

In contrast to SAC proteins, which localize to the kinetochore during mitosis, CENP-A localizes to the centromere throughout cell cycle and provides a structural and functional foundation for the kinetochore. I detail the role of CENP-A in senescence.

3.2.1 CENP-A has an impact on cell proliferation

Despite extensive studies of centromere-associated proteins, it remains unclear whether these proteins are involved in the control of cell proliferation; previous studies focused on the roles of centromere proteins in chromosome segregation, and were mainly conducted in immortalized cell lines, such as HeLa (Goshima et al., 2003). With regard to CENP-A, studies in a variety of organisms have indicated that the effects of CENP-A loss on proliferation vary widely according to the species, cell type, and methods used to delete or deplete CENP-A. *Cenpa* null mice fail to survive (Howman et al., 2000). Disruption of CID by antibody injection into *Drosophila* embryos and RNAi in cells in tissue culture exhibits a range of phenotypes affecting both cell cycle progression and mitotic chromosome segregation (Blower & Karpen, 2001). CENP-A-depleted chicken DT40 cells exhibit defects in kinetochore function and stop proliferating, although the apparent cessation of cell proliferation is caused by extensive cell death and the cells are still cycling (Régnier et al., 2005). CENP-A-depleted HeLa cells proliferate but exhibit misalignment and lagging of chromosomes during mitosis (Goshima et al., 2003). In HeLa cells, two tumor suppressor molecules, p53 and retinoblastoma protein (Rb), which have been shown to play crucial roles in cell cycle arrest in primary human cells, are inactivated due to the integration of the human papillomavirus that leads to their immortalization. Although it is essential to use primary human cells to uncover the regulatory roles of centromere proteins in cell proliferation, no such studies have yet been reported. To address whether CENP-A has an impact on cell proliferation, we examined the effects of CENP-A depletion in human primary somatic cells with functional p53 and Rb (Maehara et al., 2010). The reduction of CENP-A by retrovirally transducing CENP-A shRNA did not show growth arrest in HeLa cells, consistent with the previous results in CENP-A RNAi-mediated HeLa cells (Goshima et al., 2003). However, depletion of CENP-A in primary human TIG3 fibroblasts resulted in the immediate cessation of proliferation accompanied by increased levels of p16 and p21 expression, upregulated SAHF formation, and increased SA-β-gal activity, all of which are common markers of cellular senescence (Alcorta et al., 1996; Dimri et al., 1995; Hara et al., 1996; Narita et al., 2003; Zhang et al., 2005). Inactivation of p53 in CENP-A-depleted TIG3 cells restores proliferation leading to an increase in number of cells exhibiting aberrant chromosome behavior. These results indicate that the reduction of CENP-A drives normal human diploid fibroblasts into a senescent state in a p53-dependent manner. The senescence that arises from CENP-A depletion may be a self-defense mechanism to suppress the otherwise catastrophic impact upon genome integrity that would arise from kinetochore dysfunction following certain types of stress. It should be noted that reduction of CENP-A does not result in irreversible growth arrest in human pluripotent stem cells (Ambartsumyan et al., 2010). Ambartsumyan et al. demonstrated that CENP-A-depleted undifferentiated human pluripotent stem cells were capable of maintaining a functional centromere marks and showed no changes in morphology or proliferation rate relative to control cells, whereas CENP-A-depleted BJ fibroblasts showed arrest in G2/M and underwent apoptosis. Although the pluripotent state may cause the different phenotypes in response to CENP-A depletion, CENP-A has an impact on cell proliferation in human primary somatic cells.

3.2.2 CENP-A is downregulated in senescent human cells

Model systems with manipulation of gene expression/deletion have clearly revealed that some centromere/kinetochore-associated proteins play crucial roles in regulating commitment to the senescent state. However, the mechanisms of senescence and individual aging are presumed to be complex. To gain insights into the mechanisms that control lifespan and age-related phenotypes, Ly et al. examined mRNA abundance of more than

6000 known genes in dermal fibroblasts derived from elderly human subjects and from those with Hutchinson–Gilford Progeria Syndrome (HGPS), a rare genetic disorder characterized by accelerated aging (Ly et al., 2000). They found that genes involved in cell cycle progression, spindle assembly, and chromosome segregation, such as cyclins A, B, polo kinase, CENP-A, CENP-F, and kinesin-related proteins, were downregulated in elderly individuals and those with HGPS. We showed that CENP-A mRNA expression was reduced in both replicative and *ras*-induced senescent human TIG3 cells (Maehara et al., 2010). Another group reported a reduction in the levels of CENP-A transcripts in senescent human IMR90 fibroblasts (Narita et al., 2006). Therefore, the reduction of CENP-A mRNA levels appears to be a common feature of cellular senescence and individual aging. However, this reduction is not specific to senescence; we observed a marked reduction of CENP-A mRNA level in quiescent cells that had transiently exited from the cell cycle (Maehara et al., 2010). As CENP-A transcription is regulated by the cell cycle and occurs in G2 phase in human cells (Shelby et al., 1997), the transcription of CENP-A ceases immediately when cells are arrested regardless of whether the arrest is promoted by senescence or quiescence, even though reduction in CENP-A transcript level shows a strong association with the reduced proliferation potential of senescent cells.

In contrast to the levels of CENP-A transcript, which are reduced in both senescent and quiescent cells, CENP-A protein levels are markedly reduced in senescent cells, while quiescent cells retain similar levels of CENP-A protein to their actively growing counterparts (Maehara et al., 2010). These observations suggest that both transcriptional and posttranslational regulation are involved in the senescence-associated reduction of CENP-A protein level. CENP-A protein may be degraded via the ubiquitin – proteasome-dependent pathway in these cells. A previous study demonstrated that cullin-4A, human ring finger protein 2, and hypothetical protein FLJ23109, which have been reported or assumed to possess ubiquitin ligase activity, were coimmunoprecipitated with anti-CENP-A antibody from HeLa interphase nuclear extract (Obuse et al., 2004a). It is noteworthy that CENP-A also undergoes destruction when human cells are infected with herpes simplex virus type 1 protein ICP0 (Lomonte et al., 2001). Ubiquitin-dependent proteolysis of the yeast Cse4/CENP-A incorporated at non-centromeric regions has been reported (Collins et al., 2004). In addition to CENP-A, linker histone H1 protein level is decreased in senescent human WI38 cells, presumably because of posttranslational regulation (Funayama et al., 2006). A mitotic exit network kinase, WARTS/LATS1, was also reported to be reduced in senescent human cells (Takahashi et al., 2006). The reduction of this kinase was attenuated by addition of MG132. These results imply the presence of a senescence-associated proteolysis pathway in primary human cells. The senescence-associated proteolysis pathway may contribute to maintenance of metabolism and biosynthesis in senescent cells by recycling proteins that are no longer required for non-dividing cells and to ensure irreversible growth arrest by destruction of proteins essential for proliferation. Although the molecular mechanism of CENP-A reduction remains to be clarified, reduced levels of CENP-A protein seem to be common to cellular senescence and individual aging.

3.2.3 CENP-A reduction enhances centromeric heterochromatin formation

In our exploration of senescence-associated alterations in nuclear structure using primary human cells, we found that CENP-A levels were markedly reduced in senescent cells. In contrast to CENP-A, the levels of the other centromere proteins, CENP-B and hMis12, increased gradually, as the cells became senescent (Maehara et al., 2010). In addition, increased HP1 proteins, which are essential components of the pericentric heterochromatin region, were enriched on centromeres alongside CENP-B. These changes in the levels of centromere proteins alter the centromere chromatin structure, and are thought to represent physiologically significant phenomena associated with cellular senescence. Forced reduction of CENP-A alters the distributions of CENP-B and HP1 proteins, which are similar to those observed in replicative and *ras*-induced senescent cells, suggesting that this centromere alteration is triggered, at least in part, by the reduction of CENP-A protein level. Recent studies have demonstrated the remarkable role of CENP-B in heterochromatin formation in the centromere. In fission yeast, the disruption of CENP-B homologs, Abp1 and Cbh1, causes a reduction of Swi6, a homolog of HP1, at centromeric chromatin and a decrease in heterochromatin-specific modifications of histone H3 (Nakagawa et al., 2002). Using human artificial chromosomes (HAC) and alpha-satellite arrays integrated into chromosomal arms as models, Okada et al. demonstrated a dual role of CENP-B in CENP-A assembly and heterochromatin formation (Okada et al., 2007). Although CENP-B is required for de novo CENP-A assembly on HAC, CENP-B enhances histone H3K9 trimethylation and DNA methylation in chromosomally integrated alphoid DNA and suppresses centromere formation. Furthermore, Nakano et al. generated HAC containing both integrated alpha satellite and tet operator (tetO) sequences and tethered tet repressor (tetR) chromatin-modifying protein fusions to the HAC centromere (Nakano et al., 2008). Stimulation of the formation of a heterochromatin state by forced binding of silencers or targeted nucleation of HP1 resulted in the inactivation of a functional HAC centromere. Depletion of dimethylated histone H3K4 (H3K4me2) by tethering the lysine-specific demethylase 1 (LSD1) causes CENP-A loss from HAC kinetochores and ultimately results in inactivation of the kinetochore (Bergmann et al., 2011). These observations suggest that inactivation of the centromere occurs through epigenetic mechanisms. Thus, the loss of CENP-A and the extended heterochromatinization mediated by CENP-B and HP1 proteins on the centromere in senescent cells are assumed to promote centromere inactivation. During senescence, primary human cells alter their centromere states from a functional centromere, which is required for faithful segregation of chromosomes, to an inactivated centromere, which is likely to contribute to the establishment of the senescent state. Further qualitative and quantitative studies are needed to understand the structural and the functional changes that occur in the centromere during the senescence process.

3.2.4 How do primary human somatic cells sense centromere/kinetochore dysfunction and undergo senescence?

Forced depletion of CENP-A induces senescence-like phenotypes in the primary cells and CENP-A appears to be actively degraded in the senescent cells. This raises the question of whether CENP-A reduction is a cause or a consequence of cellular senescence. As cellular senescence is a complex trait, it is not possible to provide a clear answer to this question. There may be a positive feedback circuit between CENP-A degradation and induction of cellular senescence during senescence. I hypothesize that primary human somatic cells possess a mechanism for monitoring centromere/kinetochore integrity, which activates the p53-dependent senescence pathway in response to centromere/kinetochore defects, such as insufficient incorporation of CENP-A at the centromere (Fig. 3).

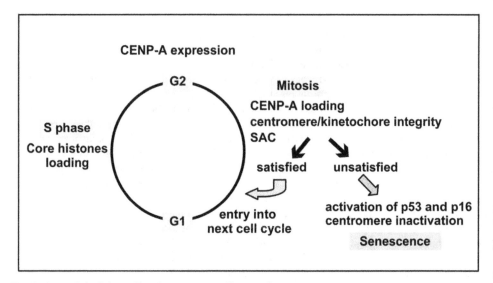

Fig. 3. A model of the roles of centromere/kinetochore proteins in senescence

How do the primary cells sense centromere/kinetochore dysfunction?

Telomere shortening triggers the DNA damage response (DDR), which is a major intrinsic factor to induce cellular senescence. Previous studies clearly demonstrated that p53 activation in oncogene-induced senescence is due to activation of the DDR (Bartkova et al., 2006; Di Micco et al., 2006; Mallette et al., 2007). We examined whether DDR plays a crucial role in activation of p53 in response to a reduction of CENP-A level. The presence of DNA damage foci (phosphorylated histone H2A.X, γ-H2AX), chk2 phosphorylated on threonine 68 and chk1 phosphorylated on serine 345, which are associated with DDR, were not detected in CENP-A-depleted-senescent cells (Maehara et al., 2010), suggesting that CENP-A depletion is not causally linked to DDR. Excess growth signals produced by oncogenes and telomere shortening seem to be sensed as DNA replication stresses, while CENP-A reduction is not. This may explain the unconventional type of senescence that does not require the activation of DNA damage signaling.

Unlike canonical core histones that are loaded into chromatin during DNA replication, newly synthesized CENP-A is incorporated into centromeric chromatin in telophase and early G1 phase (Fig. 3). Mis18 complex and HJURP/Smc3 have been implicated in the centromeric loading of CENP-A (Barnhart et al., 2011; Dunleavy et al., 2009; Foltz et al., 2009; Fujita et al., 2007; Hayashi et al, 2004). Primary cells may monitor CENP-A loading and centromere/kinetochore integrity during M and early G1 phases and immediately cease proliferation before entry into the next cell cycle in response to fatal centromere/kinetochore dysfunction under conditions in which some key centromere proteins and/or the SAC are not functioning properly. Under these conditions, senescence seems to not only prevent the cells from producing abnormal chromosomes, but also protects the organism from the potentially hazardous consequences of proliferation of cells harboring chromosomal abnormalities that arose as a consequence of defective mitosis.

4. Conclusion

Recent studies have revealed novel roles of centromere/kinetochore-associated proteins in the senescence program mainly using model systems in which target genes were manipulated. As highlighted in this chapter, while low levels of several centromere/kinetochore-associated proteins play crucial roles in regulating commitment to the senescent state, the interactions between centromere/kinetochore proteins and components of the senescence pathway remains to be determined. Further studies are required to determine the epigenetic mechanisms of centromere inactivation, particularly histone modification, and components involved in regulating the ratio of CENP-A to heterochromatin during senescence.

5. Acknowledgments

I thank Hiroshi Masumoto (Kazusa DNA Research Institute) for discussions concerning the centromere inactivation. I apologize to authors whose work could not be cited here due to space limitation. This work was supported by a Grant-in-Aid for Scientific Research (C) from the Japan Society for the Promotion of Science and Takeda Science Foundation.

6. References

Alcorta, D.A., Xiong, Y., Phelps, D., Hannon, G., Beach, D., & Barrett, J.C. (1996) Involvement of the cyclin-dependent kinase inhibitor p16 (INK4a) in replicative senescence of normal human fibroblasts. *Proc Natl Acad Sci U S A* Vol.93, No.24, (November 1996), pp. 13742-13747, ISSN 1091-6490

Ambartsumyan, G., Gill, R.K., Perez, S.D., Conway, D., Vincent, J., Dalal, Y., & Clark, A.T. (2010) Centromere protein A dynamics in human pluripotent stem cell self-renewal, differentiation and DNA damage. *Hum Mol Genet* Vol.19, No.20, (October 2010), pp. 3970-3982, ISSN 0964-6906

Baker, D.J., Jeganathan, K.B., Cameron, J.D., Thompson, M., Juneja, S., Kopecka, A., Kumar, R., Jenkins, R.B., de Groen, P.C., Roche, P., & van Deursen, J.M. (2004) BubR1 insufficiency causes early onset of aging-associated phenotypes and infertility in mice. *Nat Genet* Vol.36, No.7, (July 2004), pp. 744-749, ISSN 1061-4036

Baker, D.J., Jeganathan, K.B., Malureanu, L., Perez-Terzic, C., Terzic, A,, & van Deursen, J.M. (2006) Early aging-associated phenotypes in Bub3/Rae1 haploinsufficient mice. *J Cell Biol* Vol.172, No.4, (February 2006), pp. 529-540, ISSN 0021-9525

Baker, D.J., Perez-Terzic, C., Jin, F., Pitel, K., Niederländer, N.J., Jeganathan, K., Yamada, S., Reyes, S., Rowe, L., Hiddinga, H.J., Eberhardt, N.L., Terzic, A., & van Deursen, J.M. (2008) Opposing roles for p16Ink4a and p19Arf in senescence and ageing caused by BubR1 insufficiency. *Nat Cell Biol* Vol.10, No.7, (July 2008), pp. 825-836, ISSN 1465-7392

Barnhart, M.C., Kuich, P.H.J.L., Stellfox, M.S., Ward, J.A., Bassett, E.A., Black, B.E., & Foltz, D.R. (2011) HJURP is a CENP-A chromatin assembly factor sufficient to form a functional de novo kinetochore. *J Cell Biol* Vol.194, No.2, (July 2011), pp. 229-243, ISSN 0021-9525

Bartkova, J., Rezaei, N., Liontos, M., Karakaidos, P., Kletsas, D., Issaeva, N., Vassiliou, L.V., Kolettas, E., Niforou, K., Zoumpourlis, V.C., Takaoka, M., Nakagawa, H., Tort, F.,

Fugger, K., Johansson, F., Sehested, M., Andersen, C.L., Dyrskjot, L., Ørntoft, T., Lukas, J., Kittas, C., Helleday, T., Halazonetis, T.D., Bartek, J., & Gorgoulis, V.G. (2006) Oncogene-induced senescence is part of the tumorigenesis barrier imposed by DNA damage checkpoints. *Nature* Vol.444, No.7119, (November 2006), pp. 633-637, ISSN 0028-0836

Ben-Porath, I., & Weinberg, R.A. (2005) The signals and pathways activating cellular senescence. *Int J Biochem Cell Biol* Vol.37, No.5, (May 2005), pp. 961-976, ISSN 1357-2725

Bergmann, J.H., Rodríguez, M.G., Martins, N.M.C., Kimura, H., Kelly, D.A., Masumoto, H., Larionov, V., Lansen, L.E.T., & Earnshaw, W.C. (2011) Epigenetic engineering shows H3K4me2 is required for HJURP targeting and CENP-A assembly on a synthetic human kinetochore. *EMBO J* Vol.30, No.2, (January 2011), pp. 328-340, ISSN 0261-4189

Blower, M.D., & Karpen, G.H. (2001) The role of Drosophila CID in kinetochore formation, cell-cycle progression and heterochromatin interactions. *Nat Cell Biol* Vol.3, No.8, (August 2001), pp. 730-739, ISSN 1465-7392

Buchwitz, B.J., Ahmad, K., Moore, L.L., Roth, M.B., & Henikoff, S. (1999) A histone-H3-like protein in *C. elegans*. *Nature* Vol.401, No.6753, (October 1999), pp. 547-548, ISSN 0028-0836

Cheeseman, I.M., Chappie, J.S., Wilson-Kubalek, E.M., & Desai, A. (2006) The conserved KMN network constitutes the core microtubule-binding site of the kinetochore. *Cell* Vol.127, No.5, (December 2006), pp. 983-997, ISSN 0092-8674

Cleveland, D.W., Mao, Y., & Sullivan, K.F. (2003) Centromeres and kinetochores: from epigenetics to mitotic checkpoint signaling. *Cell* Vol.112, No.4, (February 2003), pp. 407-421, ISSN 0092-8674

Collado, M., Blasco, M.A., & Serrano, M. (2007) Cellular senescence in cancer and aging. *Cell* Vol.130, No.2, (July 2007), pp. 223-233, ISSN 0092-8674

Collins, K.A., Furuyama, S., & Biggins, S. (2004) Proteolysis contributes to the exclusive centromere localization of the yeast Cse4/CENP-A histone H3 variant. *Curr Biol* Vol.14, No.21, (November 2004), pp. 1968-1972, ISSN 0960-9822

Compton, D.A. (2011) Mechanisms of aneuploidy. *Curr Opin Cell Biol* Vol.23, No.1, (February 2011), pp. 109-113, ISSN 0955-0674

Deng, Y., Chan, S.S., & Chang, S. (2008) Telomere dysfunction and tumour suppression: the senescence connection. *Nat Rev Cancer* Vol.8, No.6, (June 2008), pp. 450-458, ISSN 1474-1768

Di Micco, R., Fumagalli, M., Cicalese, A., Piccinin, S., Gasparini, P., Luise, C., Schurra, C., Garré, M., Nuciforo, P.G., Bensimon, A., Maestro, R., Pelicci, P.G., & d'Adda di Fagagna, F. (2006) Oncogene-induced senescence is a DNA damage response triggered by DNA hyper-replication. *Nature* Vol.444, No.7119, (November 2006), pp. 638-642, ISSN 0028-0836

Dimri, G.P., Lee, X., Basile, G., Acosta, M., Scott, G., Roskelley, C., Medrano, E.E., Linskens, M., Rubelj, I., Pereira-Smith, O., Peacocke, M., & Campisi, J. (1995) A biomarker that identifies senescent human cells in culture and in aging skin in vivo. *Proc Natl Acad Sci U S A* Vol.92, No.20, (September 1995), pp. 9363-9367, ISSN 1091-6490

Dobles, M., Liberal, V., Scott, M.L., Benezra, R., & Sorger, P.K. (2000) Chromosome missegregation and apoptosis in mice lacking the mitotic checkpoint protein Mad2. *Cell* Vol.101, No.6, (June 2000), pp. 635-645, ISSN 0092-8674

Dunleavy, E.M., Roche, D., Tagami, H., Lacoste, N., Ray-Gallet, D., Nakamura, Y., Daigo, Y., Nakatani, Y., & Almouzni-Pettinotti, G. (2009) HJURP is a cell-cycle-dependent maintenance and deposition factor of CENP-A at centromeres. *Cell* Vol.137, No.3, (May 2009), pp. 485-497, ISSN 0092-8674

Earnshaw, W.C., & Rothfield, N. (1985) Identification of a family of human centromere proteins using autoimmune sera from patients with scleroderma. *Chromosoma* vol.91, No.3-4, (January 1985), pp. 313-321, ISSN 0009-5915

Earnshaw, W.C., Sullivan, K.F., Machlin, P.S., Cooke, C.A., Kaiser, D.A., Pollard, T.D., Rothfield, N.F., & Cleveland, D.W. (1987) Molecular cloning of cDNA for CENP-B, the major human centromere autoantigen. *J Cell Biol* Vol.104, No.4, (January 1987), pp. 817-829, ISSN 0021-9525

Foltz, D.R., Jansen, L.E.T., Black, B.E., Bailey, A.O., Yates, JR, 3rd, & Cleveland, D.W. (2006) The human CENP-A centromeric nucleosome-associated complex. *Nat Cell Biol* Vol.8, No.5, (May 2006), pp. 458-469, ISSN 1465-7392

Foltz, D.R., Jansen, L.E.T., Bailey, A.O., Yates, JR, 3rd, Bassett, E.A., Wood, S., Black, B.E., & Cleveland, D.W. (2009) Centromere-specific assembly of CENP-A nucleosomes is mediated by HJURP. *Cell* Vol.137, No.3, (May 2009), pp. 472-484, ISSN 0092-8674

Fujita, Y., Hayashi, T., Kiyomitsu, T., Toyoda, Y., Kokubo, A., Obuse, C., & Yanagida, M. (2007) Priming of centromere for CENP-A recruitment by human hMis18α, hMis18β, and M18BP1. *Dev Cell* Vol.12, No.1, (January 2007), pp. 17-30, ISSN 1534-5807

Funayama, R., Saito, M., Tanobe, H., & Ishikawa, F. (2006) Loss of linker histone H1 in cellular senescence. *J Cell Biol* Vol.175, No.6, (December 2006), pp. 869-880, ISSN 0021-9525

Gjoerup, O.V., Wu, J., Chandler-Militello, D., Williams, G.L., Zhao, J., Schaffhausen, B., Jat, P.S., & Roberts, T.M. (2007) Surveillance mechanism linking Bub1 loss to the p53 pathway. *Proc Natl Acad Sci U S A* Vol.104, No.20, (May 2007), pp. 8334-8339, ISSN 1091-6490

Goshima, G., Kiyomitsu, T., Yoda, K., & Yanagida, M. (2003) Human centromere chromatin protein hMis12, essential for equal segregation, is independent of CENP-A loading pathway. *J Cell Biol* Vol.160, No.1, (January 2003), pp. 25-39, ISSN 0021-9525

Hanks, S., Coleman, K., Reid, S., Plaja, A., Firth, H., FitzPatrick, D., Kidd, A., Méhes, K., Nash, R., Robin, N., Shannon, N, Tolmie, J., Swansbury, J., Irrthum, A., Douglas, J., & Rahman, N. (2004) Constitutional aneuploidy and cancer predisposition caused by biallelic mutations in *BUB1B*. *Nat Genet* Vol.36, No.11, (November 2004), pp. 1159-1161, ISSN 1061-4036

Hara, E., Smith, R., Parry, D., Tahara, H., Stone, S., & Peters, G. (1996) Regulation of p16CDKN2 expression and its implications for cell immortalization and senescence. *Mol Cell Biol* Vol.16, No.3, (March 1996), pp. 859-867, ISSN 0270-7306

Hayashi, T., Fujita, Y., Iwasaki, O., Adachi, Y., Takahashi, K., & Yanagida, M. (2004) Mis16 and Mis18 are required for CENP-A loading and histone deacetylation at centromeres. *Cell* Vol.118, No.6, (September 2004), pp. 715-729, ISSN 0092-8674

Hayflick, L., & Moorhead, P.S. (1961) The serial cultivation of human diploid cell strains. *Exp Cell Res* Vol.25, No.3, (December 1961), pp. 585-621, ISSN 0014-4827

Howman, E.V., Fowler, K.J., Newson, A.J., Redward, S., MacDonald, A.C., Kalitsis, P., & Choo, K.H.A. (2000) Early disruption of centromeric chromatin organization in centromere protein A (Cenpa) null mice. *Proc Natl Acad Sci U S A* Vol.97, No.3, (February 2000), pp. 1148-1153, ISSN 1091-6490

Hudson, D.F., Fowler, K.J., Earle, E., Saffery, R., Kalitsis, P., Trowell, H., Hill, J., Wreford, N.G., de Kretser, D.M., Cancilla, M.R., Howman, E., Hii, L., Cutts, S.M., Irvine, D.V., & Choo, K.H.A. (1998) Centromere protein B null mice are mitotically and meiotically normal but have lower body and testis weights. *J Cell Biol* Vol.141, No.2, (April 1998), pp. 309-319, ISSN 0021-9525

Izuta, H., Ikeno, M., Suzuki, N., Tomonaga, T., Nozaki, N., Obuse, C., Kisu, Y., Goshima, N., Nomura, F., Nomura, N., & Yoda, K. (2006) Comprehensive analysis of the ICEN (Interphase Centromere Complex) components enriched in the CENP-A chromatin of human cells. *Genes Cells* Vol.11, No.6, (June 2006), pp. 673-684, ISSN 1356-9597

Jeganathan, K., Malureanu, L., Baker, D.J., Abraham, S.C., & van Deursen, J.M. (2007) Bub1 mediates cell death in response to chromosome missegregation and acts to suppress spontaneous tumorigenesis. *J Cell Biol* Vol.179, No.2, (October 2007), pp. 255-267, ISSN 0021-9525

Kapoor, M., Montes, de Oca, Luna, R., Liu, G., Lozano, G., Cummings, C., Mancini, M., Ouspenski, I., Brinkley, B.R., May, G.S. (1998) The *cenpB* gene is not essential in mice. *Chromosoma* Vol.107, No.8, (December 1998), pp. 570-576, ISSN 0009-5915

Lomonte, P., Sullivan, K.F., & Everett, R.D. (2001) Degradation of nucleosome-associated centromeric histone H3-like protein CENP-A induced by herpes simplex virus type 1 protein ICP0. *J Biol Chem* Vol.276, No.8, (February 2001), pp. 5829-5835, ISSN 0021-9258

Ly, D.H., Lockhart, D.J., Lerner, R.A., & Schultz, P.G. (2000) Mitotic misregulation and human aging. *Science* Vol.287, No.5462, (March 2000), pp. 2486-2492, ISSN 0036-8075

Maehara, K., Takahashi, K., & Saitoh, S. (2010) CENP-A reduction induces a p53-dependent cellular senescence response to protect cells from executing defective mitoses. *Mol Cell Biol* Vol.30, No.9, (May 2010), pp. 2090-2104, ISSN 0270-7306

Maehara, K. (2011) Cellular senescence as a self-defense mechanism against centromere dysfunction. *Biomed Gerontol* Vol.35, No.1, (February 2011), pp. 17-23, ISSN 0912-8921

Mallette, F.A., Gaumont-Leclerc, M.F., & Ferbeyre, G. (2007) The DNA damage signaling pathway is a critical mediator of oncogene-induced senescence. *Genes Dev* Vol.21, No.1, (January 2007), pp. 43-48, ISSN 0890-9369

Marshall, O.J., Chueh, A.C., Wong, L.H., & Choo, K.H. (2008) Neocentromere: new insights into centromere structure, disease development, and karyotype evolution. *Am J Hum Genet* Vol.82, No.2, (February 2008), pp. 261-282, ISSN 0002-9297

Masumoto, H., Masukata, H., Muro, Y., Nozaki, N., & Okazaki, T. (1989) A human centromere antigen (CENP-B) interacts with a short specific sequence in alphoid DNA, a human centromeric satellite. *J Cell Biol* Vol.109, No.5, (November 1989), pp. 1963-1973, ISSN 0021-9525

Meluh, P.B., Yang, P., Glowczewski, L., Koshland, D., & Smith, M.M. (1998) Cse4p is a component of the core centromere of Saccharomyces cerevisiae. *Cell* Vol.94, No.5, (September 1998), pp. 607-613, ISSN 0092-8674

Musacchio, A., & Salmon, E.D. (2007) The spindle-assembly checkpoint in space and time. *Nat Rev Mol Cell Biol* Vol.8, No.5, (May 2007), pp. 379-393, ISSN 1471-0072

Nakagawa, H., Lee, J.K., Hurwitz, J., Allshire, R.C., Nakayama, J., Grewal, S.I., Tanaka, K., & Murakami, Y. (2002) Fission yeast CENP-B homologs nucleate centromeric heterochromatin by promoting heterochromatin-specific histone tail modifications. *Genes Dev* Vol.16, No.14, (July 2002), pp.1766-1778, ISSN 0890-9369

Nakano, M., Cardinale, S., Noskov, V.N., Gassmann, R., Vagnarelli, P., Kandels-Lewis, S., Larionov, V., Earnshaw, W.C., & Masumoto, H. (2008) Inactivation of a human kinetochore by specific targeting of chromatin modifiers. *Dev Cell* Vol.14, No.4, (April 2008), pp. 507-522, ISSN 1534-5807

Narita, M., Nuñez, S., Heard, E., Narita, M., Lin, A.W., Hearn, S.A., Spector, D.L., Hannon, G.J., & Lowe, S.W. (2003) Rb-mediated heterochromatin formation and silencing of E2F target genes during cellular senescence. *Cell* Vol.113, No.6, (June 2003), pp. 703-716, ISSN 0092-8674

Narita, M., Narita, M., Krizhanovsky, V., Nuñez, S., Chicas, A., Hearn, S.A., Myers, M.P., & Lowe, S.W. (2006) A novel role for High-Mobility Group A proteins in cellular senescence and heterochromatin formation. *Cell* Vol.126, No.3, (August 2006), pp. 503-514, ISSN 0092-8674

Obuse, C., Yang, H., Nozaki, N., Goto, S., Okazaki, T., & Yoda, K. (2004a) Proteomics analysis of the centromere complex from HeLa interphase cells: UV-damaged DNA binding protein 1 (DDB-1) is a component of the CEN-complex, while BMI-1 is transiently co-localized with the centromeric region in interphase. *Genes Cells* Vol.9, No.2, (February 2004), pp. 105-120, ISSN 1356-9597

Obuse, C., Iwasaki, O., Kiyomitsu, T., Goshima, G., Toyoda, Y., & Yanagida, M. (2004b) A conserved Mis12 centromere complex is linked to heterochromatic HP1 and outer kinetochore protein Zwint-1. *Nat Cell Biol* Vol.6, No.11, (November 2004), pp. 1135-1141, ISSN 1465-7392

Okada, M., Cheeseman, I.M., Hori, T., Okawa, K., McLeod, I.X., Yates, JR, 3rd, Desai, A., & Fukagawa, T. (2006) The CENP-H-I complex is required for the efficient incorporation of newly synthesized CENP-A into centromeres. *Nat Cell Biol* Vol.8, No.5, (May 2006), pp. 446-457, ISSN 1465-7392

Okada, T., Ohzeki, J., Nakano, M., Yoda, K., Brinkley, W.R., Larionov, V., & Masumoto, H. (2007) CENP-B controls centromere formation depending on the chromatin context. *Cell* Vol.131, No.7, (December 2007), pp. 1287-1300, ISSN 0092-8674

Palmer, D.K., O'Day, K., Wener, M.H., Andrews, B.S., & Margolis, R.L. (1987) A 17-kD centromere protein (CENP-A) copurifies with nucleosome core particles and with histones. *J Cell Biol* Vol.104, No.4, (January 1987), pp. 805-815, ISSN 0021-9525

Perez-Castro, A.V., Shamanski, F.L., Meneses, J.J., Lovato, T.L., Vogel, K.G., Moyzis, R.K., & Pedersen, R. (1998) Centromeric protein B null mice are viable with no apparent abnormalities. *Dev Biol* Vol.201, No.2, (September 1998), pp. 135-143, ISSN 0012-1606

Régnier, V., Vagnarelli, P., Fukagawa, T., Zerjal, T., Burns, E., Trouche, D., Earnshaw, W., & Brown, W. (2005) CENP-A is required for accurate chromosome segregation and sustained kinetochore association of BubR1. *Mol Cell Biol* Vol.25, No.10, (May 2005), pp. 3967-3981, ISSN 0270-7306

Ruchaud, S., Carmena, M., & Earnshaw, W.C. (2007) Chromosomal passengers: conducting cell division. *Nat Rev Mol Cell Biol* Vol.8, No.10, (October 2007), pp. 798-812, ISSN 1471-0072

Serrano, M., Lin, A.W., McCurrach, M.E., Beach, D., & Lowe, S.W. (1997) Oncogenic *ras* provokes premature cell senescence associated with accumulation of p53 and p16INK4a. *Cell* Vol.88, No.5, (March 1997), pp. 593-602, ISSN 0092-8674

Shelby, R.D., Vafa, O., & Sullivan, K.F. (1997) Assembly of CENP-A into centromeric chromatin requires a cooperative array of nucleosomal DNA contact sites. *J Cell Biol* Vol.136, No.3, (February 1997), pp. 501-513, ISSN 0021-9525

Snape, K., Hanks, S., Ruark, E., Barros-Nuñez, P., Elliott, A., Murray, A., Lane, A.H., Shannon N., Callier, P., Chitayat, D., Clayton-Smith, J., FitzPatrick, D., Gisselsson, D., Jacquemont, S., Asakura-Hay, K., Micale, M.A., Tolmie, J., Turnpenny, P.D., Wright, M., Douglas, J., & Rahman, N. (2011) Mutations in *CEP57* cause mosaic variegated aneuploidy syndrome. *Nat Genet* Vol.43, No.6, (June 2011), pp. 527-529, ISSN 1061-4036

Stoler, S., Keith, K.C., Curnick, K.E., & Fitzgerald-Hayes, M. (1995) A mutation in CSE4, an essential gene encoding a novel chromatin-associated protein in yeast, causes chromosome nondisjunction and cell cycle arrest at mitosis. *Genes Dev* Vol.9, No.5, (March 1995), pp. 573-586, ISSN 0890-9369

Takahashi, A., Ohtani, N., Yamakoshi, K., Iida, S., Tahara, H., Nakayama, K., Nakayama, K.I., Ide, T., Saya, H., & Hara, E. (2006) Mitogenic signalling and the p16INK4a-Rb pathway cooperate to enforce irreversible cellular senescence. *Nat Cell Biol* Vol.8, No.11, (November 2006), pp. 1291-1297, ISSN 1465-7392

Takahashi, K., Chen, E.S., & Yanagida, M. (2000) Requirement of Mis6 centromere connector for localizing a CENP-A-like protein in fission yeast. *Science* Vol.288, No.5474, (June 2000), pp. 2215-2219, ISSN 0036-8075

Tanaka, T.U. (2010) Kinetochore-microtubule interactions: steps towards bi-orientation. *EMBO J* Vol.29, No.24, (December 2010), pp. 4070-4082, ISSN 0261-4189

Wang, Q., Liu, T., Fang, Y., Xie, S., Huang, X., Mahmood, R., Ramaswamy, G., Sakamoto, K.M., Darzynkiewicz, Z., Xu, M., & Dai, W. (2004) BUBR1 deficiency results in abnormal megakaryopoiesis. *Blood* Vol.103, No.4, (February 2004), pp. 1278-1285, ISSN 0006-4971

Zhang, R., Poustovoitov, M.V., Ye, X., Santos, H.A., Chen, W., Daganzo, S.M., Erzberger, J.P., Serebriiskii, I.G., Canutescu, A.A., Dunbrack, R.L., Pehrson, J.R., Berger, J.M., Kaufman, P.D., & Adams, P.D. (2005) Formation of MacroH2A-containing senescence-associated heterochromatin foci and senescence driven by ASF1a and HIRA. *Dev Cell* Vol.8, No.1, (January 2005), pp. 19-30, ISSN 1534-5807

7

Central Immune Senescence, Reversal Potentials

Krisztian Kvell and Judit E. Pongracz
Department of Medical Biotechnology, University of Pecs,
Hungary

1. Introduction

1.1 Ageing in focus

Ageing is a complex process that affects all living organisms. Senescence is not only conceivable in multicellular organisms, but also in unicellulars. Unlike certain diseases that have specific morbidity rates, ageing is a physiological process that affects all individuals that live long enough (unaffected by i.e. predation or famine) to experience senescence.

A pioneer of ageing research, August Weismann has established two rather opposing concepts for aging and even today both gather numerous followers. One is the adaptive concept, according to which ageing has evolved to cleanse the population from old, non-reproductive consumers. The other, non-adaptive concept suggests that ageing is due to greater weight on early survival / reproduction rather than vigour at later ages. This latter has been reshaped by the theory of antagonistic pleiotropy (Ljubuncic et al. 2009).

Due to advances in biomedical research and care, currently an average 55-aged person is expected to live up to 85 years of age at death on average in the Western societies. This number is expected to increase if biomedical research continues to develop at the current rate and by the year 2030 an average 55-aged person is expected to live up to 115 years of age at death (according to SENS plans) (de Grey 2007). If such forecasts prove to be true, it is of extraordinary significance and will likely trigger immense social and economical conflicts.

1.1.1 Ageing and society

Ageing of the population is one of the most important challenges for the developed world to face over the next decades. The current demographic trends and consequent shrinkage of the active workforce will put enormous pressure on the financing of social protection and health systems, likely to reduce living standards. Taken together with increased migration and emergence of novel infectious diseases, broad-scale provision of immunological protection constitutes a strategic aim for longer and healthier lifespan.

At present life-span is still significantly increasing in the Western civilisations, however, this increase is not accompanied by proportional increase in life spent in overall good health referred to as 'health-span'. There are current efforts to prolong health-span within expanding life-span. This would not only extend life spent in appropriate quality of life, but also has the potential to alleviate pressure on current public health systems. This chapter focuses on central immune senescence and therefore will enumerate potential mechanisms of extending human central immune fitness in the elderly.

1.1.2 Ageing of the immune system

Impaired immunological responsiveness in the elderly poses a major difficulty. The immunological competence of an individual is determined by the presence of mature lymphocytes formed in primary lymphoid organs, and specialized secondary lymphoid tissues performing diverse immune responses. Thus at systems level the maintenance of immunological equilibrium requires steady lymphocyte output, and controlled expansion. Lymphostromal interactions in both primary and secondary lymphoid tissues play essential roles in the development and function of lymphocyte subsets in adaptive immune responses. The thymic and lymph-node stromal microenvironments thus represent key elements in the development of the adaptive immune system. Consequently, impairment of the lymphoid microenvironment will ultimately lead to insufficient primary and secondary immune responses or to the decline of thymic selection, manifesting in immune senescence accompanied by late-onset autoimmune disorders, often observed in elderly. Self-tolerant cytotoxic and helper T-lymphocytes, the crucial regulator cells in adaptive immune responses, develop in the specialized epithelial network of the thymus. The thymus, however, gradually loses its capacity to support lymphopoiesis in an involution process that results in a decline of *de novo* T-cell production.

1.1.3 Significance of thymic involution studies

In contrast to the extensive studies addressing haemopoietic cells, the in-depth analysis of determinants for stromal competence during immunological ageing is far less detailed, despite its clear significance related to immunological responsiveness in the elderly. There is literature describing quantitative changes that occur during immunological senescence in peripheral immunologically competent tissues like the spleen or lymph nodes. Probably the best characterised, significant example is that of FDCs. Compared to young counterparts the aged follicular dendritic cells express significantly less CD21 ligand and FcγRII. As a consequence aged FDCs lose their ability to trap immune complexes and present antigens to B cells. This in turn leads to impaired germinal centre reactions and antibody production (Aydar et al. 2004). However, even these well characterised quantitative changes of the peripheral lymphoid tissues are less dramatic than the adipose involution of the thymus.

The manipulation of thymic immune senescence and the restoration of de-novo T-cell production should provide direct benefits for both adult and elderly patients. Such interventions shall increase health-span within life-span significantly reducing the healthcare costs alleviating the burden on healthcare systems.

1.2 T-cell development in the thymus

T-cell progenitors migrate to the thymus from the bone marrow where they undergo an extensive differentiation and selection process. After entering the thymus, thymocytes representing different stages of development occupy distinct regions of the thymus. The earliest CD4-CD8-CD44+CD25- thymocyte progenitors, referred to as double negative 1

(DN1) cells are found near their site of entry at the cortico-medullary junction. The slightly more mature CD4-CD8-CD44+CD25+ (DN2) subset is found throughout the cortex, whereas CD4-CD8-CD44-CD25+ (DN3) subset is concentrated below the capsule. Following rearrangement of antigen receptor (TCR) genes (He et al. 2006) CD4+CD8+ (double positive or DP) thymocytes undergo positive (functional TCR) and negative (non self-reactive TCR) selection in the cortex and medulla, to finally leave the thymus for the periphery as CD4-CD8+ (cytotoxic) or CD4+CD8- (helper) single positive (SP), mature, naïve T-cells.

1.3 Thymic microenvironment in *de novo* T-cell production

Successful T-cell development requires the interaction of thymocytes with the thymic stroma, creating the special thymic microenvironment for T-cell differentiation and selection. A large proportion of the thymic stroma consists of epithelial cells that develop from the epithelial thymic anlage from the third pharyngeal pouch around embryonic day 10-11 in the mouse (Manley 2000). Following several differentiation steps, including expression of FoxN1 – a member of the forkhead transcription factor family (Mandinova et al. 2009) – that is essential for Mts24+ epithelial progenitors (Bennett et al. 2002; Gill et al. 2002) to develop into various epithelial subsets (Dooley et al. 2005) and to establish the special thymic epithelial cell phenotype (Manley 2000). FoxN1 expression in early stages of thymus organogenesis is regulated by secreted Wnt4 (Balciunaite et al. 2002) protein. The mature thymic epithelium consists of two major compartments, the cortex and the medulla, which apart from producing chemokines that attract haematopoietic stem cells to the thymus, also contribute the establishing the special thymic microenvironment. The thymic epithelial network regulates homing, intrathymic migration, and differentiation of developing T-lymphocytes through release of cytokines (e.g. interleukin-7 (Alves et al. 2009)), secretion of extracellular matrix components, and establishment of intercellular connections (Crisa L et al. 1996) (Schluns et al. 1997). Thymocytes bearing diverse TCR repertoire are selected by MHC (major-histocompatibility-complex) molecules and MHC bound-antigens presented by the thymic stroma, including epithelial cells. During T-cell development, characterised by progression through phenotypically distinct stages (Lind et al. 2001), thymocytes reside in spatially restricted domains of the mature thymus. T-cell precursors enter the thymus at the cortico-medullary junction (Blackburn et al. 2004), then migrate to the subcapsular zone of the outer cortex, back through the cortex, then to the medulla, where they finally leave to the periphery (Blackburn et al. 2004). Functional studies have shown, that the cortex is important in producing chemokines, which attract pro-thymocytes (Bleul et al. 2000) and are also essential for mediating positive selection (Anderson et al. 1994). Meanwhile the medullary epithelium has been implicated in driving the final stages of thymocyte maturation (Ge et al. 2000) and has a crucial role in tolerance induction (Farr et al. 1998; Derbinski et al. 2001). Additionally, the thymic epithelium is also the source of other secreted and cell surface proteins that regulate T-cell development. These proteins include bone morphogenic protein (BMP) (Bleul et al. 2005), Notch (Valsecchi 1997), and Wnt (Pongracz et al. 2003) family members.

1.4 Thymic involution during ageing

In comparison to other organs, ageing of the thymus is an accelerated process in all mammals. In humans, thymic senescence begins early, around late puberty and by 50 years of age 80% of the thymic stroma is converted into adipose tissue (Dixit 2010). As the thymic epithelium is replaced by adipose tissue, the whole process is called adipose involution (Marinova 2005). Due to decrease in functional thymic epithelial tissue mass, the thymus can no longer support the same output of naïve T-cell production (Ribeiro et al. 2007). T-lymphocyte composition in the periphery therefore exhibits the dominance of memory T-lymphocytes resulting in impaired responses towards novel, particularly viral infections (Chidgey et al. 2007; Gui et al. 2007; Grubeck-Loebenstein 2009). Since the thymic epithelium has also a key role in deleting auto-reactive T-cell clones, functional impairment increases the chances of developing auto-immune disease (Hsu et al. 2003). The transcription factor FoxN1, characteristic in thymus development is also affected by age. FoxN1 (Mandinova et al. 2009) is not only essential for progenitor epithelial cells of the thymic rudiment to develop into various epithelial subsets (Dooley et al. 2005) but also to maintain TEC identity in the differentiated, adult thymus. Decreased level of FoxN1 expression in adult TECs results in accelerated thymic involution (Chen et al. 2009; Cheng et al. 2010).

1.5 Thymic involution: developmental programme or senescence?

It has long been known that the thymus begins adipose involution and senescence rather early, but how early is that exactly? Recent studies have analysed the kinetics of thymic function and thymic mass versus age. It has been confirmed in both mouse and human thymic samples that the functional peek of thymic activity significantly precedes the peek of thymic mass and the first signs of adipose involution. In the mouse thymic activity is largely decreased by the age of one month compared to the newborn age (thymocyte precursor immigration at 6% and mature T cell emigration at 7% where 100% is measured at newborn age), yet the thymus reaches its largest size at one month of age (Shiraishi et al. 2003). Similar tendency has been described in humans where thymic function reaches its peak around the age of one year followed by the first signs of adipose infiltration by the age of approx. five years (Shiraishi et al. 2003). However, most studies describe significant thymic adipose involution starting around puberty / young adulthood. Therefore there is apparently significant detachment of thymic activity peak and thymic mass peak, and surprisingly activity peak significantly precedes mass peak.

The above described phenomenon raises the issue whether the early appearance of thymic involution belongs to senescence or developmental programme and how strictly these two may be separated? Similar questions are raised by the detection of miniature atherosclerotic lesions detected already at foetal age, a currently fashionable topic (Leduc et al. 2010).

1.6 Trans-differentiation of fibroblasts into adipocytes

The nuclear lamina consists of a matrix of proteins located next to the inner nuclear membrane. The lamina family of proteins makes up the matrix and that is highly conserved in evolution. The family of lamina associated polypeptides (LAP) has several members with similar functions. Studies with fibroblast cells have revealed that fibroblast to pre-adipocyte transformation is strongly connected to LAP2α, the member of the LAP2α protein family (Dorner et al. 2006). While most splice variants associate with the nuclear envelope, LAP2α is involved in several nucleoplasmic activities including cell-cycle control and differentiation (Berger et al. 1996; Hutchison et al. 2001). LAP2α is synthesized in the cytoplasm and is then transported into the nucleus by a PKC-dependent mechanism (Dreger et al. 1999). The mere over-expression of LAP2α in fibroblasts is known to directly up-regulate PPARγ expression, an acknowledged marker and key transcription factor of pre-adipocyte differentiation (Dorner et al. 2006). In pre-adipocytes PPARγ expression is followed by an increase of ADRP expression (adipose differentiation-related protein) a known direct target gene of PPARγ. Although LAP2α over-expression alone initiates pre-adipocyte differentiation in fibroblasts, it is not sufficient to complete the adipocyte differentiation programme in the absence of additional stimuli (Dorner et al. 2006).

1.7 Wnt signalling

1.7.1 Wnt signalling

The Wnt family of 19 secreted glycoproteins controls a variety of developmental processes including cell fate specification, cell proliferation, cell polarity and cell migration. There are two main signalling pathways involved in the signal transduction process from the Wnt receptor (Frizzled) complex: the canonical or β-catenin dependent, and the non-canonical pathway, which splits into the polar cell polarity (PCP) or c-Jun-N-Terminal Kinase (JNK) / Activating Protein (AP1) dependent and the Ca^{2+} or Protein kinase C (PKC) / Calmodulin Kinase (CaMKII) / Nuclear Factor of Activating T-cells (NFAT) dependent signalling pathways.

Based on their ability to activate a particular Wnt pathway, Wnt molecules have been grouped as canonical (Wnt1, Wnt3, Wnt3a, Wnt7a, Wnt7b, Wnt8) (Torres et al. 1996) and non-canonical pathway activators (Wnt5a, Wnt4, Wnt11) (Torres et al. 1996), although promiscuity is a feature of both ligands and receptors.

1.7.2 Canonical Wnt-pathway

The canonical or β-catenin / Tcf dependent Wnt pathway is extensively investigated, and has been shown to be present in the thymus both in developing thymocytes (Ioannidis et al. 2001; Staal 2001; Xu et al. 2003) as well as in the thymic epithelium (Balciunate et al. 2001, Pongracz et al. 2003). Generally, in the absence of canonical Wnt-s, glycogen synthase kinase-3β (GSK-3β) is active and phosphorylates β-catenin in the scaffolding protein complex of adenomatous polyposis coli (APC) and axin (Ikeda 1998; Yamamoto 1999). The phosporylated β-catenin is targeted for ubiquitination and 26S proteasome-mediated degradation, thereby decreasing the cytosolic level of β-catenin (Aberle 1997; Akiyama 2000). In the presence of Wnt-s, signals from the Wnt-Fz-LRP6 complex lead to the phosphorylation of three domains of Dishevelled (Dvl), a family of cytosolic signal transducer molecules (Noordermeer 1994). Activation of Dvl ultimately leads to phosphorylation and consequent inhibition of GSK-3β. Inhibition of GSK-3β results in stabilisation and finally cytosolic accumulation of β-catenin, which then translocates to the nucleus where is required to form active transcription complexes with members of the T-Cell Factor (LEF1, TCF1, TCF3, TCF4) transcription factor family (Staal et al. 2003) and transcription initiator p300 (Labalette et al. 2004). Successful assembly of the transcription complex leads to the activation of various target genes including cyclin-D1 (Shtutman et al. 1999; Tetsu et al. 1999), c-myc (He et al. 1998), c-jun (Mann et al. 1999), Fra-1 (Mann et al. 1999), VEGFR (Zhang et al. 2001).

1.7.3 Non-canonical Wnt-pathways

Generally, the two non-canonical signalling pathways are considered as regulators of canonical Wnt signalling and gene transcription. The two non-canonical Wnt pathways, the JNK/AP1 dependent, PCP (Yamanaka et al. 2002) and the PKC/CAMKII/NFAT dependent Ca^{2+} pathway (Wang et al. 2003), become activated following the formation of Wnt-Fz-LRP6 complex just like the canonical Wnt pathway. Although the non-canonical pathways differ from the canonical pathway in their dependency on the type of G-proteins (Malbon et al. 2001), activation of Dvl, downstream of Frizzled, is critical for further signal transduction in both (Boutros et al. 1998; Sheldahl et al. 2003). In further contrast to canonical Wnt signalling, phosphorylation of all three domains of Dvl, is not a requirement for transduction of non-canonical Wnt signals (Wharton Jr. 2001). Downstream of the cytosolic Dvl, the two non-canonical pathways activate different signalling cascades, which involve JNK or PKC and CaMKII, and trigger the transcription of different target genes. It has been proposed for non-canonical Wnt-signalling receptors to be linked directly to heterotrimeric G-proteins that activate phospholipase-C (PLC) isoforms, which in turn stimulate inositol lipid (i.e. Ca^{++}/PKC) signalling. Growing evidence, however, indicates that G-proteins are functionally diverse and that many of their cellular actions are independent of inositol lipid signalling (Peavy et al. 2005), indicating high levels of complexity in both the PKC dependent and independent Wnt signalling cascades. The JNK dependent PCP pathway, partly shares target genes with the canonical pathway, including cyclin-D1 (Schwabe et al. 2003) and matrix metalloproteinases (Nateri et al. 2005). Certainly, canonical Wnt signals can be rechanneled into the JNK pathway through naturally occurring, intracellular molecular switches, like the Dvl inhibitors, Naked-s (Nkd-1, Nkd-2) (Yan et al. 2001) leading to AP1 rather than TCF activation. AP1 is not a single protein, but a complex of various smaller proteins (cJun, JunB, JunD, cFos, FosB, Fra1, Fra2, ATF2, and CREB), which can form homo- and heterodimers. The composition of the AP1 complex is a decisive factor in the activation of target genes, therefore the regulation of AP1 composition is important. Two prominent members of the AP1 complex cJun and Fra1 are both targets of the canonical Wnt pathway (Mann et al. 1999), indicating strong cross-regulation between the canonical and the non-canonical JNK dependent Wnt signalling cascades (Nateri et al. 2005).

While there are shared ligands (Rosso et al. 2005; Wang et al. 2005) and target genes (Shtutman et al. 1999; Schwabe et al. 2003) in the canonical and JNK dependent Wnt pathways, Ca^{2+}/PKC dependent non-canonical signalling appears to be more independent of the other two pathways although cross-talk with both the β-catenin and the JNK pathways have been proposed (Kuhl et al. 2001). Generally, Ca^{2+} and PKC-dependent signals are frequently linked to AP1, NFkB and NFAT activation.

1.7.4 Inhibitory Wnt pathway

Besides the canonical and non-canonical Wnt pathways, inhibitory Fz pathways have also been described. Fz1 and Fz6 are, for example, able to transmit inhibitory Wnt signals. While Fz1 inhibits Wnt signal transduction via a G-protein dependent manner (Roman-Roman et al. 2004) (Zilberberg et al. 2004), Fz6 (Golan et al. 2004) inhibits Wnt dependent gene transcription by activating the transforming growth factor β–activated kinase 1 (TAK1), a member of the MAPKKK family, and nemo-like kinase (NLK) (Ishitani et al. 2003; Smit et al. 2004) via a Ca^{++} dependent signalling cascade. NLK phosphorylates TCF that consequently cannot bind to β-catenin, and the formation of active transcription complex becomes inhibited (Smit et al. 2004) .

1.7.5 Wnt-s in ageing

As Wnt-s are important regulators of stem cell survival and differentiation, recent studies have started to investigate the role of Wnt family members in ageing. Most studies confirmed that drastically reduced Wnt levels can trigger ageing as tissue specific stem cells fail to replenish mesenchymal tissues as a result of low Wnt signals. In contrast, the KLOTHO mouse, that carries a single gene mutation in KLOTHO, an endogenous Wnt antagonist also shows signs of accelerated ageing (Liu et al. 2007). It has been proposed that increased Wnt signalling leads to continuous stem cell proliferation which finally results in depletion of the stem cell pool (Brack et al. 2007).

1.7.6 Wnt-s in the thymus

The main source of Wnt glycoproteins in the thymus is the thymic epithelium, where 14 members of the Wnt family together with all 10 known Wnt receptors of the seven-loop transmembrane receptor family, Frizzleds (Fz) have been identified (Pongracz et al. 2003). That is a striking difference compared with thymocytes where developmentally regulated receptor expression is limited to Fz-5 and Fz-6 (Pongracz et al. 2003). The assembly of an active Wnt-Fz receptor complex also requires the presence of a co-receptor, the low density lipoprotein related protein 5 and 6 (LRP5/6) (Pinson 2000; Tamai 2000; Wehrli 2000), which is expressed both in thymocytes and thymic epithelial cells, indicating full ability in both cell types to respond to Wnt signals.

Initial experiments, by manipulating the level of some Wnt-s and soluble Fz-s, have shown perturbation of T-cell development (Staal 2001; Mulroy 2002), highlighting the importance of Wnt dependent signalling for T-cell proliferation and differentiation. Recent data (Pongracz et al. 2003) revealed differential expression of Wnt ligands and receptors in thymic cell types raising that T-cell development may be influenced by indirect events triggered by Wnt signalling within the thymic epithelium.

The canonical pathway has been shown to have an important role in thymocyte development regulating survival and differentiation (Ioannidis et al. 2001; Staal 2001; Pongracz et al. 2003; Xu et al. 2003). In a thymic epithelial cell study, transgenic expression of cyclin-D1, one of the principal target genes of Wnt signalling, has lead to the expansion of the entire epithelial compartment (Klug et al. 2000) suggesting that canonical Wnt signalling is involved in thymic epithelial cell proliferation, strengthening the argument, that thymic epithelial development is regulated by Wnt-s. So far, signalling studies have revealed, that Wnt4 can activate both the canonical (Lyons et al. 2004) and the non-canonical (Torres et al. 1996) (Chang et al. 2007; Kim et al. 2009) Wnt-pathways.

1.8 Steroids and ageing

Physiological steroids are implicated in the regulation of thymic ageing. For example both surgical and chemical castration have been demonstrated to decrease the progression of thymic ageing (Qiao et al. 2008) indicating that high steroid levels would accelerate the ageing process of the thymus. Still, steroids used in therapy have not been fully investigated for their effects on immune senescence. Autoimmune diseases and haematological malignancies are often treated by steroids, as they effectively promote apoptosis of leukaemia cells and trigger complex anti-inflammatory actions (Stahn et al. 2007). Apart from triggering decreased expression of cytokines and MHC class II (MHC II) molecules, glucocorticoid (GC) analogues like dexamethasone (DX) also induce apoptotic death of peripheral (Wust et al. 2008) and developing T-cells. In mouse models, GCs cause massive thymocyte depletion, especially in the CD4+CD8+ (DP) thymocyte population, (Wiegers et al. 2001; Berki et al. 2002; Jondal et al. 2004) blocking *de novo* T-cell production. Experiments have also demonstrated that high-dose GCs induce a dramatic (Blomgren et al. 1970) and apoptosis-associated (Boersma et al. 1979) involution of the thymus, and not only thymocytes but also TECs are seriously affected (Dardenne et al. 1986). Recent reports (Fletcher et al. 2009) have highlighted that TEC depletion appears reversible, and thymic epithelial stem cells play an important role in this process.

2. Thymic senescence – Current opinion

2.1 Physiological thymic senescence

2.1.1 Disintegration of epithelial network, adipose involution

Senescence exhibits characteristic histological changes in both the human and mouse thymus (Oksanen 1971; Marinova 2005). In young adult mice (at 1 month of age), histology reveals strict segregation of epithelial cell compartments by staining for medullary (EpCAM1++, Ly51-) and cortical (EpCAM1+, Ly51++) epithelial cellular subsets (Kvell et al. 2010). Thymic morphology shows high level of integrity just preceding puberty/early adulthood. However, the highly organized structure disintegrates and becomes chaotic by the age of 1 year. By this age the strict cortico-medullary delineation becomes disintegrated, degenerative vacuoles appear surrounded by areas showing strong co-staining with both epithelial markers. Also significant cellular areas appear that lack staining with either epithelial markers, a pattern completely absent at the young adult age. Staining for extracellular matrix components of fibroblast origin (ER-TR7++) identifies mesenchymal elements. The staining pattern with ER-TR7 and EpCAM1 is strikingly different at the two ages examined. In young adult thymic tissue sections, a-EpCAM1 and a-ER-TR7- show little tendency for co-localization. In stark contrast, already by the age of 9 months a-EpCAM1 and ER-TR7-staining show significant overlap within the thymic medulla. The disorganization of thymic epithelial network is followed by the emergence of adipocytes. If thymic sections of senescent mice are co-stained with neutral lipid deposit-specific stains then histology shows the presence of relatively large, inflated cells in which the cytoplasm is pushed to the periphery by red-staining neutral lipid deposits, a pattern characteristic of adipose cells (Kvell et al. 2010).

2.1.2 Gene expression changes in the thymic epithelium during ageing

To investigate the underlying molecular events of thymic epithelial senescence, the gene expression changes may be investigated in TECs purified from 1 month and 1 year old mice (Kvell et al. 2010). The expression of both Wnt4 and FoxN1 decreases in thymic epithelial cells. Highly decreased level (or total absence in some cases) of FoxN1 could be the consequence of strong Wnt4 down-regulation by the age of 1 year, indicating that TECs can down-regulate FoxN1 expression while maintaining that of epithelial cell surface markers like EpCAM1 (Balciunaite et al. 2002). At the same time, mRNA levels of pre-adipocyte differentiation markers PPARγ and ADRP rise with age. This finding is in harmony with histological data demonstrating the emergence of adipocytes in the thymic lobes of senescent mice. The expression of lamin1, a key component of the nuclear lamina remains unaffected during senescence in thymic epithelial cells; whereas, the expression of LAP2α increases significantly. This degree of dissociation between lamin1 and LAP2α expression is of note and suggests functional differences despite conventionally anticipated association of lamin1 and LAP2 molecular family members. LAP2α up-regulation associated with age-related adipose involution is, however, in perfect agreement with other literature data suggesting the pre-adipocyte differentiation-promoting effect of LAP2α in fibroblasts (Dorner et al. 2006) and the same is suggested by our reports performed, however, with epithelial cells (Kvell et al. 2010).

According to literature, EMT is associated with differential expression of E- (decrease) and N-cadherin (increase) (Seike et al. 2009). TECs were tested for these markers to investigate whether the first step towards pre-adipocyte differentiation is the EMT of epithelial cells. In purified TECs while E-cadherin mRNA levels significantly decreased, N-cadherin gene expression showed a slight increase during ageing, indicating that EMT might be the initial step in epithelial cell transition and trans-differentiation.

2.1.3 Studies of LAP2α and Wnt4 effects on TEC

The hypothesis that both LAP2α and Wnt4 play important though opposite roles in thymic senescence may be addressed using LAP2α over-expressing or Wnt4-secreting transgenic TEP1 (mouse primary-derived thymic epithelial) cell lines. The use of a primary-derived model cell line provides the advantage of absolute purity, the complete lack of other cell types that could potentially affect the gene expression profile of epithelial cells (Beardsley et al. 1983). Using such cells quantitative RT-PCR analysis revealed that LAP2α over-expression triggers an immense surge of PPARγ expression. Such an increase in mRNA level suggests that this is not a plain quantitative, but rather a qualitative change. ADRP a direct target gene of PPARγ also becomes up-regulated although to a lesser extent. On the other hand in Wnt4-secreting cells the mRNA level of both PPARγ and ADRP decreased (Kvell et al. 2010).

2.1.4 Fz-4 and Fz-6 expression and distribution are affected by age

Once the preventive role of Wnt4 was established in adipocyte-type trans-differentiation of TECs, receptor associated signalling studies have ensued to investigate what signal modifications can lead to Wnt4 effects. Initially, expression levels of the Wnt4 receptors, Fz-4 and Fz-6 were analysed in thymi of young adult and mature adult (1 month and 9 months old) mice (Varecza et al. 2011). Q-RT-PCR analysis of TECs showed increased expression of both Fz-4 and Fz-6 mRNA with age. Immune-histochemistry using Fz-4 and Fz-6 specific antibodies confirmed elevated levels of both receptor proteins. Additionally, differential expression pattern of Fz-4 and Fz-6 was also observed in the thymic medulla and cortex. While in the young thymus the medulla (EpCAM1++/Ly51-) was preferentially stained for

Fz4 and Fz6, the cortex (EpCAM1+/Ly51+) only faintly stained for these receptors. In contrast to the young tissue, the 9 month old thymus shows a different pattern as the whole section including the cortex has become increasingly positive for both receptors (Varecza et al. 2011).

2.1.5 PKCδ translocation and its relation with Wnt4 signalling

Since Wnt4 levels as well as its receptors are modulated during the ageing process, further studies were performed to investigate active receptor signalling that is invariably associated with modified level of phosphorylation of receptor associated signalling molecules. Since Fz-s associate with Dvls that are phosphorylated by the δ isoform of PKCs, PKCδ activity was in focus. To test the involvement of PKCδ in Wnt4 signal transduction, increased Wnt4 levels were achieved using the supernatant of Wnt4-transgenic cell line (Varecza et al. 2011). Wild type TEP1 cells were exposed to SNs of control and Wnt4-secreting cells for 1 hour, then cytosolic and membrane fractions were isolated from cell lysates. Similar to previous studies with Wnt-5a (Giorgione et al. 2003), Western blot analysis revealed that within one hour of Wnt4 exposure PKCδ translocated into the membrane fraction where the cleavage products (Kanthasamy et al, 2006) characteristic of PKCδ activation were detectable. Additionally, increased membrane localisation of PKCδ was also detected in the Wnt4-overexpressing cell line. As both Fz-4 and Fz-6 levels increased with age, it was assumed that active receptor signalling might require more PKCδ during ageing. Indeed, apart from localisation of PKCδ to the membrane fraction, up-regulation of PKCδ was also detected in the ageing thymi. To investigate the role of PKCδ involvement in Wnt4 signalling, PKCδ activity level was modified by either over-expressing wild type PKCδ or by silencing PKCδ translation using siRNA technology. CTGF was used as a read-out gene based on data of previous experiments (Varecza et al. 2011). Surprisingly, although over-expression of PKCδ had no radical effect on Wnt4 target gene transcription, even moderate down-regulation of PKCδ was able to significantly increase CTGF expression in the presence of Wnt4, indicating that PKCδ might be involved in a negative regulatory loop.

2.1.6 Negative regulatory loops of signalling during senescence

As Fz-6 has been implicated in previous studies as a negative regulator of β-catenin dependent signalling, it was important to determine whether PKCδ is preferentially associated with either Wnt4 receptors. Experiments demonstrated age dependent increase of both Fz-6 and PKCδ as well as co-localisation of Fz-6 and PKCδ (Varecza et al. 2011). While in the young thymus Fz-6 and PKCδ co-localisation is more pronounced in the thymic cortex, in the ageing thymus it is the medulla that exhibits stronger staining for both proteins. While increased expression and activity of the Fz-6 receptor, a suppressor of the canonical Wnt signalling pathway explains some aspects of uneven target gene transcription following manipulation of PKCδ activity, parallel changes like up-regulation of Fz-4 also occur during ageing that might add to the complexity of the signalling process. Increase in Fz-4 levels in ageing mice correlated with increased CTGF gene expression.

If Fz-6 that also increases during senescence is truly a suppressor of β-catenin signalling then CTGF expression should have decreased or remained unchanged as Fz-4 transmitted signals would have been quenched by Fz-6 signalling. To test the above hypothesis, we have considered the following: CTGF has recently been reported to negatively regulate canonical Wnt signalling by blocking β-catenin stabilisation via GSK3β activation leading to phosphorylation and consequent degradation of β-catenin (Luo et al. 2004), indicating that CTGF might be part of a negative feed-back loop. The expression of Fz-8 (Mercurio et al. 2004) a recently reported receptor for CTGF increased in ageing mice, while FoxN1 the direct target of β-catenin dependent Wnt4 signalling (Balciunaite et al. 2002) became undetectable (Kvell et al. 2010).

2.2 Thymic senescence model

2.2.1 Steroid induced accelerated thymic senescence

A commonly held view is that the thymus involutes at puberty, and this model is based primarily on studies showing that growth hormone (GH) and sex steroids can affect cell production in the thymus and that their concentrations decrease with age (Min et al. 2006). As steroids are frequently applied medications, investigations were extended to identify similarities in induced and physiological senescence and potential mechanisms that might be able to reduce adipose involution of the thymus.

Similar to physiological senescence, the level of FoxN1 transcription factor and its regulator Wnt4 decreased in TECs within 24 hours following a single dose DX injection and remained low for over 1 week (Talaber et al. 2011).

However, in clinical treatments GC analogues are widely used for extended periods of time, rather than single shots. To mimic this pattern of clinical application, mice were injected with DX repeatedly for a time course of 1 month. Both Wnt4 and FoxN1 levels were measured drastically down-regulated, while the adipocyte differentiation factor ADRP, down-stream target of PPARγ was significantly increased. The results indicate that adipocyte-type trans-differentiation is completed at the molecular level over a much shorter time period following exogenous steroid-induced senescence compared to physiological rate senescence (Talaber et al. 2011).

2.2.2 Wnt4 inhibits steroid-induced adipose trans-differentiation

To test whether Wnt4 can prevent adipocyte type trans-differentiation, Wnt4 over-expressing TEP1 cell line was exposed to DX for a week. While in the control cell line DX exposure induced up-regulation of adipose trans-differentiation markers, within the Wnt4 over-expressing cell line, none of the adipose trans-differentiation markers were up-regulated indicating that Wnt4 alone can efficiently protect TECs against exogenous steroid-induced adipose trans-differentiation (Talaber et al. 2011).

3. Conclusions

3.1 Physiological thymic epithelial senescence

There are characteristic changes in the gene expression profile of purified thymic epithelial cells during thymic epithelial senescence (Kvell et al. 2010). Of note, Wnt4 level decreases, while LAP2α level increases. Also, the expression of the transcription factor FoxN1 required for maintaining thymic epithelial identity diminishes with age. On the other hand, adipose differentiation is confirmed at the molecular level by the increased expression of PPARγ and ADRP. This process is accompanied by shift from E-cadherin to N-cadherin, typical for EMT (epithelial to mesenchymal transition). These pioneer experiments confirm in both a model cell line and purified primary cells rendered transgenic for either Wnt4 or LAP2α that their opposing effects antagonistically influence adipose trans-differentiation of thymic epithelial cells via EMT. This has lead to the establishment of a novel, confirmed theory for the source of adipose cells replacing functional thymic epithelial network during senescence (see **Figure 1**). Apparently, these cells do not differentiate from invading or resident mesenchymal cells, but rather trans-differentiate (via EMT) from thymic epithelial cells (Kvell et al. 2010).

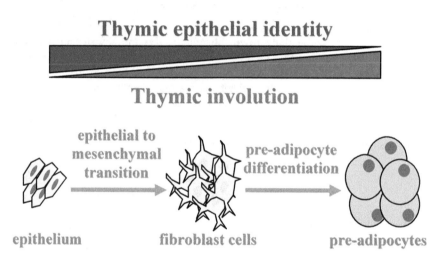

Fig. 1. Model of thymic involution process
Dedifferentiation of thymic epithelial cells triggers EMT (epithelial to mesenchymal transition) first, and then the resulting fibroblast cells undergo the conventional route of differentiation program towards adipocyte-lineage.

3.2 Signal transduction mechanisms involved in thymic epithelial senescence

While individual molecules, such as Wnt4 or LAP2α can serve as therapeutic targets to modify the ageing process, identification of complex interactions amongst signalling networks can provide further details. Investigation of Wnt signal transduction in the thymic epithelium has revealed that signalling pathways are activated or inhibited in an orderly fashion (Varecza et al. 2011). Initially, both Wnt4 receptors, Fz-4 and Fz-6 are up-regulated at young adult age. However, signals from Fz-4 and Fz-6 are different. While signals from Fz-4 initiate β-catenin dependent gene transcription, Fz-6 signals lead to suppression of β-catenin dependent signalling via increased activities of TGFβ-Activated Kinase (TAK) and Nemo-Like-Kinase (NLK). Fz-associated signals also require PKCδ to transmit Wnt signals. PKCδ associates with Fz-6 aiding suppression of β-catenin dependent signalling. Additional to Fz-6 signalling, connective tissue growth factor (CTGF, a β-catenin target gene) can also feedback on β-catenin dependent signal transduction. CTGF can interact with Fz-8 as well as LRP6, an important co-receptor of Wnt signalling and can trigger activation of GSK3β. This latter leads to accelerated proteasomal degradation of β-catenin and hence suppression of Wnt signals. Multiple signalling mechanisms that lead to suppression of Wnt signalling is summarized in **Figure 2** (Varecza et al. 2011).

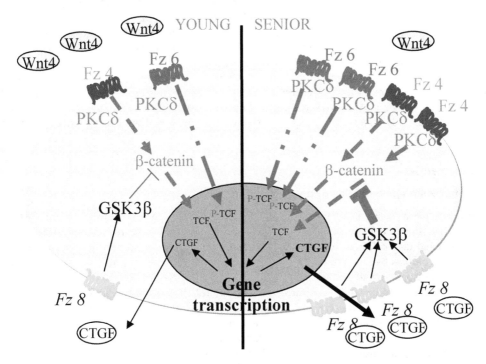

Fig. 2. Model of molecular mechanisms in thymic aging

At young age, Wnt4 levels are high and Wnt4 molecules compete for a moderate number of Fz receptors. While Fz-4 activates canonical Wnt signalling, signals from Fz-6 inhibit β-catenin dependent gene transcription keeping Wnt4 dependent signalling in balance. During the ageing process, Wnt4 levels decrease, while receptor expression increases with proportionally higher Fz-6. The β-catenin dependent Fz-4 signals lead to increased expression of CTGF. The CTGF receptor Fz8 is also up-regulated leading to enhanced activation of GSK3β. All these signalling events lead to loss of thymic epithelial cell characteristics and provide an opening for molecular events leading up to adipocyte type trans-differentiation.

3.3 Accelerated-rate, induced model of thymic epithelial senescence

Glucocorticoids are immunosuppressive drugs often used for treatment of autoimmune diseases and haematological malignancies. Although glucocorticoids can induce apoptotic cell death directly in developing thymocytes, how exogenous glucocorticoids affect the thymic epithelial network that provides the microenvironment for T cell development has been poorly characterised. The effect of DX (dexamethasone) on thymic epithelial cells has been tested both *in vitro* (model cell line) and *in vivo* (mouse model) (Talaber et al. 2011). *In vivo*, following single treatment with pharmacologically relevant dose of DX reversible changes in gene expression profile identical to physiological thymic epithelial senescence have been recorded, but occurring at a highly accelerated pace (see **Figure 1**). Specifically, the expression of Wnt4 and FoxN1 decreased, while LAP2α and PPARγ levels increased. Moreover, sustained DX treatment has induced the elevation of ADRP expression as well. The same changes of gene expression profile have been observed using the model TEP1 (thymic epithelial) cell line, however, *in vitro* studies have shown the molecular level rescue of thymic epithelial cells from adipose trans-differentiation due to the over-expression of Wnt4. These studies reveal the currently neglected effect of steroid therapy on thymic epithelial cells in patients receiving sustained or even single dose treatment and highlights novel potential side-effects appearing in the form of accelerated thymic senescence (Talaber et al. 2011).

4. Perspectives

4.1 Intervention possibilities of thymic rejuvenation

This chapter summarised current knowledge on thymic senescence, a central immune tissue that suffers significant morphological changes and functional impairment during ageing. The epithelial network is in focus that provides the niche for developing thymocytes until adipose involution begins. We have discussed physiological thymic epithelial senescence in detail with respect to the signalling pathways involved in the process (Kvell et al. 2010). It has also been shown that steroid induced accelerated rate thymic epithelial senescence quite resembles physiological rate senescence (except for its speed) at the molecular level (Talaber et al. 2011). The data presented confirm that Wnt4 can efficiently rescue thymic epithelial cells from steroid-induced adipose involution at the molecular level (Talaber et al. 2011). Since physiological and steroid-induced thymic epithelial senescence are identical at the molecular level, it is anticipated that sustained Wnt4 presence in the thymic context can efficiently prolong FoxN1 expression, maintain thymic epithelial identity and prevent trans-differentiation towards adipocyte lineage. The same works identify LAP2α as a pro-ageing molecular factor promoting the trans-differentiation of thymic epithelial cells into pre-adipocytes via EMT. The thymus selective decrease of LAP2α activity through small molecule compounds could theoretically shift the delicate molecular balance towards the same direction as increased Wnt4 presence.

However, there are also other methods that can efficiently support major functions of the thymus: T cell maturation and selection. An example is the thymus-specific enrichment of transgenic IL7 proteins using IL7-CCR9 fusion proteins that selectively home and accumulate in the thymic context to reinforce thymocyte development and maturation (Henson et al. 2005). This method has been characterised in detail and is currently being geared up towards potential human application in the form of inhalation products selectively delivering IL7 to the thymus (Aspinall et al. 2008).

The thymus-specific ablation of sex steroids also offers a target point for such interventions. Major involution in thymus mass occurs in parallel with the advance of puberty and correlation has been drawn with sex steroid levels. The use of thymus selective 11β-HSD1 inhibitor compounds could also theoretically decrease thymocyte sensitivity to steroid-induced apoptosis and steroid-induced epithelial molecular senescence, providing synergistic mechanism of action. Such artificial compounds (like the PF-00915275) have already been tested in healthy volunteers and were approved for safety (Courtney et al. 2008). However, these compounds do not specifically accumulate in the thymus and have not been tested in the thymic context.

Alternative methods for thymic rejuvenation include those targeting KGF, ghrelin and GH signal transduction pathways (Aspinall et al 2008). The ideal future thymus rejuvenation system that works selectively in the thymus at high efficiency and low side-effect ratio would likely constitute a combination of the above outlined methods and would efficiently aid restoration of immune competence.

4.2 Social and economic impact

By targeting and specifically inhibiting the molecular pathways that drive thymic adipose involution / immune senescence, it is possible to extend immune health-span within life-span, and improve health and quality of life, and also significantly decrease healthcare costs. This effect is expected to be very significant as – opposed to certain diseases – the physiological process of immune senescence affects all individuals, including currently healthy people.

It is to be evaluated whether an immune-fitness extending treatment would be predominantly useful as preventive treatment applied in younger individuals or would rather be useful as a reversal treatment in elderly individuals with various stages of thymic adipose degeneration. Both scenarios would affect vast segments of the population and would yield similarly significant economic and social benefit.

5. Acknowledgement

The authors are grateful to Prof. S. Amos (Institute of Haematology, Chaim Sheba Medical Centre, Tel-Hashomer, Israel) for providing the wild type murine LAP2α construct.

Experimental work performed by authors was funded by the following grants: OTKA PD 78310 to KK, Wellcome Trust 079415 and SROP-4.2.2/08/1/2008-0011 to JEP.

6. References

Aberle, H., Bauer, A., Stappert, J., Kispert, A., and Kemler, R. (1997). "b-Catenin is a target for the ubiquitin-proteosome pathway." EMBO J. 16: 3797-3804.
Akiyama, T. (2000). "Wnt/b-catenin signaling." Cytokine&Growth Factor Rev. 11: 273-282.
Alves NL, Richard-Le Goff O, Huntington ND, Sousa AP, Ribeiro VS, Bordack A, Vives FL, Peduto L, Chidgey A, Cumano A, Boyd R, Eberl G, Di Santo JP. (2009). "Characterization of the thymic IL-7 niche in vivo." Proc Natl Acad Sci U S A 106(5): 1512-7.

Anderson G, Owen JJ, Moore NC, Jenkinson EJ. "Thymic epithelial cells provide unique signals for positive selection of CD4+CD8+ thymocytes *in vitro*." J. Exp. Med. 179: 2027-2031.

Aspinall R, Mitchell W. Reversal of age-associated thymic atrophy: treatments, delivery, and side effects. Exp Gerontol. 2008 Jul;43(7):700-5.

Aydar Y, Balogh P, Tew JG, Szakal AK. Follicular dendritic cells in aging, a "bottle-neck" in the humoral immune response. Ageing Res Rev. 2004 Jan;3(1):15-29.

Balciunaite G, Keller MP, Balciunaite E, Piali L, Zuklys S, Mathieu YD, Gill J, Boyd R, Sussman DJ, Holländer GA. (2002). "Wnt glycoproteins regulate the expression of FoxN1, the gene defective in nude mice." Nat. Immunol. 3(11): 1102-1108.

Beardsley TR, Pierschbacher M, Wetzel GD, Hays EF. (1983). "Induction of T-Cell Maturation by a Cloned Line of Thymic Epithelium (TEPI) 10.1073/pnas.80.19.6005." Proceedings of the National Academy of Sciences 80(19): 6005-6009.

Bennett AR, Farley A, Blair NF, Gordon J, Sharp L, Blackburn CC. (2002). "Identification and characterization of thymic epithelial progenitor cells." Immunity 16(6): 803-814.

Berger R, Theodor L, Shoham J, Gokkel E, Brok-Simoni F, Avraham KB, Copeland NG, Jenkins NA, Rechavi G, Simon AJ. (1996). "The characterization and localization of the mouse thymopoietin/lamina-associated polypeptide 2 gene and its alternatively spliced products." Genome Res 6(5): 361-70.

Berki T, Pálinkás L, Boldizsár F, Németh P. (2002). "Glucocorticoid (GC) sensitivity and GC receptor expression differ in thymocyte subpopulations." Int Immunol 14(5): 463-9.

Blackburn, C. and N. Manley (2004). "Developing a new paradigm for thymus organogenesis." Nat Rev Immunol 4(4): 278-289.

Bleul, C. and T. Boehm (2000). "Chemokines define distinct microenvironments in the developing thymus." Eur. J. Immunol. 30: 3371-3379.

Bleul, C. and T. Boehm (2005). "BMP signaling is required for normal thymus development." J Immunol 175: 5213-5221.

Blomgren, H. and B. Andersson (1970). "Characteristics of the immunocompetent cells in the mouse thymus: cell population changes during cortisone-induced atrophy and subsequent regeneration." Cell Immunol 1(5): 545-60.

Boersma W, Betel I, van der Westen G. (1979). "Thymic regeneration after dexamethasone treatment as a model for subpopulation development." Eur J Immunol 9(1): 45-52.

Boutros M, Paricio N, Strutt DI, Mlodzik M. (1998). "Dishevelled activates JNK and discriminates between JNK pathways in planar polarity and wingless signaling." Cell 94: 109-118.

Brack AS, Conboy MJ, Roy S, Lee M, Kuo CJ, Keller C, Rando TA. (2007). "Increased Wnt Signaling During Aging Alters Muscle Stem Cell Fate and Increases Fibrosis." Science 317(5839): 807-810.

Chang J, Sonoyama W, Wang Z, Jin Q, Zhang C, Krebsbach PH, Giannobile W, Shi S, Wang CY. (2007). "Noncanonical Wnt-4 signaling enhances bone regeneration of mesenchymal stem cells in craniofacial defects through activation of p38 MAPK." Journal of Biological Chemistry 282(42): 30938-30948.

Chen L, Xiao S, Manley NR. (2009). "Foxn1 is required to maintain the postnatal thymic microenvironment in a dosage-sensitive manner." Blood 113(3): 567-574.

Cheng L, Guo J, Sun L, Fu J, Barnes PF, Metzger D, Chambon P, Oshima RG, Amagai T, Su DM. (2010). "Postnatal Tissue-specific Disruption of Transcription Factor FoxN1 Triggers Acute Thymic Atrophy." Journal of Biological Chemistry 285(8): 5836-5847.

Chidgey A, Dudakov J, Seach N, Boyd R.(2007). "Impact of niche aging on thymic regeneration and immune reconstitution." Semin Immunol 19(5): 331-40.

Courtney R, Stewart PM, Toh M, Ndongo MN, Calle RA, Hirshberg B. Modulation of 11beta-hydroxysteroid dehydrogenase (11betaHSD) activity biomarkers and pharmacokinetics of PF-00915275, a selective 11betaHSD1 inhibitor. J Clin Endocrinol Metab. 2008 Feb;93(2):550-6.

Crisa L, Cirulli V, Ellisman MH, Ishii JK, Elices MJ, Salomon DR. (1996). "Cell adhesion and migration are regulated at distinct stages of thymic T cell development: the roles of fibronectin, VLA4, and VLA5." J Exp Med 184(1): 21-28.

Dardenne M, Itoh T, Homo-Delarche F. (1986). "Presence of glucocorticoid receptors in cultured thymic epithelial cells." Cell Immunol 100(1): 112-8.

de Grey AD. "The natural biogerontology portfolio: "defeating aging" as a multi-stage ultra-grand challenge." Ann N Y Acad Sci. 2007 Apr;1100:409-23.

Derbinski J, Schulte A, Kyewski B, Klein L. (2001). "Promiscuous gene expression in medullary thymic epithelial cells mirrors the peripheral self." Nat. Immunol. 2(11): 1032-1039.

Dixit, V. D. (2010). "Thymic fatness and approaches to enhance thymopoietic fitness in aging." Curr Opin Immunol 22(4): 521-8.

Dooley J, Erickson M, Roelink H, Farr AG. (2005). "Nude thymic rudiment lacking functional foxn1 resembles respiratory epithelium." Dev Dyn. 233(4): 1605-1612.

Dorner D, Vlcek S, Foeger N, Gajewski A, Makolm C, Gotzmann J, Hutchison CJ, Foisner R. (2006). "Lamina-associated polypeptide 2alpha regulates cell cycle progression and differentiation via the retinoblastoma-E2F pathway." J Cell Biol 173(1): 83-93.

Dreger M, Otto H, Neubauer G, Mann M, Hucho F. (1999). "Identification of phosphorylation sites in native lamina-associated polypeptide 2 beta." Biochemistry 38(29): 9426-34.

Farr, A. and A. Rudensky (1998). "Medullary thymic epithelium: a mosaic of epithelial "self"?" J. Exp. Med. 188(1): 1-4.

Fletcher AL, Lowen TE, Sakkal S, Reiseger JJ, Hammett MV, Seach N, Scott HS, Boyd RL, Chidgey AP. (2009). "Ablation and regeneration of tolerance-inducing medullary thymic epithelial cells after cyclosporine, cyclophosphamide, and dexamethasone treatment." J Immunol 183(2): 823-31.

Ge, Q. and W. Chen (2000). "Effect of murine thymic epithelial cell line (MTEC1) on the functional expression of CD4(+)CD8(-) thymocyte subgroups." Int. Immunol. 12(8): 1127-1133.

Gill J, Malin M, Holländer GA, Boyd R. (2002). "Generation of a complete thymic microenvironment by MTS24(+) thymic epithelial cells." Nat. Immunol. 3(7): 635-642.

Giorgione J, Hysell M, Harvey DF, Newton AC. (2003). "Contribution of the C1A and C1B domains to the membrane interaction of protein kinase C." Biochemistry 42(38): 11194-11202.

Golan T, Yaniv A, Bafico A, Liu G, Gazit A. (2004). "The human frizzled 6 (HFz6) acts as a negative regulator of the canonical Wnt b-catenin signaling cascade." J. Biol. Chem. 279(15): 14879-14888.

Grubeck-Loebenstein, B. (2009). "Fading Immune Protection in Old Age: Vaccination in the Elderly." J Comp Pathol.

Gui J, Zhu X, Dohkan J, Cheng L, Barnes PF, Su DM. (2007). "The aged thymus shows normal recruitment of lymphohematopoietic progenitors but has defects in thymic epithelial cells." Int Immunol 19(10): 1201-11.

He TC, Sparks AB, Rago C, Hermeking H, Zawel L, da Costa LT, Morin PJ, Vogelstein B, Kinzler KW. (1998). "Identification of c-MYC as a target of the APC pathway." Science 281(5382): 1509-1512.

He, X. and D. J. Kappes (2006). "CD4/CD8 lineage commitment: light at the end of the tunnel?" Curr Opin Immunol 18(2): 135-42.

Henson SM, Snelgrove R, Hussell T, Wells DJ, Aspinall R. An IL-7 fusion protein that shows increased thymopoietic ability. J Immunol. 2005 Sep 15;175(6):4112-8.

Hsu, H. C. and J. D. Mountz (2003). "Origin of late-onset autoimmune disease." Immunol Allergy Clin North Am 23(1): 65-82, vi.

Hutchison CJ, Alvarez-Reyes M, Vaughan OA. (2001). "Lamins in disease: why do ubiquitously expressed nuclear envelope proteins give rise to tissue-specific disease phenotypes?" J Cell Sci 114(Pt 1): 9-19.

Ikeda, S., Kishida, S., Yamamoto, H., Murai, H., Koyama, S., Kikuchi, A. (1998). "Axin, a negative regulator of the Wnt signaling pathway, forms a complex with GSK3b and b-catenin and promotes GSK-3b-dependent phosphorylation of b-catenin." The EMBO J. 17: 1371-1384.

Ioannidis V, Beermann F, Clevers H, Held W. (2001). "The b-catenin-TCF1 pathway ensures CD4+CD8+ thymocyte survival." Nature Immunology 2: 691-697.

Ishitani T, Kishida S, Hyodo-Miura J, Ueno N, Yasuda J, Waterman M, Shibuya H, Moon RT, Ninomiya-Tsuji J, Matsumoto K. (2003). "The TAK1-NLK Mitogen-activated protein kinase cascade functions in the Wnt-5a/Ca2+ pathway to antagonize Wnt/b-catenin signaling." Mol. Cell Biol. 23(1): 131-139.

Jin EJ, Park JH, Lee SY, Chun JS, Bang OS, Kang SS. (2006). "Wnt-5a is involved in TGF-beta3-stimulated chondrogenic differentiation of chick wing bud mesenchymal cells." Int J Biochem Cell Biol 38(2): 183-95.

Jondal M, Pazirandeh A, Okret S. (2004). "Different roles for glucocorticoids in thymocyte homeostasis?" Trends Immunol 25(11): 595-600.

Kanthasamy AG, Anantharam V, Zhang D, Latchoumycandane C, Jin H, Kaul S, Kanthasamy A. A novel peptide inhibitor targeted to caspase-3 cleavage site of a proapoptotic kinase protein kinase C delta (PKCdelta) protects against dopaminergic neuronal degeneration in Parkinson's disease models. Free Radic Biol Med. 2006 Nov 15;41(10):1578-89.

Kim YC, Clark RJ, Pelegri F, Alexander CM. (2009). "Wnt4 is not sufficient to induce lobuloalveolar mammary development." BMC Dev Biol 9: 55.

Klug DB, Crouch E, Carter C, Coghlan L, Conti CJ, Richie ER. (2000). "Transgenic expression of cyclin D1 in thymic epithelial precursors promotes epithelial and T cell development." J. Immunol. 164: 1881-1888.

Kühl M, Geis K, Sheldahl LC, Pukrop T, Moon RT, Wedlich D.(2001). "Antagonistic regulation of convergent extension movements in Xenopus by Wnt/beta-catenin and Wnt/Ca2+ signalling." Mech. Dev. 106: 61-76.

Kvell K, Varecza Z, Bartis D, Hesse S, Parnell S, Anderson G, Jenkinson EJ, Pongracz JE. Wnt4 and LAP2alpha as pacemakers of thymic epithelial senescence. PLoS One. 2010 May 18;5(5):e10701.

Labalette C, Renard CA, Neuveut C, Buendia MA, Wei Y. (2004). "Interaction and functional cooperation between the LIM protein FHL2, CBP/p300, and beta-catenin." Mol. Cell Biol. 24(24): 10689-10702.

Leduc L, Levy E, Bouity-Voubou M, Delvin E. Fetal programming of atherosclerosis: possible role of the mitochondria. Eur J Obstet Gynecol Reprod Biol. 2010 Apr;149(2):127-30.

Lind EF, Prockop SE, Porritt HE, Petrie HT.(2001). "Mapping precursor movement through the postnatal thymus reveals specific microenvironments supporting defined stages of early lymphoid development." J. Exp. Med. 194(2): 127-134.

Liu H, Fergusson MM, Castilho RM, Liu J, Cao L, Chen J, Malide D, Rovira II, Schimel D, Kuo CJ, Gutkind JS, Hwang PM, Finkel T. (2007). "Augmented Wnt Signaling in a Mammalian Model of Accelerated Aging." Science 317(5839): 803-806.

Ljubuncic P, Reznick AZ. "The evolutionary theories of aging revisited—a mini-review." Gerontology. 2009;55(2):205-16.

Luo Q, Kang Q, Si W, Jiang W, Park JK, Peng Y, Li X, Luu HH, Luo J, Montag AG, Haydon RC, He TC. (2004). "Connective tissue growth factor (CTGF) is regulated by Wnt and bone morphogenetic proteins signaling in osteoblast differentiation of mesenchymal stem cells." J Biol Chem 279(53): 55958-68.

Lyons JP, Mueller UW, Ji H, Everett C, Fang X, Hsieh JC, Barth AM, McCrea PD. (2004). "Wnt-4 activates the canonical beta-catenin-mediated Wnt pathway and binds Frizzled-6 CRD: functional implications of Wnt/beta-catenin activity in kidney epithelial cells." Exp Cell Res 298(2): 369-87.

Malbon CC, Wang H, Moon RT. (2001). "Wnt signaling and heterotrimeric G-proteins: strange bedfellows or a classic romance?" Biochem. Biophys. Res. Commun. 287(3): 589-93.

Mandinova A, Kolev V, Neel V, Hu B, Stonely W, Lieb J, Wu X, Colli C, Han R, Pazin MJ, Ostano P, Dummer R, Brissette JL, Dotto GP. (2009). "A positive FGFR3/FOXN1 feedback loop underlies benign skin keratosis versus squamous cell carcinoma formation in humans." J Clin Invest 119(10): 3127-37.

Manley, N. R. (2000). "Thymus organogenesis and molecular mechanisms of thymic epithelial cell differentiation." Sem. Immunol. 12: 421-428.

Mann B, Gelos M, Siedow A, Hanski ML, Gratchev A, Ilyas M, Bodmer WF, Moyer MP, Riecken EO, Buhr HJ, Hanski C. (1999). "Target genes of beta-catenin-T cell-factor/lymphoid-enhancer-factor signaling in human colorectal carcinomas." Proc. Natl. Acad. Sci U.S.A. 96(4): 1603-1608.

Marinova, T. T. (2005). "Epithelial framework reorganization during human thymus involution." Gerontology 51(1): 14-8.

Mercurio S, Latinkic B, Itasaki N, Krumlauf R, Smith JC. (2004). "Connective-tissue growth factor modulates WNT signalling and interacts with the WNT receptor complex." Development 131(9): 2137-47.

Michie AM, Soh JW, Hawley RG, Weinstein IB, Zuniga-Pflucker JC. (2001). "Allelic exclusion and differentiation by protein kinase C-mediated signals in immature thymocytes." Proc. Natl. Acad. Sci U.S.A. 98(2): 609-614.

Min H, Montecino-Rodriguez E, Dorshkind K. (2006). "Reassessing the role of growth hormone and sex steroids in thymic involution." Clinical Immunology 118(1): 117-123.

Mulroy, T., McMahon, J.A., Burakoff, S.J., McMahon, A.P., and Sen, J. (2002). "Wnt-1 and Wnt-4 regulated thymic cellularity." Eur. J. Immunol. 32: 967-971.

Nateri AS, Spencer-Dene B, Behrens A. (2005). "Interaction of phosphorylated c-Jun with TCF4 regulates intestinal cancer development." Nature July, Epub.

Newton, A. (2001). "Protein kinase C: structural and spatial regulation by phosphorylation, cofactors, and macromolecular interactions." Chem. Rev. 101: 2353-2364.

Noordermeer, J. K., J., Perrimon, N., and Nusse, R. (1994). "Dishevelled and armadillo act in the wingless signalling pathway in Drosophila." Nature 367: 80-83.

Oksanen, A. (1971). "Multilocular fat in thymuses of rats and mice associated with thymus involution: a light- and electron-microscope and histochemical study." J Pathol 105(3): 223-6.

Peavy RD, Hubbard KB, Lau A, Fields RB, Xu K, Lee CJ, Lee TT, Gernert K, Murphy TJ, Hepler JR. (2005). "Differential effects of Gq alpha, G14 alpha, and G15 alpha on vascular smooth muscle cell survival and gene expression profiles." Mol Pharmacol. 67(6): 2102-21014.

Pinson, K. I., Brennan, J., Monkley, S., Avery, B.J., Skarnes, W.C. (2000). "An LDL-receptor-related protein mediates Wnt signalling in mice." Nature 407(6803): 535-538.

Pongracz J, Hare K, Harman B, Anderson G, Jenkinson EJ. (2003). "Thymic epithelial cells provide Wnt signals to developing thymocytes." Eur. J. Immunol. 33: 1949-1956.

Qiao S, Chen L, Okret S, Jondal M. (2008). "Age-related synthesis of glucocorticoids in thymocytes." Exp Cell Res 314(16): 3027-35.

Ribeiro, R. M. and A. S. Perelson (2007). "Determining thymic output quantitatively: using models to interpret experimental T-cell receptor excision circle (TREC) data." Immunol Rev 216: 21-34.

Roman-Roman S, Shi DL, Stiot V, Haÿ E, Vayssière B, Garcia T, Baron R, Rawadi G. (2004). "Murine Frizzled-1 behaves as an antagonist of the canonical Wnt/beta-catenin signaling." J Biol Chem. 279: 5725-5733.

Rosso, S., D. Sussman, et al. (2005). "Wnt signaling through Dishevelled, Rac and JNK regulates dendritic development." Nat. Neuroscience 8: 34-42.

Schluns KS, Cook JE, Le PT. (1997). "TGF-beta differentially modulates epidermal growth factor-mediated increases in leukemia-inhibitory factor, IL-6, IL-1 alpha, and IL-1 beta in human thymic epithelial cells." J Immunol 158(6): 2704-12.

Schwabe RF, Bradham CA, Uehara T, Hatano E, Bennett BL, Schoonhoven R, Brenner DA. (2003). "c-Jun-N-terminal kinase drives cyclin D1 expression and proliferation during liver regeneration." Hepatology 37(4): 824-832.

Seike M, Mizutani H, Sudoh J, Gemma A. (2009). "Epithelial to mesenchymal transition of lung cancer cells." J Nippon Med Sch 76(4): 181.

Sheldahl LC, Slusarski DC, Pandur P, Miller JR, Kühl M, Moon RT. (2003). "Dishevelled activates Ca2+ flux, PKC, and CamKII in vertebrate embryos." J. Cell Biol. 161(4): 767-777.

Shiraishi J, Utsuyama M, Seki S, Akamatsu H, Sunamori M, Kasai M, Hirokawa K. Essential microenvironment for thymopoiesis is preserved in human adult and aged thymus. Clin Dev Immunol. 2003 Mar;10(1):53-9.

Shtutman M, Zhurinsky J, Simcha I, Albanese C, D'Amico M, Pestell R, Ben-Ze'ev A. (1999). "The cyclin D1 gene is a target of the beta-catenin/LEF-1 pathway." Proc. Natl. Acad. Sci U.S.A. 96(10): 5522-5527.

Smit L, Baas A, Kuipers J, Korswagen H, van de Wetering M, Clevers H. (2004). "Wnt activates the Tak1/Nemo-like kinase pathway." J Biol Chem 279(17): 17232-40.

Staal, F. J. and H. Clevers (2003). "Wnt signaling in the thymus." Curr. Opin. Immunol. 15(2): 204-208.

Staal, F. J. T., Meeldijk, J., Moerer, P., Jay, P., van de Weerdt, B.C.M., Vainio, S., Nolan, G.P., Clevers, H. (2001). "Wnt signaling is required for thymocyte development and activates Tcf-1 mediated transcription." Eur. J. Immunol. 31.: 285-293.

Stahn C, Löwenberg M, Hommes DW, Buttgereit F. (2007). "Molecular mechanisms of glucocorticoid action and selective glucocorticoid receptor agonists." Mol Cell Endocrinol 275(1-2): 71-8.

Talaber G, Kvell K, Varecza Z, Boldizsar F, Parnell SM, Jenkinson EJ, Anderson G, Berki T, Pongracz JE. Wnt-4 protects thymic epithelial cells against dexamethasone-induced enescence. Rejuvenation Res. 2011 Jun;14(3):241-8.

Tamai, K., Semenov, M., Kato, Y., Spkony, r., Chumming, L., Katsuyama, Y., Hess, F., Saint-jeannet, J.-P., He, X. (2000). "LDL-receptor-related proteins in Wnt signal transduction." Nature 407: 530-535.

Tanaka Y, Mamalaki C, Stockinger B, Kioussis D. (1993). "In vitro negative selection of alpha beta T cell receptor transgenic thymocytes by conditionally immortalized thymic cortical epithelial cell lines and dendritic cells." Eur. J. Immunol. 23(10): 2614-2621.

Tetsu, O. and F. McCormick (1999). "Beta-catenin regulates expression of cyclin D1 in colon carcinoma cells." Nature 398(6726): 422-426.

Torres MA, Yang-Snyder JA, Purcell SM, DeMarais AA, McGrew LL, Moon RT. (1996). "Activities of the Wnt-1 class of secreted signaling factors are antagonized by the Wnt-5A class and by a dominant negative cadherin in early Xenopus development." J. Cell Biol. 133(5): 1123-1137.

Valsecchi C, G. C., Ballabio A, Rugarli EI. (1997). "JAGGED2: a putative Notch ligand expressed in the apical ectodermal ridge and in sites of epithelial-mesenchymal interactions." Mech. Dev. 69(1-2): 203-207.

Varecza Z, Kvell K, Talabér G, Miskei G, Csongei V, Bartis D, Anderson G, Jenkinson EJ, Pongracz JE. Multiple suppression pathways of canonical Wnt signalling control thymic epithelial senescence. Mech Ageing Dev. 2011 May;132(5):249-56.

Wang, H. and C. Malbon (2003). "Wnt signaling, Ca2+, and cyclic GMP: visualizing frizzled functions." Science 300: 1529-1530.

Wang Z, Shu W, Lu MM, Morrisey EE.(2005). "Wnt7b activates canonical signaling in epithelial and vascular smooth muscle cells through interactions with Fzd1, Fzd10, and LRP5." Mol. Cell Biol. 25: 5022-5030.

Wehrli, M., Dougan, S.T., Caldwell, K., O'Keefe, L., Schwartz, S., Vaizel-Ohayon, D., Schejter, E., Tomlinson, A., DiNardo, S. (2000). "Arrow encodes an LDL-receptor-related protein essential for Wingless signalling." Nature 407(6803): 527-530.

Wharton Jr., K. A., Zimmermann, G., Rousset, R., Scott, M.P. (2001). "Vertebrate proteins related to Drosophila naked cuticle bind dishevelled and antagonize Wnt signaling." Developmental Biology 234: 93-106.

Wiegers GJ, Knoflach M, Böck G, Niederegger H, Dietrich H, Falus A, Boyd R, Wick G. (2001). "CD4(+)CD8(+)TCR(low) thymocytes express low levels of glucocorticoid receptors while being sensitive to glucocorticoid-induced apoptosis." Eur J Immunol 31(8): 2293-301.

Wüst S, van den Brandt J, Tischner D, Kleiman A, Tuckermann JP, Gold R, Lühder F, Reichardt HM. (2008). "Peripheral T cells are the therapeutic targets of glucocorticoids in experimental autoimmune encephalomyelitis." J Immunol 180(12): 8434-43.

Xu Y, Banerjee D, Huelsken J, Birchmeier W, Sen JM. (2003). "Deletion of b-catenin impairs T cell development." Nature Immunol. 4: 1177-1182.

Yamamoto, H., Kishida, S., Kishida, M., Ikeda, S., Takada, S., and Kikuchi, A. (1999). "Phosphorylation of axin, a Wnt signal negative regulator, by glycogen synthase kinase-3beta regulates its stability." EMBO J. 274: 10681-10684.

Yamanaka H, Moriguchi T, Masuyama N, Kusakabe M, Hanafusa H, Takada R, Takada S, Nishida E. (2002). "JNK functions in the non-canonical Wnt pathway to regulate convergent extension movements in vertebrates." EMBO Rep. 3(1): 69-75.

Yan D, Wallingford JB, Sun TQ, Nelson AM, Sakanaka C, Reinhard C, Harland RM, Fantl WJ, Williams LT. (2001). "Cell autonomous regulation of multiple Dishevelled-dependent pathways by mammalian Nkd." Proc. Natl. Acad. Sci U.S.A. 98(7): 3802-3807.

Zhang X, Gaspard JP, Chung DC. (2001). "Regulation of vascular endothelial growth factor by the Wnt and K-ras pathways in colonic neoplasia." Cancer Res. 61(16): 6050-6054.

Zilberberg A, Yaniv A, Gazit A. (2004). "The low density lipoprotein receptor-1, LRP1, iteracts with the human Frizzled-1 (HFz1) and down-regulates the canonical Wnt signaling pathway." J Biol Chem. 279: 17535-17542.

8

Imagistic Noninvasive Assessment of Skin Ageing and Anti-Ageing Therapies

Maria Crisan[1], Radu Badea[1], Carlo Cattani[2] and Diana Crisan[1]
[1]University of Medicine and Pharmacy "Iuliu Hatieganu",
[2]University of Salerno,
[1]Romania
[2]Italy

1. Introduction

The significant increase in life expectancy and the process of population ageing are aspects that generate important social and economic changes and influence the health and research policies throughout the world. The ageing phenomenon represents a natural, slow and irreversible process, which affects all body tissues, being determined by a multitude of factors that contribute in different proportions to the characteristic molecular, cellular, tissular and clinical changes. (Kohl et al, 2011)

The skin is a bio-membrane situated at the interface with the external environment. It reflects the state of health of the body, the human personality and has numerous psycho-social implications. Ageing is a complex process that implies external and internal factors. Chronological skin ageing comprises those changes in the skin that occur as a result of passage of time alone. Photo-ageing comprises those changes in the skin that are result of chronic sun exposure superimposed on chronological skin ageing. Several scientific theories on the ageing regulation at molecular, cellular and systemic levels have been postulated in order to define and control this process.

1.1 Ageing theories

For decades, researchers concluded that ageing comes as a consequence of both genetic and environmental influences. Among the environmental factors, solar UV radiation is the most important cause of premature aging.

1.1.1 The genetic theory

The theory of error accumulation, described by Orgel in 1963, interferes with the DNA replication, RNA transcription and translation into proteins. The risk for errors increases with age, determining at a certain moment a critical change in the genome that triggers senescence, apoptosis and cellular death.

1.1.2 The free radical theory

Postulated by Harman in 1956, this theory prooved that the reactive oxygen species interfere with the cellular and subcellular systems, inducing molecular degradations.

1.1.3 The mitochondrial theory

It is based on the fact that the mitochondria are the main source of free radicals within the cell. Skin exposure to sun determines the accumulation of mutations in the mitochondrial DNA, with implications in senescence (Shy et al, 2010).

1.1.4 The telomerase theory

Telomeres are sequences of nucleic acids extending from the ends of chromosomes. Every time our cells divide, telomeres are shortened, leading to cellular damage and cellular death associated with ageing. Telomerase, the "immortalizing" enzyme, appears to repair telomeres, manipulating the "clock" mechanism that controls the life span of dividing cells. The telomerase controls the telomere length and could be involved in the prevention of the ageing process. Recent studies have shown that specific molecules, applied topically or generally can activate the telomerase, preventing thus the shortening of the telomeres. (Han et al , 2009)

1.1.5 The theory of glycation

Maillard's theory is widely recognized as a general intrinsic ageing mechanism, focused on another potentially destructive agent, glucose. Glycation is the non-enzimatic reaction of a sugar and a protein forming multiple chemicals called advanced glycation end products (AGEs) (Pageon et al, 2010). The reaction products accumulate during the ageing process, and seem to also be involved in different pathologies associated with diabetes, atherosclerosis, Altzheimers disease, arthrytis (Bos et al, 2011). Proteins with a long biological half-life (collagen, elastin) are more affected. Glycation has different side effects on extracellular matrix fibers, leading to stiffer and more brittle collagen. In addition, elastin is easily glycated. Denaturated elastin is associated with slackened skin. It is important to mention that AGEs have cellular receptors that initiate inflammatory reactions when they are activated by an AGE complex. These reactions are associated with metabolic disorders, arterial diseases, the premature ageing process and the whole associated pathology. It has been reported that glycation affects the precise aggregation of collagen monomers into fibers, aspect that may be correlated with the different amplitude of the pixels when performing high-frequency ultrasound. Literature data as well as our own studies show that the echogenicity of the pixels and their density, correlated with the classical histological aspect of the integument, offer important information regarding age, cutaneous phototype, anti-ageing therapies, cutaneous pathologies etc. The extension of glycation in the skin can be measured with an instrument that measures a fluorometric chemical named pentosidine. Pentosidine is a fluorescent crosslinker that accumulates in a linear fashion in the collagen of all tissues. The fluorescence degree is correlated with the amount of accumulated pentosidine, age, risc of developing a certain pathology etc.

According to literature, collagen may be considered the key protein that allows the noninvasive assessment (fluorimetry, ultrasonography) of the cutaneous senescence process as well as the efficacy of various anti-ageing therapies. The non-invasive assessment of the cutaneous structure opens a new era of skin care and anti-ageing treatment.

Photoaging occurs as a result of cumulative damage from ultraviolet (UV) radiation. The UV rays induce and accelerate the glycation process, interact with cells and extracellular matrix, induce the synthesis and release of cytokines, stimulate the metalloproteinase synthesis, especially collagenase and elastase, and represent the major aggression factor on the cellular DNA. In photoaged skin collagen fibrils are disorganized, and abnormal elastin accumulates.

Despite the progress in aging research, these have yet to be an unanimous vote on one specific theory of ageing. Most of these theories have been disputed by researchers over and over again and many of them, as Dr.Hans Kugler editor of the Journal of Longevity Research, said, "...are dying of old age." Age-related changes do not occur uniformly in individuals because they are under genetic and environmental control. What is certain is that we are all involved in a global-ageing phenomenon.

1.2 Signs of skin ageing

The skin is the only organ completely displayed at the body surface and represents the ideal system for the study of both the intrinsic and extrinsic ageing process. Changes of the skin structure, such as wrinkles, irregularities of pigmentation, in contrast to ageing of other organs are visible and provide social clues to estimate the individual age.

1.2.1 Intrinsic aged skin occurs as a result of passage of time alone

Clinical manifestations include xerosis, laxity, wrinkles, slackness, benign tumors (cherry angiomas, seborrheic keratoses). Chronological ageing is affected by the changes of hormones and growth factors that appear with age. **Histological features** involve the epidermis, dermis and appendages. The hallmarks of intrinsic ageing are the thinning of the epidermis, flattening of the dermo-epidermal junction and reduction of extracellular matrix components.

1.2.2 Photoaged skin occurs as a result of cumulative damage from UV radiation

Clinical manifestations include roughness, irregular pigmentation, wrinkles, pseudo-scars, fine nodularity (elastotic material), telangiectasia, sebaceous hyperplasia,etc. **Histological manifestations** involve irregular epidermal thickness, nodular aggregations of elastotic material in the papillary dermis. The most obvious histological aspect is solar elastosis along with an increased amount of ground substance consisting of glycozaminoglycans and proteoglycans, and a decreased number of collagen fibers. Solar elastosis may correspond to the subepidermal low echogenicity band (SLEB) a specific imagistic parameter that appears on photoagressed areas. See Figure 3

In the reticular dermis collagen fibers appear degraded, clumped and fragmented. In addition, an inflammatory infiltrate can be identified. The elastosis process, collagen degenerescence, inflammatory infiltrates (histological aspects identified in usual or special stains) represent the morphological substrate of the sonograms. The amplitude and density of the pixels, correlated with the histological aspect, quantify different mollecular, cellular, biochemical and structural reactions that govern the ageing process.

The response to UV- induced damage is correlated with the individuals' skin type. Thus, subjects with skin type II show an atrophic and dysplastic response to UV rays, present fewer wrinkles, smoother skin, actinic keratoses, and epidermal malignancies (carcinoma, melanoma). The individuals with skin type III or IV show hyperplastic responses, present thick skin with coarse wrinkles. Our observations on 140 subjects have shown that individuals belonging to phototype class II (70 subjects) have a different imagistic pattern on photoexposed and photoprotected sites, in comparison to subjects belonging to phototype class III (70 subjects). The dynamics of the pixels on the studied areas indicate significant variations according to the phototype class and are correlated to different clinical aspects of the ageing process. The specific ageing features of phototype class II and III are shown in Table I.

	Phototype classs II	Phototype class III
Clinical aspects	Smooth skin, less superficial wrinkles, numerous pre-malignant lesions and cutanous carcinomas	Thick, pigmented, deeply wrinkled skin
Imagistic aspect: Photoexposed area (zigomatic area)	Thicker epidermis Thicker dermis Increased number of LEP, MEP, HEP	Thinner epidermis Thinner dermis Lower number of LEP, MEP, HEP
Imagistic aspect: Photoprotected area (medial arm)	Thiner epidermis Thicker dermis Higher amount of LEP Lower amount of MEP, MEP, LEPs/LEPi higher	Thicker epidermis Thinner dermis Lower amount of LEP Higher amount of MEP and HEP LEPs/LEPi lower

Table 1. Clinical and imagistic characteristics of skin phototype II and III.

It is well known that phenotypical and functional skin differences of individuals belonging to different ethnic backrounds are related to genetic factors, pigmentary system, life-time UV exposure, life-time style. In contrast to the studies on the pigmentary system, we can appreciate that there are other histological, biochemical differences, which govern the different shades of color. Despite the interest in finding objective markers for phototype classification, more complex studies comparing ageing and phototype between different ethnic groups remain to be published. Our observations suggest a complex interrelationship between the histological structure and the individual pigmentary system. The identification of certain measurable objective markers for every phototype will allow an optimisation of the phototherapy protocols and will reduce the photoinduced side-effects.

1.3 Histology of the skin

From histological point of view, skin consists of two layers of different origin, structure and function.

The epidermis (0,07 to 0,12 mm) is the outermost structure, derived from ectoderm, consisting of cells organized into five layeres. Stratum basale (germinativum), suported by a basement membrane consists of a single layer of mitotically active cells. Stratum spinosum, the thickest layer of the epidermis consists of polyhedral to flattened cells, attached to each other by unstable desmosomes, confering it a prickly appearance. Stratum granulosum consists of cells that contain lipid-rich granules that act as a waterproof barrier. Stratum lucidum is present only in thick skin. Stratum corneum is the most superficial layer, composed of numerous layers of flattened, keratinized cells.

The dermis (corium), lying directly beneath the epidermis is derived from the mesoderm and is subdivided into two layers: the superficial, loosely woven papillary layer and the deeper, much denser reticular layer.The dermis ranges in thickness from 0,6mm to 3mm. Histologically, the dermis is a dense, irregular collagenous connective tissue, containing mostly type I collagen fibers and networks of elastic fibers, which support the epidermis and bind the skin to the underlying hypodermis.

The papillary dermis is a loose connective tissue consisting of: type III collagen fibers, elastic fibers, fibroblasts, mast cells etc. The reticular layer is composed of dense, irregular collagenous connective tissue, displaying thick type I collagen fibers, closely packed into large bundles lying mostly parallel to the skin surface. Thick elastic fibers form networks that are more abundant around sebaceous and sweat glands. Proteoglycans fill the interstices of the reticular dermis. Cells are sparser and include fibroblasts, mast cells, macrophages, lymphocytes and fat cells. The hypodermis (subcutaneous adipose tissue) is considered a diffuse organ, normally well represented. From structural and functional point of view it is well integrated with the dermis and epidermis thru vessels and nerve structures. It lies underneath the reticulary dermis, being composed of adipose cells, disposed in adipose lobules, separated by conjunctive septa that contain blood vessels, lymphatics, nerve fibres and numerous mastocytes. The architecture of the adipose tissue differs for men and women. The subcutis is not part of the cutaneous structure, but is studied together with it, due to the associated pathology.

1.3.1 Epidermis

Epidermis is a stratified squamous nonkeratinized epithelium that covers the body on its surface. It consists of cells organized into five rows. Among keratinocytes that represent the most important cellular population, other cells, such as melanocytes, Langerhans cells and Merkel cells are found.

Imagistically, the epidermis appears as a hyperechogenic band, displayed parallel to the cutanous surface, having a thickness that can be assessed in mm. The thickness of the epidermis changes in relationship to the ageing process, applied therapy, associated pathology.

1.3.2 Dermo-epidermal junction

The junction between epidermis and dermis is a special ondulated basement membrane rich in collagen type IV filaments, collagen type III, collagen type VII, glycoproteins. The morphofunctional integrity of this barrier is essential for the skin protection function.

Imagistically, it is visualised as an extremly thin band situated at the limit of the hyperechogenic epidermis and the underlying dermis.

1.3.3 Dermis

Is a dense connective tissue, situated between the epidermis and hypodermis. The dermis consists of cells and extracellular matrix, composed of collagen and elastic fibres, proteoglycans, glycoproteins, tissular fluid. The limit with the hypodermis is a straight or sometimes ondulated line, because of the underlying adipose lobules, that are prominent in the lower dermis (visible aspect in „orange skin appearance" cellulitis). (Crisan, 2007)

Imagistically, we can assess the thickness of the dermis in mm as well as the number of pixels with different amplitudes, each codifying different structural, physiological or pathological aspects.

1.3.3.1 Extracellular matrix

Collagen is an important protein for the skin as it is essential for the structure and function of the extracellular matrix in the dermis. Thinner and wrinkled skin are typical signs of normal ageing and are consequences of reduced collagen. Collagen may be considered a "gold protein"for the assessment of the ageing process (fluorometrical, imagistical, clinical) and risc prediction of associated pathology. Collagen type I and II are the main types of collagen in the skin. In young subjects, there is a prominence of collagen type III or reticular fibres that are organised in fibrills and disposed at the level of papillary demis. In adults there is a prominence of type I collagen, organised in fibres disposed in parallel bands in the reticular or profound dermis. Collagen type IV forms filaments and is situated at the dermo-epidermal junction. In elderly subjects all types of collagen are diminuished. Collagen type I is the most common form (80%) in the dermis and is responsable for the cutaneous resistance. It is continuously produced and recycled throughout lifetime. In young subjects the synthesis process is prominent whereas in subjects over the age of 40, degradation processess are more common. (Uitto et al, 2008)

The key cell that forms and maintains the extracellular matrix is the fibroblast. The synthesis of the connective tissue fibres is initiated intracellularly, whereas the formation of filaments, fibrills or fibres are extracellular processes, controlled by several factors of the extracellular matrix. The main source of dermal echogenicity is represented by collagen fibres, disposed in an organized manner. Collagen and elastin are important proteins in maintaining the cutaneous architecture and ensuring the biostructural qualities of the integumentary system.

Proteoglycans beside glycoproteins and fibers are important components of the extracellular matrix. They consist of a protein core to which different glycozaminoglycans are linked. Hyaluronic acid binds uncovalently the proteoglycans, forming macromolecules that attract water, resulting in a true hydrating capsule with great importance for the hydration of the skin. With age, the amount of proteoglycans decreases and consequently the cutaneous hydration degree as well. The degree of hydration can be asssessed ultrasonographically by establishing the amount of low echogenic pixels in the skin.

1.4 Ultrasonography in dermatology

The imaging techniques have imposed themselves as usefull non-invasive methods for skin examination and diagnostic tools for skin conditions. During the past years conventional and high resolution ultrasonography (US) have extended their utility in the field of clinical dermatology (Schmid-Wendtner& Burgdorf W, 2005). The procedure involving ultrasound is a non invasive method allowing "in vivo" and "in real time" histological assessment of the cutaneous structure as well as its specific conditions. Several studies have prooven the similarities between sonograms and histological sections. (Jasaitiene et al, 2011)

The inclusion of this method among the procedures used for the diagnosis of skin diseases is an attempt to replace as much as possible the invasive procedures, especially biopsy, with non invasive ones. The motivation for the extensive use of US derives from its ability to reveal in detail the skin components, up to 1.5 cm in depth, to assess the axial and lateral tumoral extension, the inflammatory and degenerative processes, as well as the efficacy of different topical and general therapies.

1.4.1 High-frequency ultrasound

High-frequency ultrasound is a new, noninvasive method that allows an "in vivo assessment" of the physiological and pathological aspects of the integumentary system. It represents a more desirable and less emotionally-involving alternative to skin biopsy that is routinely used in the dermatological field. It also represents an important research tool for the characterization of skin properties on different intervals of age, allowing the establishment of an imagistic ageing model of the integumentary system. (Badea et al, 2010)

The use of high-frequency ultrasound in dermatology allows a clear identification of the skin layers and thus tissue assessment. At frequencies above 10 MHz, it was prooven that the technology provides enough resolution to characterize microstructures. High-frequency ultrasound allows, as the senescence process progresses, the identification of variations both in skin thickness and echogenicity, offering specific, ultrasonographic markers that allow an objective assessment of the skin ageing process. The changes of the extracellular matrix, consisting in variations of the dermal density and echogenicity throughout the physiological senescence process can be easily identified with the use of high-frequency ultrasound.

The ultrasonographic assessment of the integument can be performed with a 20 MHz high-frequency Dermascan device (Dermascan C, Cortex Technology, Denmark), as seen in Figure 1, that allows the "in vivo" acquirement of cross-sectional images of the skin (B mode) up to 2.5 cm in depth.

Fig. 1. Ultrasonographic equipment (Dermascan C, Cortex Technology, Dennmark).

The device consists of three major parts: a transducer, an elaboration system and a data storing system. The ultrasonic wave is partially reflected at the boundary between adjacent structures and generates echoes of different amplitudes. The intensity of the reflected echoes is evaluated by a microprocessor and visualized as a colored two-dimensional image. The color scale of echogenicity is: white- yellow – red – green – blue – black. On a normal cutaneous image, the epidermal echogenicity appears as a white band, the dermis is expressed as a 2 color composition: yellow and/or red, and the subcutaneous layer appears either green or black, as displayed in Figure 2.

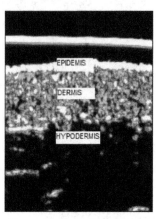

Fig. 2. Ultrasound image of the skin: epidermis, dermis, hypodermis

The ultrasonographic images are saved and processed with a specific image analysis software (Dermavision, Cortex Technology), that has a certain property: the amplitudes of the echoes of the pixels are given as a value on a numerical scale that ranges from 0-255. On this scale, the low echogenity pixel area corresponds to the 0-30 interval, the medium echogenity pixels to 50-150, and the high echogenity pixels to the 200-255 interval.

During the study, we can adjust the gain curve as well as the speed of ultrasound at tissular level. Ultrasonographic gel is applied on the aperture of the ring of the transducer, which is then placed perpendicularly to the skin surface for the acquirement of the cross-sectional image. There are several parameters that can be assessed by using Dermascan device and Dermavision analysis software as illustrated in Table 2

The thickness of the epidermis can be obtained by establishing the mean of three measurements performed in A-mode at three different sites of each image (the 2 extremities and the center of the analyzed image). The thickness of the dermis is obtained in B-mode, by measuring the distance between the dermo-epidermal and the dermo-hypodermic junction at the same three different sites and by establishing the mean of the three values. By selecting a certain interval from the 0-255 pixel scale, we obtain values corresponding to the low, medium and high echogenic pixels, present in the analyzed image.

Additionally, the LEP can be quantified separately in the upper (LEPs) and lower (LEPi) dermis. To separate the 2 areas, we draw a parallel line to the epidermal entrance echo, dividing the dermis into 2 equally thick parts. The ratio of LEP number in the upper and lower dermis (LEPs/LEPi) can be calculated.

PARAMETER	Description
Thickness of the epidermis Thickness of dermis	Given in mm
The number of LEP (low echogenic pixels):	Quantifiy the degree of cutaneous hydration, inflammatory processes, solar elastosis, collagen degeneration
The number of MEP (medium echogenic pixels)	Quantify the protein structures, the collagen and ellastin precursors (different assembly degrees)
The number of HEP (high echogenic pixels):	Quantify mature collagen assembled in fibres and disposed in parallel bands – marker of intrinsic ageing
SLEB - subepidermal low echogenicity band	A well delimited, subepidermal low echogenicity band (0-30), situated in the upper dermis, mainly present on photoexposed sites – marker of extrinsic ageing, shown in Figure 3
LEPs / LEPi ratio: number of low echogenic pixels in the upper dermis/number of low echogenic pixels in the lower dermis	Allows an appreciation of the density and integrity of the extracellular matrix, both from the upper and lower dermis, which varies according to age, UV-rays exposures, therapy – photoageing marker

Table 2. Ultrasonographic parameters

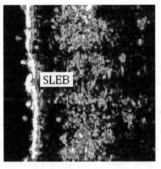

Fig. 3. Subepidermal low echogenicity band (SLEB)

The ultrasonographic skin examination with a high frequency transducer offers a 80 micrometer axial resolution, a 200 micrometer lateral resolution and a 1-2,5 cm depth [8,9]. According to the literature data and our experience, high frequency US is a non invasive instrument for skin examination having multiple applications both in the clinical and the research setting. We mention some of the most important contributions of the method in Table 3:

	Applications of high-frequency ultrasound
1.	Histologic skin evaluation and identification of each skin component (epidermis, dermis, dermo-epidermic and dermo-hypodermic junction, hypodermis); it is worth mentioning that the skin thickness is measured in"mm", while the density of the dermis is measured in number of pixels of different amplitude.
2.	Assessment and description of pigmentary and non-pigmentary tumoral structures: pigmented nevi, melanomas, carcinomas, dermoid cysts, sclerodermia, etc.
3.	Non invasive monitoring, both qualitative and quantitative, of the cutaneous alterations induced by senescence
4.	Monitoring of chronic inflammatory conditions
5.	Monitoring the efficacy of various therapies
6.	Objective markers for determining the phototype

Table 3. Applications of high-frequency ultrasonography in Dermatology

1.4.2 Conventional ultrasound

The use of conventional US has gained greatly in importance in clinical dermatology starting with the 70's. It proved to be a valuable diagnostic method and to have several indications, such as: a. identification and description of visible and palpable tumors, including melanoma; b. preoperative and postoperative assessment of periferal lymph nodes in all patients with malignant skin tumors; c. monitoring of metastases, especially during chemotherapy. (Wortsman et al, 2010) The main applications of US in dermatology in the present times are conventional cutaneous US and examination with the high frequency transducer.

2. High-frequency ultrasound study of the skin

2.1 High frequency ultrasound study of the skin aging process

High frequency ultrasound allows the "in vivo" appreciation of certain histological parameters and offers new characteristic markers, which may quantify the severity of the cutaneous senescence process. Moreover, it may differentiate between the chronological aging process and photoaging. It evaluates the physio-chemical properties of the integument, epidermis, dermis and subcutis that induce acoustical variations, expressed through certain changes of tissue echogenicity. Our study focused on measuring the changes in skin thickness and dermis echogenicity, as part of the complex ageing process, on different intervals of age.

The study was performed on 40 Caucasian patients, 12 men, 28 women, aged 4 -75 years and divided into four age categories: 4-20, 21-40, 41-60, >60. For each subject, cutaneous ultrasound images were taken from 3 different sites: dorsal forearm (DF), medial arm (MA) and zygomatic area (ZA). The data we obtained was statistically assessed, based on the ANOVA and Student T test, using the EPIINFO program. We evaluated the differences between values referring to different intervals of age at the 3 examined sites. A p value <0.05 was considered significant.

The **thickness of the epidermis** remains at approximately similar values on all examined sites, for all age intervals, with no statistically significant differences. The **thickness of the**

dermis shows certain variations. A growth of the dermal thickness at facial level can be noticed with aging. From a mean of approximately 1,320 mm on the 4-20 age interval, the dermis reached a value of 1,614 mm for the subjects taking part of the >60 age interval.

At dorsal forearm and medial arm level, we noticed that the dermis thickness varied in the same way: a decrease of the dermis thickness for the 20-40 age interval, followed by an increase for the 41-60 age interval. The 20-40 age interval, corresponding to the maturity period, is characterized by active synthesis processes, which lead to the thickening of the extracellular matrix.The degenerative processes that lead to the thinning of the dermis appear slowly after the age of 60. The variation pattern of the dermal and epidermal thickness with age can be observed in Table 4.

Area	Interval of age			
	4-20	21-40	41-60	>60
Epidermis DF (mm)	0,19675	0,182333	0,182667	0,186444
Epidermis MA (mm)	0,165375	0,163778	0,184889	0,169333
Epidermis ZA(mm)	0,175875	0,173333	0,155222	0,162333
Dermis DF (mm)	1,211875	1,191222	1,311556	1,168889
Dermis MA (mm)	0,856375	0,772889	0,838333	0,861
Dermis ZA (mm)	1,320375	1,45	1,448289	1,614

Table 4. Mean of dermis and epidermis thickness on 4 age intervals, at dorsal forearm (DF), medium arm (MA) and zygomatic area (ZA) level

Generally, considering the **total thickness of the integument** (dermis and epidermis), a significant increase may be noticed especially at the facial site, which prooves that the integument thickness increase is dependant on the severity of UV photoexposure. Also, comparing young subjects (aged 4-20) with elderly ones (>60), it is noticeable that the integument is thinner in the second group at dorsal forearm level, has similar values on the medial arm and increases at facial level.

The number of hypoechogenic pixels shows a significant variation in case of the dorsal forearm and medial arm of the patients taken into study, as follows: hypoechogenic pixels significantly decrease on the dorsal forearm in the 20-40 age interval compared to the 4-20 interval (p= 0.038018, p<0.05) and increase significantly in the >60 age interval in comparison to the 41-60 interval (p= 0.00777, p<0.05); on the medial arm, hypoechogenic pixels increase significantly in the 41-60 age interval, compared to 20-40 interval (p= 0.018056, p<0.05). The significant increase of hypoechogenic pixels after the age of 40, both on photoexposed and photoprotected sites, is correlated with the degenerative changes which are typical for the ageing process in general. Initially, elastic and reticular fibres from the papillary dermis are altered. Generally, we noticed that hypoechigenic pixels are more numerous in the upper dermis in elderly subjects on all studied areas, being correlated with the elastosis and cutaneous degenerescence processes.

Intermediate echogenic (50-100, 100-150) pixels increase significantly (p<0.05) on photoexposed sites in the 20-40 age interval (synthesis processes; assembly to filaments, microfibrils) and decrease after the age of 40 (decreased synthesis and degenerative processes). The repartition dynamics of the intermediate echogenic pixels in case of the 20-40 age interval indicates the presence of intense metabolic processes that continue on to the next intervals of age, but in a much slower rhythm. We consider this interval as a"critical age interval" that represents the optimal timing to inititate the prophylaxis of the senescence process and associated pathology.

Hyperechogenic pixels also display statistically significant variations on the three analyzed regions: on the dorsal forearm, high echogenic pixels increase significantly in the 20-40 interval of age, compared to the 4-20 interval (p=0.025154, p<0.05, and slightly decrease after the age of 40; on the medial arm, they decrease in the 40-60 age interval compared to the 20-40 age interval (p= 0.038523, p<0.05) and at facial site, high echogenic pixels increase in the 21-40 interval (p= 0.025405, p< 0.05) and decrease between 41-60 (p= 0.048694, p<0.05). The highest amount of hypoechogenic pixels was identified at facial level, an intensely photoagressed site, whereas the highest amount of hyperechogenic pixels was found at the medial arm site, a less photoexposed area. High echogenic pixels (200-255) are poorly expressed in patients belonging to the 4-20 age interval, and much better expressed in the 20-40 age interval on all studied areas. According to Table 5, the mean of hyperechogenic pixels is higher on photoprotected areas compared to the photoagressed ones for all intervals of age. Thus, we may consider hyperechogenic pixels as ultrasonographic markers of the chronological ageing process.

	0-30	50-100	100-150	200-255
DF				
0-20	11656.50	2184.12	627.62	470.50
20-40	7657.00	3581.55	1314.11	1221.55
40-60	12613.11	2413.44	740.555	701.33
>60	11007.44	2093.00	708.444	737.55
MA				
0-20	4792.50	2407.25	1036.50	1960.00
20-40	3371.55	2187.22	981.55	2687.22
40-60	5716.33	2225.88	916.11	1644.88
>60	5584.88	2204.66	822.22	1450.66
ZA				
0-20	15263.63	1568.87	348.25	120.75
20-40	13979.89	2670.00	821.77	602.22
40-60	17047.11	1561.88	394.77	213.22
>60	19055.56	1823.44	504.77	301.22

Table 5. Mean of 0-30, 50-100, 100-150, 200-255 pixels measured on subjects divided into 4 age intervals, at the examined sites: dorsal forearm (DF), medial arm (MA) and face level (ZA)

Subepidermal low echogenic band (SLEB) was identified in case of the subjects part of the 41-60 and >60 age intervals, and appeared especially on photoexposed sites (dorsal forearm, face) [10]. In some patients though, especially the younger ones, we were able to identify SLEB at medial arm level as well. On photo-aggressed sites, it may be noticed that the echogenicity of the upper dermis decreases with age.

SLEB may be considered a specific ultrasonographic parameter that allows a noninvasive quantification of the elastosis degree and actinic collagen degeneration. (Lacarrubba et al, 2008) SLEB varies in thickness and localization according to age and UV exposure. In young subjects, SLEB is present in the lower dermis and quantifies the degree of cutaneous hydration, since the extracellular matrix is rich in proteoglycans and hyaluronic acid.

Hyaluronic acid binds uncovalently the proteoglycans, forming macromolecules that attract water, forming a true hydrating capsule. In elderly subjects, SLEB quantifies the elastosis process and basophilic degenerescence of collagen, common aspects of the senescence process, but increased by UV. Thus, we may consider SLEB as a qualitative marker of the photoagression process.

The ultrasound study shows different echogenicity degrees for the **upper (LEPs) and lower (LEPi) dermis**. For the upper dermis, the study revealed an increase of hypoechogenic pixels (0-30), in comparison to the lower dermis, for all 4 age intervals studied. According to Figure 4, the hypoechogenicity degree is higher on photoexposed sites, both for the upper and the lower dermis. **LEPs/LEPi ratio** showed a statistically significant increase ($p < 0.05$) for the 20-40 and 40-60 age intervals on photoexposed sites, especially at facial level ($p = 0.000999$, $p < 0.05$).

On the medial arm , a progressive decrease was noticed till the age of 60, followed by a light increase in people >60 years. This aspect may be explained by the increase of hypoechogenic pixels in the upper dermis. Unlike the upper dermis, in the lower dermis, an increase of echogenicity may be noticed with ageing as visible in Figure 4. The ratio between the echogenicity of the upper and lower dermis represents an objective marker of the photoageing process.

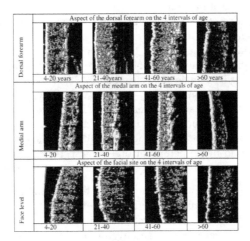

Fig. 4. Ultrasonographic aspect of the dorsal arm, medial arm and facial site on the 4 studied age intervals

The thickness of the integument, SLEB, as well as the dermal echogenicity are parameters that evaluate, with high accuracy the cutaneous senescence process at a microscopical level. The ratio between the echogenicity of the upper and lower dermis represents an objective marker of the photoaging process. SLEB is an ultrasonographic marker of the collagen degenerescence process and photoinduced cutaneous elastosis.

2.2 Non-invasive imagistic assessment by "in vivo" histological sections, of the efficacy of anti-ageing therapies

Taking into consideration that nowadays we assist a general ageing tendency of the world population, the antiaging therapy is a priority and a continuous challenge for researchers. The identification of the mechanisms involved in the cutaneous aging process and their impact on certain age categories, correlated with the hormonal and neurogenetic constellation of the subject would be highly desirable since it is estimated that about 31% of the population is over the age of 60 (US Census Bureau, online database.www.census.gov).

The increase of life expectancy, the psychosocial impact of the cutaneous aspect justifies the high amount of research studies of the ageing mechanisms as well as of the efficacy of certain anti-aging therapies. (Vaupel, 2010) The purpose of the 2 studies to be presented was the assessment with the help of high-frequency ultrasound of the cutaneous changes induced by topical use of Viniferol-containing products as well as by topical anti-ageing product (Interactive P63).

2.2.1 Efficacy of Viniferol as anti-ageing therapy

The first study assessed, with the help of high-frequency ultrasound, the cutaneous changes induced by topical use of products containing Viniferol. As far as the anti-ageing therapy is concerned, Viniferol ® (Resveratrol), an extract from Bordeaux vine stalks is one of the newest and more efficient anti-wrinkle and anti-ageing agents. Having a direct action upon the protein expression of the genes involved in the proliferation and differentiation of integumentary cells, Viniferol profoundly restructures and regenerates the skin. Due to its antioxidant properties, it reestablishes the metabolic balance of the cutaneous cells, slowing down the tissular degeneration and disorganization process (Vranesic-Bender, 2010). Even though the general anti-ageing effects of the flavonoids are well known, until now no scientific studies investigated the action of Viniferol at cutaneous level by using high-frequency ultrasound. Eighty female subjects, aged 22-75, who presented themselves to the practice for prophylaxis and anti-ageing therapy with flavonoids, were prospectively included in the study. 50% of the subjects belonged to Fitzpatrick phototype class II and 50% to phototype class III. The study excluded patients with known allergies to topical flavonoids, cutaneous facial lesions, resurfacing or other anti-ageing therapies in the last 2 months, or those who used phototherapy or oral contraception. The subjects taken into the study were divided into 2 categories: a study group and a control group.

The study group followed the proposed antiaging therapy for 12 weeks, according to a standard protocol. In the morning, a hydrating emollient cream, based on occlusive hydrating agents, was applied at facial level (including zygomatic area), lightly massaging the area for 2 minutes. In the evening, an anti-ageing product containing Viniferol, extracted from grapevine, was applied in the same manner. No other cosmetic products were used by the subjects during the 12 weeks of the study. The control group followed a placebo therapy for 12 weeks, using only moisturising cream in the morning and evening, applied at facial level. For every subject, ultrasonographic images were taken from zygomatic level initially and 12 weeks after local application of the emollient, hydrating product and anti-ageing, Viniferol-based cream. The data we obtained was analyzed, calculating the mean and standard deviation for the quantitative variables of every group and the proportions for the qualitative variables. The difference of means before and after treatment was tested using T-test for paired samples and the relationship between different parameters was assessed thru Spearmann correlation coefficients. A p-value <0.05 was considered significant.

All subjects involved in the study tolerated well the therapy, without evoking adverse effects (erythema, pruritus, ocular disturbance). Subjectively, post flavonoid-therapy a significant hydration of the skin throughout the day and an increase of the cutaneous tonicity was noticed.

After therapy, an increase of the mean **thickness of the epidermis** (0.129 ± 0.237 mm vs 0.150 ± 0.323 mm, p<0.000), and of the dermis (1.434 ± 0.241 mm vs 1.569 ± 0.219 mm, p<0.0001) was observed. (fig 3, 4)The **thickness of the dermis** increased mainly in the 40-60 age interval (1.413 ± 0.280 mm vs 1.569 ± 0.279 mm, p=0.001), and less, but still significantly < 40 years (1.416 ± 0.266 mm vs 1.585 ± 0.150 mm, p=0.015), while >60 years the increase was not statistically significant (1.480 ± 0.157 mm vs 1.554 ± 0.204 mm, p=0.097). At the same time, at dermal level, the **number of low echogenic pixels** decreased (15153.53 ± 3589.86 vs. 12958.48 ± 3628.35, p<0.0001), but this aspect was only noticed in the lower dermis (6949.75 ± 1966.93 vs 6257.62 ± 2224.88, p=0.016), not in the upper dermis (7290.55 ± 1794.60 vs 6940.65 ± 2150.30, p=0.168). Overall, the **LEPs/LEPi ratio** increased significantly after flavonoid therapy (1.092 ± 0.330 vs 1.259 ± 0.631, p=0.011). We also noticed an increase of **medium echogenic pixels** (3359.72 ± 1457.36 vs 3983.47 ± 1401.24, p=0.013) and **high echogenic pixels** (460.27 ± 323.93 vs 750.90 ± 493.82, p<0.0001) after therapy.The general variation pattern of the quantifiable ultrasonographic parameters after flavonoid therapy is illustrated in Table 6.

	Before treatment	After treatment	P
Thickness of epidermis (mm)	0.129 ± 0.237	0.150 ± 0.323	<0.0001
Thickness of dermis (mm)	1.434 ± 0.241	1.569 ± 0.219	<0.0001
LEP	15153.53 ± 3589.86	12958.48 ± 3628.35	<0.0001
MEP	3359.72 ± 1457.36	3983.47 ± 1401.24	0.013
HEP	460.27 ± 323.93	750.90 ± 493.82	<0.0001
LEPs	7290.55 ± 1794.60	6940.65 ± 2150.30	0.168
LEPi	6949.75 ± 1966.93	6257.62 ± 2224.88	0.016
LEPs/LEPi	1.092 ± 0.330	1.259 ± 0.631	0.011

Table 6. Cutaneous parameters quantified by high-frequency ultrasound before and after treatment

If we consider the variation of the ultrasonographic parameters after topical flavonoid therapy according to the phototype class of the subjects, it can be noticed that after therapy, there is a significant increase of the LEPs/LEPi ratio in the subjects belonging to phototype class II, not III, as shown in Table 7.

	Phototype 3			Phototype 2		
	Before treatment	After treatment	P	Before treatment	After treatment	P
Epidermis (mm)	0.1288±.026	0.151±0.027	<0.0001	0.129±0.021	0.148±0.037	0.020
Dermis (mm)	1.441±0.270	1.570±0.263	0.001	1.427±0.214	1.568±0.172	0.003
LEP	15059.25±4063.97	13864.75±3824.33	0.015	15247.8±3162.14	12052.2±730.34	<0.0001
MEP	3120.10±1725.95	3850.95±1487.32	0.046	3599.35±1122.42	4116.0±1334.62	0.151
HEP	379.35±280.94	645.50±373.59	<0.0001	541.20±350.25	856.3±581.03	0.004
LEPs	7071.65±1754.22	6961.1±2236.66	0.725	7509.45±1852.70	6920.20±2118.53	0.148
LEPi	7007.65±2150.80	6737.15±2205.91	0.480	6891.85±1818.86	5778.10±2193.29	0.010
LEPs/LEPi	1.035±0.262	1.096±0.300	0.200	1.1499±0.384	1.4227±0.820	**0.026**

Table 7. Variation of the cutaneous parameters quantified by high-frequency ultrasound before and after treatment, according to phototype

In the placebo group, we noticed no significant increase of the epidermis and a slight increase of the dermis after therapy (1.433 ± 0.34 mm vs. 1.486 ± 0.14 mm). The number of low echogenic pixels at dermal level also show a slight increase (13213 ± 1284 vs. 15374 ±2318 , p=0.1) due to an optimal hydration of the skin and a discrete decrease of high echogenity pixels (421,8 ± 121.18 vs 368. 3 ±104.03, p=0.07). The LEPs/LEPi ratio showed no particular display according to the age or phototype of the subjects.

Previous studies have shown that the thickness of the epidermis and dermis, as well as the dermal density are important parameters that assess the cutaneous regeneration process (Crisan M et al, 2009). The neosynthesis of the proteic structures induces an increase of the dermal echogenicity and density, local cell architecture changes and implicitly there is an increase of the dermis and epidermis thickness. It has been prooved that certain ultrasonographic markers, such as SLEB (subepidermal low echogenity band) or the LEPs/LEPi ratio can quantify the cutaneous senescence process, as well as the efficacy of various antiaging therapies.

The obtained results are in accordance with the data published in literature. Thus, locally applied flavonoids induce the neosynthesis of the fibrillary structures, but also of glicosaminoglycans, intense hydrophil molecules, favouring the cutaneous hydration. It is well known that flavonoids have important antiaging properties not only at cutaneous level, but at the level of the entire organism. Viniferol, a molecule with proven anti-ageing and antioxidant properties, exhibits a complex action at cutaneous level: it interacts with fibroblastic receptors, amplifies the interrelation fibrocyte-extracellular matrix, modulates the adhesivity molecules and interferes with the oxidative stress process and non-enzymatic glycation, with regenerative effect at cutaneous level.

After therapy, a significant increase of the mean thickness of the epidermis and dermis was noticed, fact that once again confirms the presence of a complex, regenerative dermal process, induced by flavonoids. The dermal thickness increased the most in the 40-60 age interval, to a lesser extent, but still significant under the age of 40, and insignificantly over 60 years. We can affirm that topical flavonoid products have the best efficacy on mature integument, with specific structural and hormonal characteristics. In young subjects (<40) the thickness of the dermis increases discretely as the dermis is a young connective tissue, rich in glicosaminoglycans and thus, properly hydrated. After the age of 60, interval characterized by the presence of degenerative changes of the extracellular matrix the flavonoid-based anti-ageing therapy induces less intense regenerative changes that could be amplified by the association of products able to interfere the characteristic age-related aging mechanisms. At the same time, concomitantly with the change in dermal thickness, the number of low echogenic pixels (LEP) decreased in the lower dermis, not in the upper part. The LEPs/LEPi ratio also increased significantly after therapy. The decrease of the number of low echogenic pixels in the lower dermis is proportionate with the significant increase of medium and high echogenic pixels that quantify proteic neosynthesis, as well as cytoarchitectural reorganizations of the extracellular matrix.

Our data shows important ultrasonographic changes at cutaneous level after anti-ageing therapy, as visible in Figure 5. Flavonoids have a complex action at the dermal level, interfering with several mechanisms involved in the senescence process. They act at the level of fibrocytes, on specific receptors, turning inactive mature cells into young, metabolically active ones.

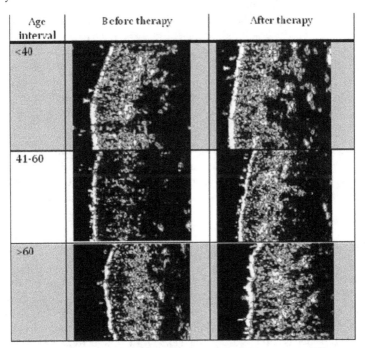

Fig. 5. Ultrasonographic skin aspect before and after topical therapy, on different age intervals

The synthesis of the proteic structures is initiated at intracellular level. The trophocollagen molecules, ellastin, the glicosaminoglycans are extracellularly assembled into microfibrills, fibers or proteoglycans. Depending on the biochemical structure, the level of organisation, architectural orientation and quantity, the proteins show a certain cutaneous echogenity degree. The low echogenity pixels that quantify the hydration degree of the extracellular matrix especially in the lower dermis are replaced by medium and high echogenity pixels, quantifying proteic synthesis. We can consider that MEP codify elastic and collagen precursors that are to be assembled into mature connective tissue fibers, codified by HEP.

The increase of the LEPs/LEPi ratio quantifies the replacement of the hypoechogenic pixels from the lower dermis with medium and high echogenic pixels as a result of protein neosynthesis. Type I collagen, that is prominant at the dermal level (punctiform hyperechogenic pixels) is organised in fibers, visible as hyperechogenic bands, having a parallel display in the lower dermis. These hyperechogenic bands, visible especially on photoprotected sites represent an ultrasonographic marker of the intrinsic aging process.

If we consider the significant changes of the ultrasonographic parameters after anti-ageing therapy depending on the phototype of the subject, a significant increase of the LEPs/LEPi ratio is present in the subjects in phototype II class, but not class III. This observation would justify the correlation of the anti-ageing therapy with the cutaneous phototype. Further studies are necessary to confirm the different reactivity of the phototype classes to local therapies.

Flavonoids, through complex mechanisms, interfere with the reactions involved in the senescence process, and induce the synthesis of the extracellular matrix. According to our data, Viniferol-based products are more efficient in the 40-60 age interval, characterized by complex biological changes at cutaneous level. Viniferol shows real and important anti-ageing properties, since it interferes concomitantly with the genetic, oxidative, immunologic, metabolic mechanisms that are involved in the cutaneous aging process. The prophylaxis of the ageing process should start before the age of 40, preferable in the "critical age interval" (20-40 years), that is characterized by important changes at tissular, cellular and molecular levels, (Crisan et al, 2010)

The optimization of the anti-ageing therapy, according to special studies requires targeted, personalized therapies, adapted to the hormonal, genetic, oxidative, immunologic and metabolic status of the subject, capable of interfering with deficient mechanisms on certain age intervals. Viniferol-based products have a higher efficacy in phototype II subjects compared to phototype III ones.

2.2.2 Efficacy of INTERACTIVE PEEL P63 as anti-ageing therapy

Interactive P63 is a metabolical dynamiser, capable of interacting simultaneously at different cutaneous levels, both on anabolic and catabolic mechanisms. It contains 8 active principles, among which: alfahidroxiacids, retinoids, a complex derived from growth factors, gluconolactone incapsulated in liposomes etc. This anti-ageing complex has a simulataneous action on three levels, epidermis, dermis and dermoepidermic junction. It has been tested in vitro on human fibroblasts cell cultures (Line Hs27) for cytotoxicity, apoptosis, proliferation index, collagen synthesys, matrix metaloproteinases activity.

This study included fifty female subjects aged 40-75, who addressed themselves to the practice for anti-ageing therapy. The subjects were divided into 2 groups of 30 and 20 patients. From the study group (30 subjects), 16 subjects belonged to Phototype class II, 14 subjects to phototype class III, wheareas from the placebo goup, 10 subjects were phototype II and the rest of 10 phototype III. The subjects were divided into 3 age categories: 40-50, 51-60, >60. The subjects from the study group underwent topical therapy with Interactive P63 product, whereas the rest of 20 subjects from the control group used a placebo product. The subjects taken into the study followed the proposed antiaging therapy for 12 weeks, according to a standard protocol. The Interactive P63 and placebo product were applied twice a week for 30 minutes at facial level for 12 weeks. During this period, no other treatments apart from moisturising cream were applied. For every subject, ultrasonographic images were taken from zygomatic area, initially and after 12 weeks of treatment. The data we obtained was analyzed, calculating the mean and standard deviation for all quantitative variables. The difference of means before and after treatment was tested using T test for paired samples. A p-value <0.05 was considered significant.

All subjects involved in the study tollerated well the therapy, without evoking adverse effects (erithema, pruritus, ocular disturbance) after 30 minutes of contact. In the Interactive P63 group, after therapy, an increase of the mean thickness of the epidermis (0.117 ± 0.021 mm vs 0.135 ± 0.023 mm, p=0.0024), and of the dermis (1.537 ± 0.23 mm vs 1.710 ± 0.244 mm, p=0.0076) was observed. At dermal level, the number of low echogenity pixels decreased in a significant manner after topical therapy (18484.4 ± 4666.5 mm vs. 14138.97 ± 3779.5 mm, p=0.00021) whereas the number of the medium (3118.63 ± 974.4 mm vs. 4608.93 ± 1105.6 mm, p=0.001) and high echogenic pixels (379.6 ± 274.17 mm vs. 1004.9 ± 458.78 mm, p<0.0001) increased significantly as displayed in Figure 6.

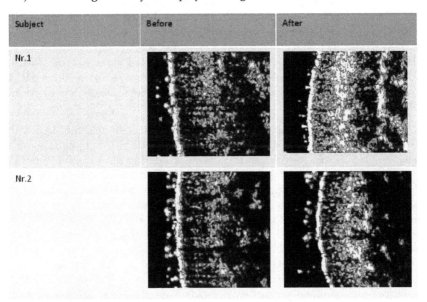

Fig. 6. Ultrasonographic evaluation of the zygomatic area, before and after topical P63 therapy

The LEPs/LEPi ratio increased significantly after therapy (1.149 ± 0.251 mm vs 1.574 ± 0.317 mm) especially due to the significant decrease of the number of low echogenic pixels (LEPi) in the lower dermis (8740.4±2711.01 vs 4921±2373.6, p=0.016).

If we consider the variation of the ultrasonographic parameters after topical therapy according to the phototype class of the subjects, it can be noticed in Figure 7 that after therapy, there is a significant increase of the LEPs/LEPi ratio in the subjects belonging to phototype class III, not II.

Considering the LEPs/LEPi parameter, we noticed a significant increase of the ratio in all subjects part of the study, especially in the 51-60 age interval (40-50 age interval: p=0.01, 51-60 age interval: p=0.0009, >60 age interval: p=0.02) as visible in Figure 8:

In the placebo group, we noticed a slight increase of the dermis (1.496 ± 0.14 mm vs. 1.571 ± 0.174, p=0.07) and of the dermal low echogenic pixels (13812 ± 2070 vs. 14787 ±2218 , p=0.08) due to an optimal hydration of the skin and a discret tendency of the high echogenity pixels to decrease (379,8 ± 137.18 vs 316.3 ±163.43, p=0.11). The LEPs/LEPi ratio showed no particular display according to the age or phototype of the subjects

INTERACTIVE P63 complex interacts concomitantly different mechanisms involved in the cutaneous aging process, conferring from imagistical point of view, a characteristic display of the pixels at cutaneous level. (Rouabhia et al, 2002) The increase of the epidermal/dermal thickness represents the morphological exppression of the changes induced by INTERACTIVE P63 complex at fibroblastic and extracellular matrix level. The activation of the fibroblasts as well as the inductive effect upon stem cells, associated with the inhibition of the mechanisms responsible for the distruction of the fibrillary structures, induce an increase of the dermal density. Thus, we noticed a general, significant decrease of the mean number of low echogenic pixels (LEP) at dermal level, more pronounced in the lower dermis (LEPi) than the upper one (LEPs), suggesting important structural, biochemical, mollecular and architectural changes that vary according to certain particular properties of the upper and lower dermis. Parallel to the decrease of LEP after therapy, a statistically significant increase of the mean number of medium (MEP) and high ecogenic pixels (HEP) was noticed, quantifying the increase of dermal density and thus, collagen neosynthesis.

The LEPs/ LEPi ratio, an essential imagistic marker that quantifies the dermal density, increased in a significant manner, due to the important decrease of the number of low echogenity pixels from the lower dermis (LEPi). Considering the LEPs/LEPi ratio on the three age categories: 40-50,51-60, >60 , a significant increase was noticed in all three age intervals. The fact that the most significant increase of the dermal density occured in the 51-60 age interval, may be correlated with the post-menopausal status as well as with the estrogen-like activity of INTERACTIVE P63 complex (El-Alfy et al, 2010). During menopause, due to a decrease of estrogen and cutaneous estrogen receptors, a progressive decrease of dermal collagen occurs, with a loss of collagen content of 1-2% every menopausal year. Several studies certify the fact that topical estrogen therapy in menopausal women induces an increase of almost 5.1% of dermal collagen. It is also a fact that the efficacy of hormone therapy is dependant on the basal collagen status at the beginning of the therapy. The initiation of a precoucious therapy in menopause has a prophylactic role, while a delayed therapy has a therapeutic purpose.

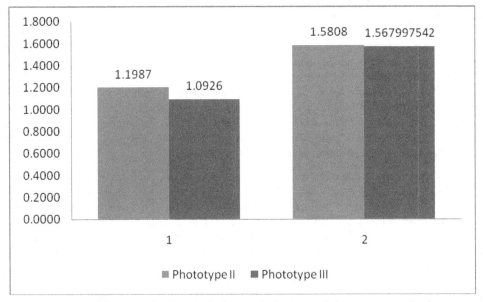

Fig. 7. Variation of the LEPs/LEPi ratio before and after topical therapy, according to phototype.

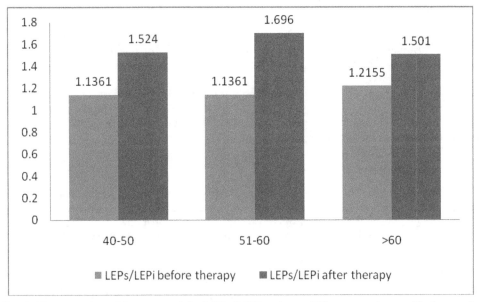

Fig. 8. The variation of the mean of the LEPs/LEPi ratio on different age interval before and after therapy

INTERACTIVE P63 complex acts at cellular level, interffering the cutaneous estrogenic receptors (α and β) that, even though with structural and functional similarities, have different expression conditions and act differently in menopause, explaining the changes regarding dermal density on age categories. Regarding the phototype of the subjects, a certain, particular reactivity is to be mentioned: a more significant growth of the dermal density in subjects belonging to phototype class III, compared to phototype class II. The LEPs/LEPi ratio showed no significant variation neither with the age or the phototype of the subjects. Interactive P 63 product acts on specific sensitive receptors may interfere with estrogen-like receptors, activating the fibroblast "key cell" and increases the synthesis of collagen. It has real and important anti-ageing properties on large age intervals, since it interferes concomitantly the oxidative, genetic, immunologic, hormonal and metabolic mechanisms. It is highly efficient especially in the 51-60 age interval and in phototype III patients.

In the past years, many advances in the diagnosis of skin ageing have become available for an earlier and more specific diagnosis. Our studies show the importance of high-frequency ultrasound as a noninvasive method for the assessment of the cutaneous senescence process. The correlations between the histological and imagistic parameters allow the establishment of noninvasive diagnosis and treatment protocols. Our observations require further development and review to determine the diagnostic accuracy. Some show great promise in assessing, with less invasive methodes, histological features required for an earlier diagnosis, or for establishing the efficacy of various therapies. The need to develop new strategies on how to prevent and how to accommodate the ageing society requires the elaboration of mathematical models in order to predict the evolution of the ageing phenomenon. (Crisan et al, 2010) The anti-ageing medicine may discover there is no limit to human life span.

3. Conclusions

High-frequency ultrasound is a non-invasive histological tool that allows the visualisation of "in vivo" histological sections, offering information with microscopical correspondence and also characteristic ultrasonographic markers. The efficacy of anti-ageing therapies varies with the age interval, according both to the applied product but also the cutaneous reactivity, phototype, hormonal and metabolical status. The prevention of the cutaneous ageing process should begin in the "critical age interval" and the improvement of the clinical aspect requires precocious, personalized therapies, using efficient substances, previously tested on cell-cultures.

4. Acknowledgment

Special thanks to all of my colleagues who contributed with usefull advice and pertinent observations to the elaboration of this chapter. I would also like to thank Cortex Technology for allowing us to use the Dermascan equipment, whitout which our study would not have been possible. This study is part of CNCSIS national research grant nr. 2624.

5. References

Badea R, Crişan M, Lupşor M, Fodor L.(2010). Diagnosis and characterization of cutaneous tumors using combined ultrasonographic procedures (conventional and high resolution ultrasonography). *Med Ultrason*. (4):317-22.

Bos DC, de Ranitz-Greven WL, de Valk HW. (2011) Advanced glycation end products, measured as skin autofluorescence and diabetes complications: a systematic review.*Diabetes Technol Ther.* Jul;13(7):773-9.

Crisan M, Cattani C, Badea R, Mitrea P et al. (2010) Modelling Cutaneous Senescence Process. *Lecture Notes in Computer Science,* (6017): 215-224

Crisan M, Cattani C, Badea R, Cosgarea R, Dudea S, Mitrea P, Lupsor M. (2009). Complex histological, genetical, ultrasonography and clinical studies in early noninvasive diagnosis of the photoinduced cutaneous senescence and in the photoinduced skin cancers, using computerized imaging, modern biotechnology and mathematical modelling methods." *Automatic computers applied mathematics*, vol.18.No.2, pg. 231-255 Scientific Journal ISSN 1221- 437X MEDIAMIRA SCIENCE PUBLISHER - ACAM CNCSIS - B +

Crisan M. (2007) Histology of the integumentary system with practical aplications. *Casa Cartii de Stiinta.* ISBN 973-686-935-0

El-Alfy, M., Deloche, C., Azzi, L., Bernard, B., Bernerd, F., Coutet, J., Chaussade, V., Martel, C., Leclaire, J. and Labrie, F. (2010). Skin responses to topical dehydroepiandrosterone: implications in antiageing treatment.. *British Journal of Dermatology*, 163: 968–976.

Han J, Quresh AA, Prescott J et al.(2009). Prospective study of telomere length and the risk of skin cancer. *J Invest Dermatol*, 129, 415-421

Jasaitiene D, Valiukeviciene S, Linkeviciute G, Raisutis R, Jasiuniene E, Kazys R. (2011) Principles of high-frequency ultrasonography for investigation of skin pathology.*J Eur Acad Dermatol Venereol;* 25(4): 375-382.

Kohl, E., Steinbauer, J., Landthaler, M. and Szeimies, R.-M. (2011), Skin ageing. *Journal of the European Academy of Dermatology and Venereology*, 25: no. doi: 10.1111/j.1468-3083.2010.03963.x

Lacarrubba F, Tedeschi A, Nardone B, Micali G. (2008) Mesotherapy for skin rejuvenation: assessment of the subepidermal low-echogenic band by ultrasound evaluation with cross-sectional B-mode scanning.. *Dermatol Ther. Suppl* 3:S1-5.

Pageon H. 2010. Reaction of glycation and human skin: the effects on the skin and its components, reconstructed skin as a model. (2009) *Pathol Biol (Paris)*. Jun;58(3):226-31. Epub . Review.

Rouabhia and al. (2002). G factor P63 increases P63 protein and activates the differentiation of stem cells. (2009) *Arch Dermatol Res* ; 301(4):301-6.

Schmid-Wendtner MH, Burgdorf W (2005). Ultrasound scanning in dermatology. *Arch Dermatol.* Feb; 141(2):217-24.

Shi Y, Buffenstein R, Pulliam DA, Van Remmen H. (2010). Comparative studies of oxidative stress and mitochondrial function in aging. *Integr Comp Biol.* (5):869-79. Epub 2010 Jul2.

Uitto J. The role of elastin and collagen in cutaneous aging: intrinsic aging versus photoexposure. (2008). *J Drugs Dermatol.* (2 Suppl):s12-6

Vaupel JW. (2010). Biodemography of human ageing *Nature;* 464: 536-542.

Vranesić-Bender D. (2010) The role of nutraceuticals in anti-aging medicine. .*Acta Clin Croat.*;49(4):537-44.

Wortsman X, Wortsman J. (2010) Clinical usefulness of variable-frequency ultrasound in localized lesions of the skin. *J Am Acad Dermatol*; 62(2): 247-256.

9

Pharmacologic Inhibition
of Cardiac Stem Cell Senescence

Daniela Cesselli[1], Angela Caragnano[1],
Natascha Bergamin[1], Veronica Zanon[1], Nicoletta Finato[1],
Ugolino Livi[2], Carlo Alberto Beltrami[1] and Antonio Paolo Beltrami[1]
[1]Department of Medical and Biological Sciences,
University of Udine, Udine,
[2]Department of Experimental and Clinical Medical Sciences,
University of Udine, Udine,
Italy

1. Introduction

Mammalian aging may be viewed as a reduction in the capacity to adequately maintain tissue homeostasis or to repair tissues after injury(Sharpless and DePinho, 2007). When homeostatic control diminishes to the point at which tissue/organ integrity and function are no longer sufficiently maintained, physiological decline develops, and aging becomes apparent. Cells that express senescence markers accumulate at sites of chronic age-related pathology, such as osteoarthritis, atherosclerosis and chronic heart failure(Blasco, 2007; Campisi and d'Adda di Fagagna, 2007; Chimenti, et al., 2003; Deng, et al., 2008; Jeyapalan and Sedivy, 2008; Minamino and Komuro, 2008; Sharpless and DePinho, 2007; Shawi and Autexier, 2008; Torella, et al., 2004; Urbanek, et al., 2005). Thus, senescent cells are associated with aging and age-related diseases in vivo(Campisi, 2011).

The discovery of tissue-resident stem and progenitor cells has suggested that these cells are responsible for tissue homeostasis and regeneration(Hosoda, et al., 2009; Hsieh, et al., 2007; Li and Clevers, 2010). For this reason, pathological and patho-physiological conditions characterized by altered tissue homeostasis and impaired regenerative capacity can be viewed as a consequence of the reduction in stem cell number and/or function. Following the evolutionary theory of antagonistic pleiotropy, stem cell senescence can be considered a double edged-sword that exerts both a tumor-suppressor effect, by preventing the expansion of injured self-renewing cells, and detrimental effects, contributing to tumor invasiveness in a paracrine fashion or to aging by causing stem cell arrest or attrition (cancer-ageing hypothesis)(Campisi, 2005; Sharpless and DePinho, 2007). In line with this, stem cell aging has been demonstrated in hematopoietic stem cells, as well as in other self-renewing compartments(Beltrami, et al., 2011a).

The recognition that the heart possesses a pool of primitive, clonogenic, self-renewing, and multipotent cells responsible for tissue homeostasis has opened a new era of research aimed at harvesting, expanding and utilizing these cells for cardiac repair(Beltrami, et al., 2003).

However, experimental studies have demonstrated that, although the cardiac stem cell (CSC) pool is expanded acutely after myocardial infarction, this response is attenuated in chronic heart failure(Urbanek, et al., 2005). In addition, a significant accumulation of senescent CSC in cardiac tissue both in pathological settings and with aging has been described(Cesselli, et al., 2011; Chimenti, et al., 2003; Rota, et al., 2006). More recently, our group has demonstrated that both age and pathology exert detrimental effects on human CSC (hCSC). Specifically, they attenuate CSC telomerase activity, reduce telomeric length, determine telomere erosion, are associated with the presence of telomere induced dysfunction foci and impair CSC function(Cesselli, et al., 2011). Importantly, comparing the gene expression profile of CSC obtained from normal and pathological tissues we identified several possible molecular targets for pharmacological interventions aimed at reverting or attenuating the senescence processes.

Aims of this chapter will be to review the knowledge on the impact that CSC senescence exerts on cardiac function, to discuss interventions aimed at reverting it and to focus on original results investigating the effects of Rapamycin, Resveratrol and DETA/NO on CSC senescence.

2. Cellular senescence

In 1961 Hayflick applied the term *cellular senescence* to cells that ceased to divide in culture despite favorable growth conditions, based on the speculation that their behavior recapitulated organism aging(Hayflick and Moorhead, 1961). Since then, this phenomenon was proposed to be either a detrimental cause of aging or a beneficial tumor suppression mechanism. In fact, cell senescence plays both these roles, supporting the evolutionary theory of antagonistic pleiotropy that postulates that cellular processes, selected to benefit young organisms, may have unselected deleterious effects in older organisms(Campisi and d'Adda di Fagagna, 2007).

Cellular senescence is currently defined as a specialized form of growth arrest, confined to mitotic cells, induced by various stressful stimuli and characterized by several, although not specific, markers(Sharpless and DePinho, 2007). Specifically, senescent cells are characterized by a permanent growth arrest, resistance to apoptosis, an altered pattern of gene expression, and the expression of proteins that are characteristic of, although not exclusive to, the senescent state(Beltrami, et al., 2011a), such as the senescence-associated β-galactosidase (SA-βgal) (Dimri, et al., 1995).

Recently identified markers of cellular senescence are p16, DEC1, p15, and DCR2(Collado, et al., 2005), and the cytological markers: senescence-associated heterochromatin foci (SAHFs), and senescence associated DNA-damage foci (SDFs)(Di Micco, et al., 2008; Narita, et al., 2003). SDFs are present in senescent cells from mice and humans and contain proteins that are associated with DNA damage. When these foci result from dysfunctional telomeres they are defined as telomere-induced dysfunctional foci (TIFs) (Campisi and d'Adda di Fagagna, 2007; Jeyapalan and Sedivy, 2008; Sharpless and DePinho, 2007).

Regarding the molecular mechanisms responsible for cellular senescence, intrinsic and extrinsic pathways have been described. While the first one is initiated by intracellular damages/stimuli, the second one is related to extracellular molecules. Importantly, these two mechanisms are strictly interconnected since senescent cells are characterized by the production of molecules able to alter the microenvironment thus inducing senescence on the neighborhood cells through a paracrine mechanism(Campisi, 2005).

Intrinsic inductors of cellular senescence are either the progressive telomere erosion that is associated with cell proliferation (i.e. replicative senescence) (Deng, et al., 2008; Shawi and Autexier, 2008) or the formation of irreparable DNA lesions that induce a persistent DNA damage response (DDR) which keeps the cells alive, but arrests their proliferation (i.e. telomere independent, stress-induced premature senescence)(Beltrami, et al., 2011a). In this latter case, DDR is induced by activated oncogenes, and DNA double strand break-inducing agents, such as reactive oxygen species (ROS). Many proteins participate in the DDR, including protein kinases (e.g. ataxia telangiectasia mutated -ATM- and checkpoint-2 - CHK2-), adaptor proteins (e.g. 53BP1 and MDC1 -mediator of DNA damage checkpoint protein-1-) and chromatin modifiers (for example, γ-H2AX)(von Zglinicki, et al., 2005). Therefore, intrinsic inductors of cellular senescence initiate a DDR, consisting of the activation of ATM and ataxia telangiectasia- and Rad3-related (ATR), and downstream kinases CHK1 and CHK2, and phosphorylation of p53. Phosphorylated p53 transcriptionally up-regulates genes, such as p21, that mediate cellular senescence and/or apoptosis to inhibit tumorigenesis. Although less well-understood, telomere dysfunction could also activate the p16^{INK4A}-RB pathway and inhibit cellular proliferation(Campisi and d'Adda di Fagagna, 2007; Deng, et al., 2008; Sharpless and DePinho, 2007; Shawi and Autexier, 2008).

Recent reports have demonstrated that autophagy plays a crucial role in the induction of cellular senescence(Adams, 2009), either replicative (Young and Narita, 2010) or stress-induced premature senescence (Patschan and Goligorsky, 2008; Young and Narita, 2010). However, several interventions that extended lifespan in various species (e.g. caloric restriction, and negative regulation of insulin and mTOR pathways) are associated with the activation of autophagy(Vellai, 2009). To reconcile this observation with the observed accumulation of senescent cells in aged tissues and organs, Authors hypothesize that autophagy may play a beneficial role in mild but long-term stress conditions, counteracting the accumulated damage, while it contributes to senescence establishment in more severely damaged cells(Young and Narita, 2010). Until recently, the general notion was that once senescence was established, cells were locked into a senescent phenotype through a global induction of heterochromatin, which results in the formation of Senescence Associated Heterochromatin Foci (SAHF). In this process, the cyclin dependent kinase inhibitor p16INK4A seemed to play a primary role. However, recently it was shown that SAHF are induced mainly in response to activated oncogenes in a cell type- and insult- dependent manner(Kosar, et al., 2011).

Several *extrinsic inductors of cellular senescence* have also been described so far(Beltrami, et al., 2011a). Specifically, it has been demonstrated that: Advanced Glycation End-products (AGE) (Patschan and Goligorsky, 2008), Angiotensin II (Fukuda and Sata, 2008; Imanishi, et al., 2005; Kunieda, et al., 2006), IGFBP7, IL-6, IL-8(Kuilman, et al., 2008; Orjalo, et al., 2009), GROα, urokinase- or tissue-type plasminogen activators (uPA or tPA), the uPA receptor (uPAR), and inhibitors of these serine proteases (PAI-1 and -2) (Blasi and Carmeliet, 2002; Kortlever, et al., 2006), can induce cellular senescence in different cell types. In this regard, a special role is played by the altered secretome of senescent cells (e.g. Senescence Associated Secretory Phenotype - SASP-). In fact, it has been demonstrated that senescent cells may alter profoundly their microenvironment, by inducing cellular senescence in neighboring cells in a paracrine fashion, by remodeling the extracellular matrix and by stimulating inflammation(Acosta, et al., 2008; Coppe, et al., 2008; Wajapeyee, et al., 2008).

3. Cardiac stem cell senescence

The recognition that the human adult heart possesses a pool of resident cardiac progenitor cells (hCSC), which are self-renewing, clonogenic, and multipotent(Bearzi, et al., 2009; Bearzi, et al., 2007; Beltrami, et al., 2007; Castaldo, et al., 2008; Messina, et al., 2004; Smith, et al., 2007), changed the dogma of the heart as a terminally differentiated organ, offered new hints in the understanding of the pathophysiology of heart diseases and opened a new area of research focused on the use of stem cells for cardiac repair(Beltrami, et al., 2011b; Dimmeler and Leri, 2008). Several different hCSC populations have been identified and characterized on the basis of the expression of specific markers, i.e. c-Kit(Bearzi, et al., 2009; Bearzi, et al., 2007; Castaldo, et al., 2008), ABCG2(Meissner, et al., 2006) and Islet-1(Bu, et al., 2009), or utilizing selective culture conditions, i.e. cardio-spheres(Messina, et al., 2004; Smith, et al., 2007) and multipotent adult stem cells(Beltrami, et al., 2007). Whether these cells are distinct populations or whether they represent different stages of maturation of the same cell type is still a debated question (Beltrami, et al., 2011b; Laflamme and Murry, 2011). Nonetheless, moving from the robust evidence of the efficacy of cardiac stem cell therapy in animal models(Bearzi, et al., 2007; Smith, et al., 2007), the feasibility, safety and some hints on the efficacy of autologous CSC therapy in patients suffering from cardiac pathology is currently under investigation in several clinical trials (ClinicalTrials.gov identifier NCT00474461, NCT00893360, and NCT00981006). Autologous CSC represent a population of cells intrinsically committed to cardiac lineages and would offer the advantage to avoid immunological issues(Dimmeler and Leri, 2008). Nonetheless, it would be important to identify whether and at which extent cardiac diseases can affect this resident stem cell reservoir.

3.1 CSC senescence in cardiac pathologies

The first evidence that hCSC could undergo cellular senescence was given by Anversa's group showing that aged diseased hearts were characterized, at tissue level, by an accumulation of p16^{INK4a}-positive/c-Kit-positive CSC(Chimenti, et al., 2003). Later it was shown that chronic heart failure was associated, in human heart tissues, with an increase in the number of p16^{INK4a}-p53-positive senescent hCSC, further characterized by short telomeres(Urbanek, et al., 2005). More recently, our group provided a direct demonstration of the impact that both aging and pathology exert on hCSC function(Cesselli, et al., 2011). Specifically, we observed that age and pathological state are both associated with: a reduction in telomerase activity, telomeric shortening, and an increased frequency of CSC with telomere induced dysfunction foci, and eventually expressing p16^{INK4a} and p21^{CIP}. These pathologic alterations were coupled with a reduced hCSC function; in fact, hCSC obtained from failing hearts showed, with respect to those obtained from healthy hearts, a significant reduction in clonogenic, proliferative, and migratory potential. Moreover, senescent hCSC displayed an altered gene expression profile, enriched in transcripts of proteins involved in the senescence associates secretory phenotype (SASP), such as IL6 and IGFBP7(Cesselli, et al., 2011). Of note, the underlying diseases of the patients enrolled in this study were different, ranging from ischemic cardiomyopathy to hypertrophic and dilated cardiomyopathy(Cesselli, et al., 2011), suggesting that, independently from the etiology, end stage heart failure is characterized by a progressive loss of the compartment of hCSC with high regenerative potential, paralleled by an increase in the pool of stem cells with minimal or no ability to divide and acquire cardiac cell lineages. Moreover, animal models showed an involvement of CSC senescence in other pathologies such as the diabetic cardiomyopathy (Rota, et al., 2006) and the antracyclin-induced cardiomyopathy(De Angelis, et al., 2010).

3.2 Pathways involved in hCSC senescence

Whith regard to the mechanisms responsible for the replication, differentiation, senescence, and death of hCSC, different growth-factor receptor systems have been shown to play a key role: IGF-1-IGF1R, IGF-2-IGF2R, HGF-c-Met and the renin angiotensin system (RAS)(Dimmeler and Leri, 2008). While IGF-1-IGF1R and HGF-c-Met seemed to exert a protective effect, IGF-2-IGF2R and the RAS up-regulation is associated with CSC senescence.

Specifically, the expression of IGF-1R and the production of IGF-1 are attenuated in aging CSC, and this negatively interferes with oxidative damage and telomere shortening(D'Amario, et al., 2011a; Torella, et al., 2004). In fact, IGF-1 – IGF-1R induces CSC division, upregulates telomerase activity, maintains telomere length, hinders replicative senescence, and preserves the population of functionally competent cardiac stem cells in animals(Torella, et al., 2004) and in humans(D'Amario, et al., 2011a).

Ageing is also associated with a reduction in HGF production, thus impairing the migratory ability of CSC in response to tissue damage(Gonzalez, et al., 2008; Khan, et al., 2011); importantly, CSC dysfunction was shown to be partially restored by HGF injection(Gonzalez, et al., 2008). Regarding IGF-2-IGF2R, it has been recently demonstrated that hCSC expressing IGF-2R are characterized, with respect to IGF-1R positive hCSC, by a more senescent phenotype and by a reduced in vivo regenerative capacity(Gonzalez, et al., 2008).

Similarly, it has been documented that a local RAS is present on hCSC and that the formation of Angiotensin II (Ang II), together with the expression of AT1R, increases with age in hCSC(D'Amario, et al., 2011a). Ang II generates ROS possibly contributing to the age-dependent accumulation of oxidative damage in the heart(Fiordaliso, et al., 2001; Smith, et al., 2007). In fact, the use of ACE-inhibitors positively interferes with heart failure and prolongs life in failing patients(McMurray and Pfeffer, 2005). Moreover, sustained oxidative stress can trigger telomere shortening and uncapping initiating a permanent DNA-damage response(von Zglinicki, et al., 2005). The importance of oxidative stress has been confirmed in a murine model of diabetes, where it has been shown the association of cardiomyopathy with the premature senescence and apoptosis of CSC; importantly, in this model the deletion of p66shc could prevent CSC senescence and was associated with the preservation of myocyte number and cardiac function(Rota, et al., 2006).

Despite these important data, it remains to be determined whether other pathways, that are involved in the senescence of other cell compartments, could contribute to hCSC senescence as well. For example, data on the role played by autophagy, mitochondrial dysfunction, nucleolar dysfunction and epigenetic changes are still missing(Beltrami, et al., 2011a). However, comparing the gene expression profile of hCSC isolated from end-stage failing hearts with that of hCSC isolated from normal hearts, it was possible to demonstrate changes in the expression of genes strictly related to these senescence associated pathways(Cesselli, et al., 2011). Importantly, the analysis identified several possible molecular targets for pharmacological interventions aimed at reverting or attenuating the senescence processes. Interestingly, some of them were very well known target of drugs commonly used in clinical practice, such as beta-blockers and ACE-inhibitors(Cesselli, et al., 2011).

4. How to interfere with cardiac stem cell senescence

Cell therapy is a promising option for treating ischemic disease and heart failure(Dimmeler and Leri, 2008). In fact, various experimental studies documented that tissue-resident primitive cells improve recovery after ischemia(Beltrami, et al., 2011b). Moreover, different groups have demonstrated the feasibility of isolating and expanding hCSC even from end-stage failing hearts(Bearzi, et al., 2007; Beltrami, et al., 2007; Itzhaki-Alfia, et al., 2009; Smith, et al., 2007). However, accumulated evidences indicate that both ageing and pathology are associated with hCSC senescence and functional impairment (Cesselli, et al., 2011; D'Amario, et al., 2011a; D'Amario, et al., 2011b; Itzhaki-Alfia, et al., 2009). In fact, hCSC obtained from failing hearts present reduced migration, proliferation and differentiation(Cesselli, et al., 2011), features considered to be crucial for the regenerative potential of this autologous cell source. Moreover, these cells are characterized by a gene expression profile enriched in elements that are part of the senescence associated secretory phenotype (SASP). Therefore, senescent hCSC can contribute to create a microenvironment favoring, through a paracrine mechanism, senescence on neighbor cells, inflammation and extracellular matrix remodeling, thus creating a vicious circle hampering regenerative purposes.

For this reason, it would be extremely intriguing any attempt aimed at "improving" the quality of the expanded cells, selecting the fraction of cells with the highest regenerative potential or devoid of senescent cells. Conversely, we can hypothesize an intervention aimed at attenuating/reverting the molecular pathways characterizing senescent cells. In this regard, three main strategies can be envisioned: a sorting-based strategy, a function-based strategy and a drug-based strategy.

4.1 Sorting-based strategy to enrich in non-senescent cells

The sorting-based strategy would consist in the physical selection of the cells of interest. Sorting can be achieved, for example, utilizing a fluorescence-activated cell sorting (FACS) or a magnetic activated cell sorting (MACS). The selection strategy can be either positive (we choose and sort "young" cells on the basis of specific surface antigens) or negative (we enrich in non-senescent cells removing from the un-fractioned population those cells we believe to be senescent). Both approaches require the knowledge of specific surface antigens able to recognize the right population to sort. D'Amario et al have recently given examples of positive selection, utilizing antibodies recognizing insulin-like growth factor (IGF)-1 receptors to select, within hCSC isolated from end-stage failing patients, a population of young cells characterized by high telomerase activity, intact telomere length and endowed with a high regenerative ability, being able to restore a large quantity of infarcted myocardium, thus representing a potent cell population for cardiac repair(D'Amario, et al.,

2011a). Although not yet utilized for a negative selection, AT-2(D'Amario, et al., 2011a), IGF-2R and CD49a(Cesselli, et al., 2011) are surface markers that, being more expressed in senescent cells, could be utilized to deplete hCSC culture of the most senescent cells.

The major drawback of the sorting-based approach is the fact that it adds a further grade of complexity to the procedure aimed at producing clinical grade hCSC, since it requires Good Manufacturing Practice-compliant cell sorting and large-scale expansion starting from a reduced number of cells.

4.2 Function-based strategy to enrich in non-senescent cells

It is possible to take advantage of the fact that non-senescent cells are functionally impaired to select cells whose stem cell properties are still preserved. For example, we showed that single-cell derived clones, obtained from hCSC isolated from end-stage failing patients, are less senescent than the overall population(Cesselli, et al., 2011). Again, this approach would be hard to transfer to clinical practice, since it requires Good Manufacturing Practice-compliant cell sorting and large-scale expansion starting from very few cells. In fact, we have recently shown that only 0.7% of the hCSC obtained from end stage failing hearts gives rise to highly proliferating clones(Cesselli, et al., 2011) and, since hCSC are finite cell lines(Beltrami, et al., 2007; Cesselli, et al., 2011), hCSC-derived clones could undergo replicative senescence as a consequence of the high number of population doublings that are required to obtain a number of cells suitable for clinical purposes.

Whether selecting cells on the basis of the ability to actively extrude Hoechst 33342 (side population) could enrich in less senescent cells is still unknown(Hierlihy, et al., 2002; Martin, et al., 2004). However, also in this case a Good Manufacturing Practice-compliant cell sorting would be required.

4.3 Drug-based strategy to enrich in non-senescent cells

Several molecular pathways have been either associated with the development of cell senescence or, on the contrary, with organism longevity (Beltrami, et al., 2011a). Interestingly, the key elements of these two are common and the possibility to act on them can be explored to interfere with stem cell senescence and dysfunction, ameliorating stem cell regenerative approaches and organ pathology.

Briefly, as shown in Figure 1, the pathways main involved are: Insulin/Insulin-like Growth Factor Signaling (IIS), mTOR, AMPK/Autophagy, Nitric Oxide/Estrogen/Telomerase, Sirtuins, and p38MAPK(Beltrami, et al., 2011a).

Although the *Insulin/Insulin-like Growth Factor Signaling (IIS)* is critical for nutrient homeostasis, growth and survival, experimental evidences show that reduced IIS signaling in animals is associated with life extension(Beltrami, et al., 2011a). Insulin like Growth Factors and insulin inhibit the FoxO family of transcription factors through a pathway involving Insulin Receptor Substrate (IRS), PI3K and Akt. FoxO transcription factors promote a variety of cellular responses that include apoptosis, cell cycle arrest, differentiation, resistance to oxidative stress, and autophagy(Ronnebaum and Patterson, 2010; Salih and Brunet, 2008). The transcriptional activities and biological effects of FoxO depend on post-translational modifications and, in this regard, Sirt1 is believed to increase the ability of FoxO to respond to stress through cell cycle arrest and other adaptations but inhibits FoxO transcription of apoptotic genes. Last, FoxO is required for preventing Akt-mediated cardiac hypertrophy(Ronnebaum and Patterson, 2010). Regarding the possibility to interfere with this pathway, we have previously reported that hCSC expressing IGF1-R represent a subset of young and fully functional cells(D'Amario, et al., 2011a), and that IGF1 was able to support proliferation and differentiation of IGF-1R-positive hCSC.

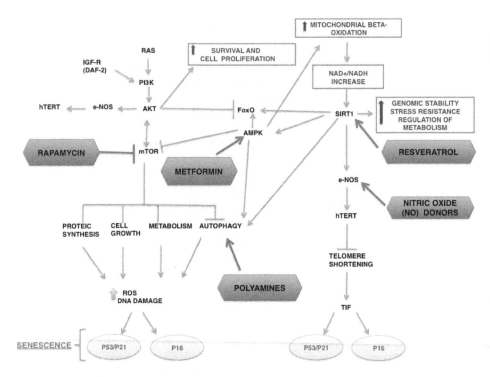

Fig. 1. Senescence pathways and possible pharmacological targets.

In aging, the *mammalian Target Of Rapamycin (mTOR)* plays a prominent role, which is, at least in part, mediated through IIS (Bhaskar and Hay, 2007). In fact, mTOR is activated by insulin, growth factors, nutrients and, indirectly, by Akt. mTOR forms two protein complexes; the Rapamycin-sensitive mTORC1, when bound to Raptor, and the Rapamycin insensitive mTORC2, when bound to Rictor. mTORC1 phosphorylates S6 kinase (S6K), eukariotic translation initiation factor 4E (eIF4E), and other factors involved in protein synthesis and hypertrophy. S6K, in turn, inhibits IRS by phosphorylation, while mTORC2 has a positive feedback on Akt. Nutrients and energy balance can regulate mTORC1, where aminoacids can activate it through the Rag family of GTPases, while AMPK, which is activated by ATP depletion, inhibits it (Bhaskar and Hay, 2007). Because of the central role of mTOR in ageing, Rapamycin has emerged as a very promising drug able to interfere with aging and, possibly, cell senescence(Blagosklonny, 2010). Importantly, Rapamycin is already used in clinical practice for its immunosuppressant and antiproliferative effects. Moreover, accumulated evidences display a possible role of Rapamycin in ageing and cell senescence. In fact, Rapamycin can extend the maximum lifespan of mice, when given late in life, restore self-renewal of hematopoietic stem cells of aged mice, and prevent epidermal stem cell exhaustion induced by Wnt-1 in mouse skin(Blagosklonny, 2010). Last, it has recently been shown that Rapamycin, the mTOR inhibitor PP242 or the IGF1R inhibitor PQ401, are able to increase the efficiency of iPS generation(Chen, et al., 2011). Despite the fact that rapamyicin is utilized in heart transplanted patients to avoid immunrejection, the effects of Rapamycin on hCSC senescence are still unexplored.

Nitric Oxide (NO) and estrogen signaling have been shown to counteract endothelial progenitor cell senescence through the catalytic subunit of human telomerase (hTERT)(Farsetti, et al., 2009). Estrogens' action is mediated either via genomic or nongenomic signaling pathways. The first ones follow the binding of estrogens to nuclear hormone receptors, which are capable of regulating transcription of a number of genes involved in development, metabolism, and differentiation following interaction with a hormone molecule. Therefore, estrogen receptors are ligand-dependent transcription factors. In addition, estrogens can trigger nongenomic signaling pathways through membrane associated estrogen receptors (mER) that activate both the PI3K and the Mitogen Activated Protein Kinase (MAPK) pathways. Estrogens can also activate Adenylate Cyclase and c-Src through the G-protein coupled estrogen receptor (GPER)(Meyer, et al., 2009). NO, on the other hand, is a free radical and an ancestral regulator of biological functions that include endothelial function, vasodilation, inflammation, and heart and muscle organogenesis(Farsetti, et al., 2009). NO is produced by a family of NO synthases (NOS) starting from L-arginine: neuronal NOS (n-NOS), endothelial NOS (e-NOS), and inducible NOS (i-NOS). Despite their names, the distribution of these enzymes is ubiquitous, but, while e-NOS and n-NOS are activated following an increase of intracellular calcium levels, i-NOS is calcium insensitive and is activated by inflammatory cytokines(Farsetti, et al., 2009). Importantly, it has been shown that VEGF-induced angiogenesis is mediated by NO and relies on hTERT activity. Estrogens, on the other hand, exert a beneficial role on the cardiovascular system which is, at least in part, mediated through the induction of e-NOS and hTERT(Farsetti, et al., 2009). Last, it has been recently shown that e-NOS and estrogen receptor (ERα) physically interact and cooperate in regulating hTERT and possibly other genes, thus delaying vascular senescence. Although NO production and endothelial nitric oxide synthase have been shown to be greater in longer living rodents, NO donors do not seem to influence animal maximum lifespan(Csiszar, et al., 2007). However, it has been shown both that NO can regulate telomerase activity(Farsetti, et al., 2009) and that it has a profound impact on mouse embryonic stem cell differentiation towards a cardiovascular fate(Spallotta, et al., 2010). Data on the effects of NO on hCSC are still missing.

Mammalian Sirtuins (Sirt) are yeast Sir2 orthologs possessing both NAD+ dependent- protein deacetylase and ADP-ribosyltransferase activity(Beltrami, et al., 2011a). Although Sir proteins are key regulators of *S. Cerevisiae, Drosophila,* and *C. Elegans* lifespan, the effect of Sirtuins on mammalian lifespan is less dramatic. Nonetheless, in mice lacking Sirt1, caloric restriction is unable to extend lifespan. Mammalians, in fact, possess at least 7 sirtuins, that act as metabolic sensors directly linking environmental signals to metabolic homeostasis and stress response. Sirt1, the most studied mammalian Sirtuin, controls gene expression, metabolism and aging, through a continuously growing list of substrates, that include: p53, members of the FoxO family, HES1 (hairy and enhancer of split 1), HEY2 (hairy/enhancer-of-split related with YRPW motif 2), PPARγ (peroxisome proliferator-activated receptor gamma), p300, PGC-1α (PPARγ coactivator), and NF- κB (nuclear factor kappaB)(Rahman and Islam, 2011). Although aging has been associated with stem cell senescence and dysfunction, the molecular mechanisms through which Sirt1 could protect primitive cells have not been completely delineated yet. However, the most prominent ones are: the positive regulation of telomeric length(Palacios, et al., 2010), the reduction of ROS production, the inhibition of p53(Rahman and Islam, 2011) and the induction of autophagy(Lee, et al., 2008). In this regard, Resveratrol is emerging as a potent drug able to delay age-related deteriorations and in mediating cardio-protection, conceivably by activating Sirt1(Petrovski, et al., 2011). In fact this polyphenolic compound has the ability to mimic the effects of caloric restriction by activating sirtuins and therefore acting modulating cell cycle, inhibiting apoptosis, increasing resistance to stress, and, finally, interfering with mTOR (Petrovski, et al., 2011). Accordingly, Resveratrol has shown beneficial effects against most degenerative and cardiovascular diseases from atherosclerosis, hypertension, ischemia/reperfusion, and heart failure to diabetes, obesity, and aging (Petrovski, et al., 2011). Importantly, pretreatment of either the infarcted heart or of cardiac stem cells with Resveratrol prior to cell injection results in an improvement of the regenerative capacities of the injected cells that eventually leads to improved heart function(Gorbunov, et al., 2011). However, in this specific case it was only evaluated the ability of Resveratrol to increase the engraftment of "normal" donor cells. In fact, up to now, the effects of Resveratrol on hCSC senescence remain to be elucidated. In addition, Resveratrol has been shown to be the most potent drug able to enhance iPS generation(Petrovski, et al., 2011).

p38MAPK is rapidly and transiently activated, by phosphorylation, following acute cellular stress. It is involved in senescence growth arrest by activating both p53 and pRb/p16INK4A pathways. Additionally, p38MAPK activity is required for the oncogene-induced premature senescence caused by oncogenic RAS, while its inhibition is able to delay replicative senescence, and to reverse the accelerated aging phenotype of fibroblasts obtained from Werner syndrome patients(Freund, et al., 2011). Further, p38MAPK is necessary and sufficient for the development of SASP in cells undergoing cellular senescence as a result of direct DNA damage or by oncogenic RAS(Freund, et al., 2011). Last, it was recently shown that p38MAPK inhibits Sirt1 by inducing its proteasomal degradation(Hong, et al., 2010). Although P38MAPK inhibitors have been successfully used to counteract in vitro the accelerated senescence phenotype seen in Werner syndrome progeria, it is still unclear whether this effect could be generalized to more physiological aging conditions. Importantly, it has been shown that p38MAPK inhibition can maintain hematopoietic stem cell quiescence, inhibiting the exhaustion of the hematopoietic stem cell pool(Ito, et al., 2006). In addition, p38MAPK inhibition can reduce cellular senescence in EPC exposed to doxorubicin (Spallarossa, et al., 2010). No data are available regarding hCSC. However, a role played by p38MAPK inhibition in inducing myocyte differentiation of embryonic stem cells has been reported(Gaur, et al., 2010).

Altogether, we can conclude that, although extremely interesting, the possibility to pharmacologically interfere with hCSC senescence has not yet been exploited.

5. Experimental data

In order to establish whether drugs known to interfere with the ageing processes could positively interfere with hCSC senescence and rescue their functional competence, hCSC obtained from failing hearts were cultured in the presence of increasing concentration of Rapamycin (1nM, 10nM, 100nM), Resveratrol (0.2μM, 0.5μM, 1μM) and DETA/NO (5μM, 10μM, 50μM). To reduce cell line variability, we selected hCSC obtained from ≈60 year old, male patients affected by end stage ischemic cardiomyopathy. After a 3-day treatment, cell lines (n=8) were analyzed both in terms of stem cell marker expression and cell proliferation, death and senescence. The ability of hCSC to differentiate and migrate was further assessed.

5.1 Methods

5.1.1 hCSC isolation and culture

Human atrial specimens, weighing 3-6 g, were collected over a period of five years from explanted hearts of patients in NHYA class 4 undergoing cardiac transplantation at the Cardiac Surgery Unit of the University Hospital of Udine, Italy. Informed consent was obtained in accordance with the Declaration of Helsinki and with approval by the Independent Ethics Committee of the University of Udine. Samples were employed for the isolation and expansion of c-kit-positive human cardiac stem cells (hCSC), as previously described (Bearzi, et al., 2009; Bearzi, et al., 2007; Beltrami, et al., 2007; Cesselli, et al., 2011). Specifically, two protocols were employed for the isolation of hCSC: enzymatic dissociation of the samples with collagenase and primary explant technique(Cesselli, et al., 2011). These two methodologies yielded comparable results up to 20-25 population doublings; efficiency and viability of hCSC were superimposable. Collagenase treatment was not found to affect these variables.

5.1.2 Pharmacological treatment of hCSC

After about 20 population doublings, growing cultures of hCSC were exposed to Rapamycin (1-100 nM, Sigma-Aldrich), 1-[N-(2-Aminoethyl)-N-(2-ammonioethyl)amino]diazen-1-ium-1,2-diolate or DETA/NO (5-50 μM, Sigma-Aldrich) and Resveratrol (0.2-1 μM, Sigma-Aldrich) for three days. At the end of the treatment, part of the cells was analyzed in terms of immunophenotype and to quantify cellular senescence, cell proliferation, and cell death. Part of the vehicle-treated and drug-treated cells was switched for n=2 days to a drug-free medium and subsequently assayed in terms of growth kinetic, differentiation and migration ability (see below).

5.1.3 Cell growth kinetic

Cells were seeded at a density of 2,000 cells/cm^2 in expansion medium. Cells were detached and counted at 1-2-5-9-12 and 14 days.

5.1.4 Cell differentiation assay

Muscle cell differentiation was achieved plating 0.5 to 1x10^4/cm^2 cells in expansion medium containing 5% FCS (Sigma-Aldrich, st. Louis, MO, USA), 10 ng/mL bFGF, 10 ng/mL VEGF, and 10 ng/mL IGF-1 (all from Peprotech EC, London, UK), but not EGF. Cells were allowed to become confluent and cultured for up to 4 weeks with medium exchanges every 4 days(Beltrami, et al., 2007; Cesselli, et al., 2011). Endothelial cell differentiation was obtained plating 0.5 to $1 \times 10^4/cm^2$ hCSC in *EGM®-2* Endothelial Cell Growth Medium-2 (Lonza, Switzerland) for 2 weeks.

5.1.5 Migration assay

In order to evaluate in vitro cell migration of drug treated or untreated hCSC, a scratch assay was performed(Liang, et al., 2007). In 33mm-plates at high confluence, scratches were created utilizing 200µl tips. Phase contrast images of the scratches were acquired at 3-hour intervals, until their complete closure, utilizing Leica DMI6000B. Images were then compared and quantified by ImageJ in order to calculate the rate of cell migration. The mean scratch width did not differ significantly in the different culture conditions (p>0.05).

5.1.6 Flow cytometry

Proliferating cells were detached with 0.25% trypsin-EDTA (Sigma-Aldrich) and, after a 20 minutes recovery phase, were incubated with either properly conjugated primary antibodies: CD13, CD29, CD49a, CD49b, CD49d, CD90, CD73, CD44, CD59, CD45, HLA-DR, CD117, CD271, CD34, (BD Biosciences), CD105, CD66e (Serotech), CD133 (Miltenyi Biotec), E-cadherin (Santa Cruz Biotechnology), ABCG-2 (Chemicon International), or with an unconjugated primary antibody: N-cadherin (Sigma-Aldrich). Unconjugated antibody was revealed using PE or FITC conjugated secondary antibodies (DakoCytomation). Properly conjugated isotype matched antibodies were used as a negative control.

Apoptosis and necrosis were evaluated utilizing the Annexin V-FITC Apoptosis Detection Kit (Bender MedSystem), following manufacturer's instructions.

The analysis was performed either by FACS-Calibur (BD Biosciences) or by CyAn (Dako Cytomation).

5.1.7 Immunofluorescence and fluorescence microscopy

Cells cultured either in expansion or in differentiation medium were fixed in 4% buffered paraformaldehyde for 20 minutes at room temperature (R.T.). For intracellular stainings, fixed cells were permeabilized for 8 minutes at R.T. with 0.1% Triton X-100 (Sigma-Aldrich) before exposing them to primary antibodies. Primary antibody incubation was performed over-night at 4°C using following dilutions: Oct-4 (Abcam, 1:150); Sox-2 (Chemicon, 1:150); Nanog (Abcam, 1:150), cKit (R&D; 1:100), p21 (Santa Cruz; 1:40), p16[INK4A] (CIN-TEK, pre-diluted), γH2A.X (Upstate, 1:500), Ki67 (Novocastra, 1:1000); α-Sarcomeric Actin (Sigma, 1:100) and CD31 (Dako, 1:50). To detect primary antibodies, A488 and A555 dyes labeled secondary antibodies, diluted 1:800, were employed (Molecular Probe, Invitrogen). Finally, 0.1 µg/ml DAPI (Sigma) was used to identify nuclei. Vectashield (Vector) was used as mounting medium. Confocal image acquisition was carried out by a Confocal Laser Microscope (Leica TCS-SP2, Leica Microsystems) utilizing either a 63x oil immersion objective (numerical aperture: 1.40) or a 40x oil immersion objective (numerical aperture: 1.25). Epifluorescence and phase contrast images were obtained utilizing a live cell imaging dedicated system consisting of a Leica DMI 6000B microscope connected to a Leica

DFC350FX camera (Leica Microsystems, Wetzlar, Germany). 10X (numerical aperture: 0.25), 40X oil immersion (numerical aperture: 1.25), and 63X oil immersion (numerical aperture: 1.40) objectives were employed for this purpose. Bright field images were captured utilizing a Leica DMD108 microscope (Leica Microsystems). 10X (numerical aperture: 0.40), 20X (numerical aperture: 0.70), and 40x (numerical aperture: 0.95) objectives were employed. Adobe Photoshop software was utilized to compose, overlay the images and adjust the contrast (Adobe, USA).

5.1.8 Statistics

Two-tailed unpaired- Student t- test and one-way Anova followed by Bonferroni post-test were utilized to compare means between two or more groups, respectively (Prism, version 4.0c). Results are expressed as mean±standard deviation. P values less than 0.05 were considered significant.

5.2 Results

We evaluated the effects of Rapamycin, Resveratrol and DETA/NO on hCSC stem cell marker expression, proliferation, senescence, death and function.

5.2.1 Effects of drugs on hCSC stem cell marker expression

As previously mentioned, hCSC obtained from failing hearts presented a mesenchymal immunophenotype and largely expressed the pluripotent state specific transcription factors Oct-4, Nanog and Sox-2. Drug treatment did not alter the mesenchymal immunophenotype and left unchanged the faction of cells expressing the pluripotent state specific transcription factors (data not shown).

5.2.2 Effects of drugs on hCSC proliferation, senescence and death

The effects exerted by Rapamyicin, Resveratrol and DETA/NO on hCSC proliferation, senescence and death resulted to be drug- and concentration- dependent; therefore, the effects exerted by each drug will be presented separately.

Rapamycin

As shown in Figure 1, Rapamycin mainly acts inhibiting mTOR-related pathway, thus inhibiting cell growth, autophagy and reducing oxidative stress. However, the anti-proliferative effect of the drugs is partially counteracted by a positive effect on Akt.

Accordingly, after a 3-day treatment hCSC, with respect to vehicle-treated cells, did not display changes in proliferation, as testified both by Ki67 expression and nuclear density (Figure 2).

Rapamycin was instead effective in reducing the fraction of senescent cells acting primarily on the fraction of cells expressing p16 that, at a 10nM concentration, resulted to be halved. No changes in the fraction of cells with DNA-damage foci were observed (Figure 2). DNA-damage foci positive cells were identified by the presence of the historic protein γH2AX in the absence of Ki67 expression (Lawless, et al., 2010).

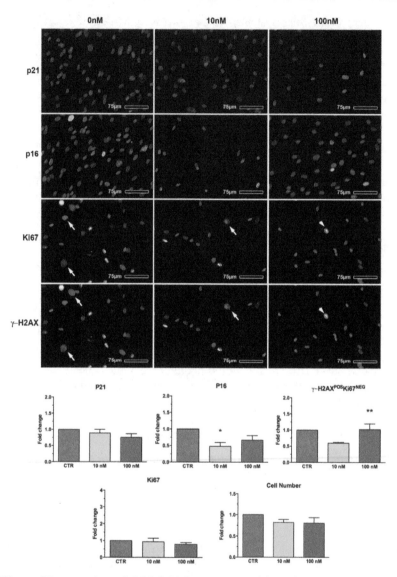

Fig. 2. **Effects of Rapamycin on hCSC.** hCSC were exposed for 3 days to 0nM (left panels), 1nM (central panels) and 10nM (right panels) Rapamyicin. Cells were then stained for p21 (red fluorescence), p16 (green fluorescence), Ki67 (yellow fluorescence) and γH2AX (magenta fluorescence). DNA-damage foci positive cells (arrows) were recognized as cells positive for γH2AX (magenta fluorescence) but negative for Ki67 (yellow fluorescence). Cells positive for both Ki67 and γH2AX (arrowheads) were excluded from the count. Histograms represent the fold changes in the fraction of cells expressing senescence (p21, p16, γH2AX+Ki67-) and proliferation markers (Ki67) and in hCSC number of treated cells with respect to vehicle-treated cells (CTR). *, **, p<0.05 with respect to CTR and 10nM treated cells, respectively.

Interestingly, Rapamyicin at both concentrations tested increased the fraction of cells undergoing cell death through apoptosis (Figure 3).

In conclusion, Rapamyicin seemed to act reducing the fraction of p16-positive senescent cells, without affecting cell proliferation.

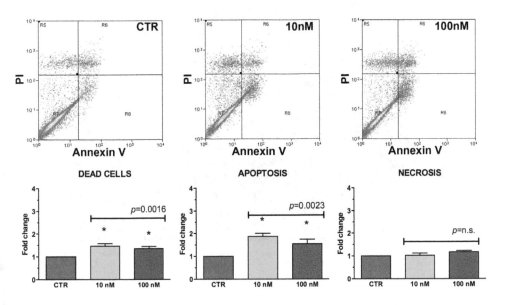

Fig. 3. **Effects of Rapamycin on hCSC death.** In the upper panel dot-plots graphically represent hCSC cultured in the presence of different concentrations of Rapamycin, stained for PI and AnnexinV and analyzed by FACS. Apoptotic cells were defined as AnnexinV+PI+/- cells, necrotic cells as AnnexinV-PI+cells. In the lower panels, histograms represent the quantitative analysis of the fold change in the fraction of dead cells. *, p<0.05 with respect to vehicle-treated cells (CTR).

Resveratrol

As displayed in figure 1, Resveratrol has the ability to mimic the effects of caloric restriction by activating sirtuins and therefore acting modulating cell cycle, inhibiting apoptosis, increasing resistance to stress, and, finally, interfering with mTOR (Petrovski, et al., 2011).

Accordingly, Resveratrol-treated cells presented a larger fraction of Ki67-positive cells and an increased nuclear density (Figure 4). Importantly, the fraction of senescent cells resulted to be significantly reduced at both drug concentration used. Differently from Rapamyicin, acting on p16-positive cells, Resveratrol was effective in reducing the fraction of cells presenting DNA-damage foci and expressing p21 (Figure 4).

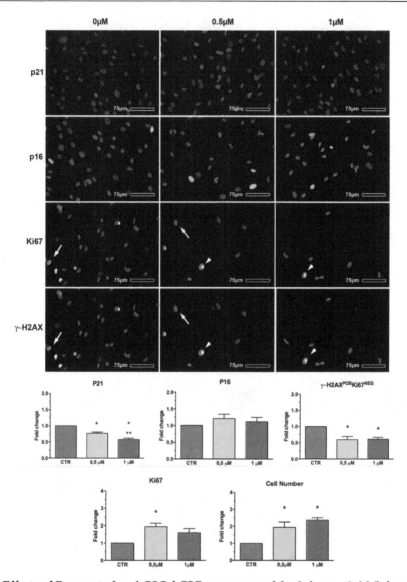

Fig. 4. **Effects of Resveratrol on hCSC.** hCSC were exposed for 3 days to 0μM (left panels), 0,5μM (central panels) and 1μM (right panels) Resveratrol. Cells were then stained for p21 (red fluorescence), p16 (green fluorescence), Ki67 (yellow fluorescence) and γH2AX (magenta fluorescence). DNA-damage foci positive cells (arrows) were recognized as cells positive for γH2AX (magenta fluorescence) but negative for Ki67 (yellow fluorescence). Cells positive for both Ki67 and γH2AX (arrowheads) were excluded from the count. Histograms represent the fold changes in the fraction of cells expressing senescence (p21, p16, γH2AX+Ki67-) and proliferation markers (Ki67) and in hCSC number of treated cells with respect to vehicle-treated cells (CTR). *, p<0.05 with respect to CTR.

Moreover, Resveratrol significantly reduced the fraction of cells dying by necrosis (Figure 5).

Altogether these results indicate that Resveratrol presented beneficial effects on hCSC stimulating cell proliferation, reducing DNA-damage induced senescence and cell death.

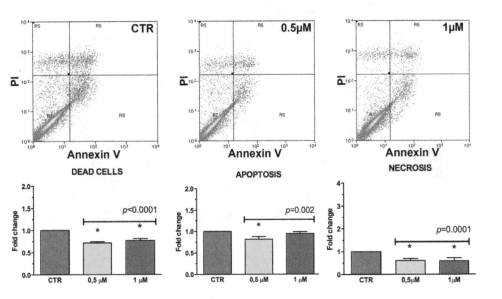

Fig. 5. **Effects of Resveratrol on hCSC death.** In the upper panel dot-plots graphically represent hCSC cultured in the presence of different concentrations of drug, stained for PI and AnnexinV and analyzed by FACS. Apoptotic cells were defined as AnnexinV+PI+/- cells, necrotic cells as AnnexinV-PI+cells. In the lower panels, histograms represent the quantitative analysis of the fold change in the fraction of dead cells. *, p<0.05 with respect to vehicle-treated cells (CTR).

DETA/NO

As displayed in Figure 1, NO donors regulate telomerase activity. Moreover, it has been shown that it has a profound impact on stem cell differentiation towards a cardiovascular fate (Farsetti, et al., 2009).

After a 3-day treatment, DETA/NO-treated cells did not differ, with respect to vehicle-treated cells, in terms of nuclear density, while Ki67 resulted to be increased only in 10μM-treated cells (Figure 6). Importantly, all DETA/NO used concentrations significantly decreased the fraction of γH2AX-positive cells, with a trend to reduce the fraction of p21 positive cells only at 10 μM (Figure 6).

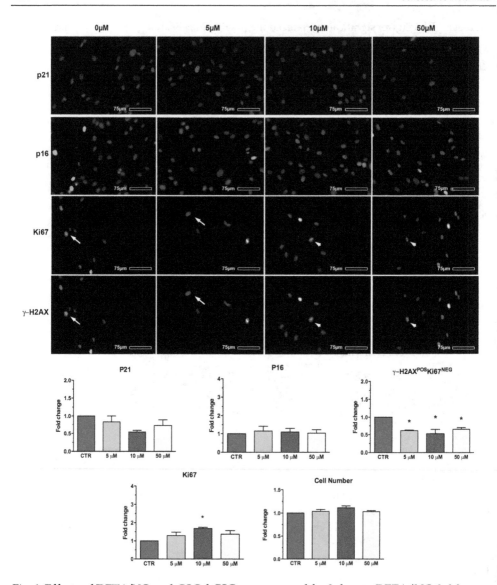

Fig. 6. **Effects of DETA/NO on hCSC.** hCSC were exposed for 3 days to DETA/NO 0μM, 5μM, 10μM and 50μM. Cells were then stained for p21 (red fluorescence), p16 (green fluorescence), Ki67 (yellow fluorescence) and γH2AX (magenta fluorescence). DNA-damage foci positive cells (arrows) were recognized as cells positive for γH2AX (magenta fluorescence) but negative for Ki67 (yellow fluorescence). Cells positive for both Ki67 and γH2AX (arrowheads) were excluded from the count. Histograms represent the fold changes in the fraction of cells expressing senescence (p21, p16, γH2AX+Ki67-) and proliferation markers (Ki67) and in hCSC number of treated cells with respect to vehicle-treated cells (CTR). *, p<0.05 with respect to CTR.

No significantly changes in the fraction of dying cells was assessed (Figure 7).

In conclusion, DETA/NO seemed to act specifically by reducing the fraction of cells with DNA-damage foci eliciting a DNA-damage response. This is in line with its ability to activate telomerase activity.

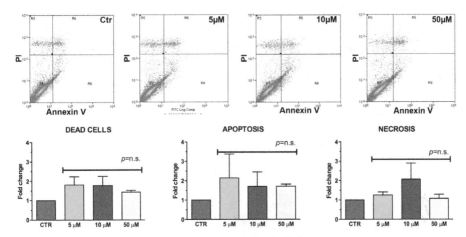

Fig. 7. **Effects of DETA/NO on hCSC death.** In the upper panel dot-plots graphically represent hCSC cultured in the presence of different concentrations of drug, stained for PI and AnnexinV and analyzed by FACS. Apoptotic cells were defined as AnnexinV+PI+/- cells, necrotic cells as AnnexinV-PI+cells. In the lower panels, histograms represent the quantitative analysis of the fold change in the fraction of dead cells.

Altogether, the analysis of the effects of a three-day drug treatment of senescent hCSC, showed that, although all the utilized drugs exerted a beneficial effect in reducing the fraction of senescent cells, they differed not only in the pathway of cell senescence specifically targeted (p16 vs γH2AX/p21), but also in their ability to interfere with other key-processes such as cell proliferation and cell death. Table 1 summarizes the effects of the drugs on the principal cell processes and indicates the identified optimal drug concentration.

	Variable	Rapamycin	Resveratrol	DETA/NO
Senescence	P16	↓		
	P21		↓	↓
	γH2AX		↓	↓
Proliferation	Ki67		↑	↑ (10 μM)
	Nuclear density		↑	
Cell Death	Apoptosis	↑		
	Necrosis		↓	
Optimal drug concentration		10 nM	0.5 μM	10 μM

Table 1. Summary of drug effects on hCSC senescence, proliferation and death.

5.2.3 Effects of drugs on hCSC function

In order to verify whether the beneficial effects exerted by drugs on hCSC senescence were paralleled by an improvement in hCSC function, hCSC treated for three days with the optimal drug concentration were assayed, after two days of recovery, for: growth kinetic, differentiation capacity and migration abilities.

Growth kinetic

Despite the fact that during the three-day treatment only Resveratrol-treated cells resulted to increase their number (Figure 3, 5 and 7), all the drugs resulted to be effective in significantly reduce the population doubling time (Figure 8, p=0.002), suggesting that the reduction in the fraction of senescent cells was afterward associated with an increased proliferation rate.

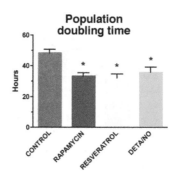

Fig. 8. Effects of drugs on hCSC population doubling time. *, p<0.05 with respect to vehicle-treated cells (CONTROL).

Differentiation ability

We investigated the ability of drug-treated hCSC to differentiate along the endothelial and myogenic fate. CD31 was utilized as endothelial marker, while alpha-sarcomeric actin as myogenic markers.

Interestingly, we have seen that cells treated for three days with drugs and then exposed to endothelial-differentiation inducing conditions displayed different behaviour. Specifically, while Rapamycin-treated cells significantly improved their ability to differentiate into endothelial cells expressing CD31, Resveratrol and DETA/NO did not (Figure 9).

Regarding, myocyte differentiation capacity, we noticed that cell cultures differed not only in the percentage of alpha-sarcomeric actin (ASA) positive cells, but also in the level of organization of the filaments. Therefore, we decided to use a score able to taking into account these two factors and defined as the product of the fraction of ASA-positive cells and an index expressing ASA organization, which ranged from 1 (not-organized) to 3 (well defined filaments)(Cesselli, et al., 2011). Applying these criteria, we established that Rapamycin did not interfere with the differentiation ability of hCSC, while the other two, especially DETA/NO, improved the myogenic potential of hCSC (Figure 10).

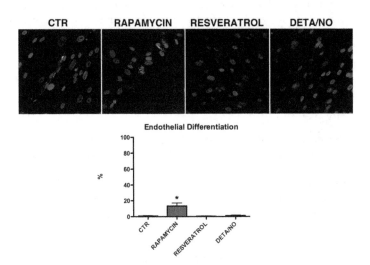

Fig. 9. **Effects of drugs on hCSC endothelial differentiation ability.** Green fluorescence represent CD31 expression on hCSC exposed to endothelial differentiation medium. Nuclei are depicted by the blue fluorescence of DAPI staining. Histograms represent the quantitative analysis of the fraction of CD31-positive cells in the cells treated with different drugs. *, $p < 0.05$ vs vehicle-treated cells (CTR).

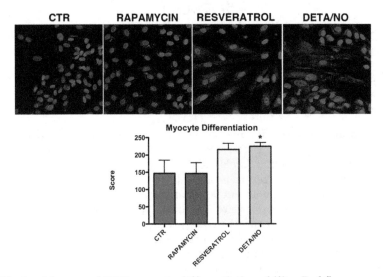

Fig. 10. **Effects of drugs on hCSC myocyte differentiation ability.** Red fluorescence represent alpha-sarcomeric actin expression on hCSC exposed to myocyte differentiation medium. Nuclei are depicted by the blue fluorescence of DAPI staining. Histograms represent the quantitative analysis of the level of myocyte differentiation of the cells treated with different drugs. See text for score meaning. *, $p < 0.05$ vs vehicle-treated cells (CTR).

It remains to be demonstrated whether the improved differentiation ability of drug-treated cells is a consequence of the benefical effects of the drug on senescence or if it is due to a direct effect of the drug on differentiation pathways. In fact, the ability of DETA/NO to favor stem cell differentiation towards a cardiovascular fate has already been demonstrated in mouse embryonic stem cells (Farsetti, et al., 2009). Moreover, oxytocin, a hormone present also in the heart, induces embryonic and cardiac somatic stem cells to differentiate into cardiomyocytes, possibly through nitric oxide(Danalache, et al., 2007).

Migration capacity

In order to establish the migration speed of hCSC, a scratch assay was performed. With respect to vehicle treated cells, only DETA/NO-treated cells showed a trend to increase their migration ability (Figure 11), while Rapamycin and Resveratrol treated cells did not. Even in this case, it is difficult to establish whether DETA/NO would act directly on the migration ability of the cells, since it has already demonstrated a role of NO on SDF-1/CXCR4-mediated bone-marrow cell migration(Cui, et al., 2007).

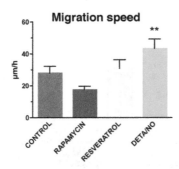

Fig. 11. **Effects of drugs on hCSC migration ability.** Histograms represent the quantitative analysis of migration speed of the cells treated with different drugs. **, p<0.05 vs Rapamycin-treated cells.

As a whole, drug treatment did not modify hCSC phenotype and stem cell marker expression. However, different effects were observed with respect to cell death, where Rapamycin increased of about 1.5 fold hCSC apoptosis (p=0.002), whereas Resveratrol showed a protective effect on cell necrosis, reducing it by 50% (p=0.0001). Although all drugs were associated with a significant decrease in the fraction of senescent cells, different pathways of cellular senescence were involved. Specifically, while Resveratrol and DETA/NO treatment were associated with a significant reduction by half of cells with DDR and p21 expression, Rapamycin treatment was mainly associated with a ≈60% reduction in p16 expression (p<0.05). Importantly, although all drug-treated cells showed, with respect to vehicle, an increase in cell proliferation, the effects on hCSC differentiation and migration ability were different. Specifically, Rapamycin treated cells displayed an improved endothelial differentiation capacity, while Resveratrol seemed to positively affect only the myogenic potential of hCSC. Finally, DETA/NO improved both the myocyte differentiation capacity and the migration ability of hCSC, without effects on endothelial differentiation capacity.

6. Conclusions

Severe heart failure is characterized by the loss of the growth reserve of the adult heart, dictated by a progressive decrease in the number of functionally-competent hCSC(Cesselli, et al., 2011). Despite these limitations, autologous CSC therapy is feasible and can be considered a therapeutic option for the large population of patients affected by severe heart failure(Beltrami, et al., 2011b; Segers and Lee, 2008). In fact, even in patients with advanced cardiomyopathies hCSC can be isolated from small myocardial biopsies and expanded in vitro(Cesselli, et al., 2011; D'Amario, et al., 2011b; Smith, et al., 2007). For this reason, it would be extremely intriguing any attempt aimed at "improving" the quality of the expanded cells, selecting the fraction of cells with the highest regenerative potential.

In this regard, Anversa's group showed that different membrane receptors influence the regenerative ability of hCSC and that IGF-1 receptor-positive hCSC are endowed with a high regenerative ability, representing a potent cell population for cardiac repair(D'Amario, et al., 2011a). However, this approach would require the sorting of cells expressing specific surface antigens, thus adding a further grade of complexity to the procedure aimed at producing clinical grade hCSC.

The strategy we wanted to undertake in this project was slightly different, since we decided to treat hCSC with drugs in culture. The results we obtained indicate that, although hCSC isolated from failing hearts are senescent and functionally impaired, it is possible to interfere pharmacologically, at least in vitro, with the senescence processes, rescuing the properties of the primitive cells. Specifically, we have shown that a three-day treatment with Rapamycin, Resveratrol or DETA/NO was able to reduce the fraction of senescent cells, improving their proliferative capacity. Importantly, the tested drugs seemed to exert their effects on different subpopulations of senescent cells; in fact, while Rapamycin mainly reduced the p16-positive fraction, DETA/NO and Resveratrol principally acted on the pool of cells characterized by DNA-damage foci and expressing p21. Similarly, different drugs showed different effects on hCSC function. In fact, while Rapamycin increased endothelial differentiation ability, DETA/NO improved hCSC myogenic and migration capacity.

These results represent the first demonstration that hCSC senescence can be attenuated in vitro, and that this is associated with an improved proliferative capacity.

Future research will be aimed: 1) at understanding more in depth the mechanism through which drugs exert their effects on cellular senescence, e.g. removal of senescent cells, modulation of SASP-mediated pathways; 2) establishing whether drug-treated cells possess an increased in vivo regenerative potential; 3) establishing criteria to define which is the best drug to use.

7. References

Acosta, J. C., O'Loghlen, A., Banito, A., Guijarro, M. V., Augert, A., Raguz, S., Fumagalli, M., Da Costa, M., Brown, C., Popov, N., Takatsu, Y., Melamed, J., d'Adda di Fagagna, F., Bernard, D., Hernando, E., & Gil, J. (2008). Chemokine signaling via the CXCR2 receptor reinforces senescence. Cell, 133(6), 1006-1018.

Adams, P. D. (2009). Healing and hurting: molecular mechanisms, functions, and pathologies of cellular senescence. Mol Cell, 36(1), 2-14.

Bearzi, C., Rota, M., Hosoda, T., Tillmanns, J., Nascimbene, A., De Angelis, A., Yasuzawa-Amano, S., Trofimova, I., Siggins, R. W., Lecapitaine, N., Cascapera, S., Beltrami, A. P., D'Alessandro, D. A., Zias, E., Quaini, F., Urbanek, K., Michler, R. E., Bolli, R., Kajstura, J., Leri, A., & Anversa, P. (2007). Human cardiac stem cells. Proc Natl Acad Sci U S A, 104(35), 14068-14073.

Bearzi, C., Leri, A., Lo Monaco, F., Rota, M., Gonzalez, A., Hosoda, T., Pepe, M., Qanud, K., Ojaimi, C., Bardelli, S., D'Amario, D., D'Alessandro, D. A., Michler, R. E., Dimmeler, S., Zeiher, A. M., Urbanek, K., Hintze, T. H., Kajstura, J., & Anversa, P. (2009). Identification of a coronary vascular progenitor cell in the human heart. Proc Natl Acad Sci U S A, 106(37), 15885-15890.

Beltrami, A. P., Barlucchi, L., Torella, D., Baker, M., Limana, F., Chimenti, S., Kasahara, H., Rota, M., Musso, E., Urbanek, K., Leri, A., Kajstura, J., Nadal-Ginard, B., & Anversa, P. (2003). Adult cardiac stem cells are multipotent and support myocardial regeneration. Cell, 114(6), 763-776.

Beltrami, A. P., Cesselli, D., Bergamin, N., Marcon, P., Rigo, S., Puppato, E., D'Aurizio, F., Verardo, R., Piazza, S., Pignatelli, A., Poz, A., Baccarani, U., Damiani, D., Fanin, R., Mariuzzi, L., Finato, N., Masolini, P., Burelli, S., Belluzzi, O., Schneider, C., & Beltrami, C. A. (2007). Multipotent cells can be generated in vitro from several adult human organs (heart, liver, and bone marrow). Blood, 110(9), 3438-3446.

Beltrami, A. P., Cesselli, D., & Beltrami, C. A. (2011a). At the stem of youth and health. Pharmacol Ther, 129(1), 3-20.

Beltrami, A. P., Cesselli, D., Beltrami, C. A., Cohen, I. S., & Gaudette, G. R. (2011b). Multiple Sources for Cardiac Stem Cells and Their Cardiogenic Potential, In: Regenerating the Heart, Cohen, I. S., & Gaudette, G. R., pp. 149-171, Humana Press.

Bhaskar, P. T., & Hay, N. (2007). The two TORCs and Akt. Dev Cell, 12(4), 487-502.

Blagosklonny, M. V. (2010). Rapamycin and quasi-programmed aging: Four years later. Cell Cycle, 9(10).

Blasco, M. A. (2007). Telomere length, stem cells and aging. Nat Chem Biol, 3(10), 640-649.

Blasi, F., & Carmeliet, P. (2002). uPAR: a versatile signalling orchestrator. Nat Rev Mol Cell Biol, 3(12), 932-943.

Bu, L., Jiang, X., Martin-Puig, S., Caron, L., Zhu, S., Shao, Y., Roberts, D. J., Huang, P. L., Domian, I. J., & Chien, K. R. (2009). Human ISL1 heart progenitors generate diverse multipotent cardiovascular cell lineages. Nature, 460(7251), 113-117.

Campisi, J. (2005). Senescent cells, tumor suppression, and organismal aging: good citizens, bad neighbors. Cell, 120(4), 513-522.

Campisi, J., & d'Adda di Fagagna, F. (2007). Cellular senescence: when bad things happen to good cells. Nat Rev Mol Cell Biol, 8(9), 729-740.

Campisi, J. (2011). Cellular senescence: putting the paradoxes in perspective. Curr Opin Genet Dev, 21(1), 107-112.

Castaldo, C., Di Meglio, F., Nurzynska, D., Romano, G., Maiello, C., Bancone, C., Muller, P., Bohm, M., Cotrufo, M., & Montagnani, S. (2008). CD117-positive cells in adult human heart are localized in the subepicardium, and their activation is associated with laminin-1 and alpha6 integrin expression. Stem Cells, 26(7), 1723-1731.

Cesselli, D., Beltrami, A. P., D'Aurizio, F., Marcon, P., Bergamin, N., Toffoletto, B., Pandolfi, M., Puppato, E., Marino, L., Signore, S., Livi, U., Verardo, R., Piazza, S., Marchionni, L., Fiorini, C., Schneider, C., Hosoda, T., Rota, M., Kajstura, J., Anversa, P., Beltrami, C. A., & Leri, A. (2011). Effects of age and heart failure on human cardiac stem cell function. Am J Pathol, 179(1), 349-366.

Chen, T., Shen, L., Yu, J., Wan, H., Guo, A., Chen, J., Long, Y., Zhao, J., & Pei, G. (2011). Rapamycin and other longevity-promoting compounds enhance the generation of mouse induced pluripotent stem cells. Aging Cell, 10(5), 908-911.

Chimenti, C., Kajstura, J., Torella, D., Urbanek, K., Heleniak, H., Colussi, C., Di Meglio, F., Nadal-Ginard, B., Frustaci, A., Leri, A., Maseri, A., & Anversa, P. (2003). Senescence and death of primitive cells and myocytes lead to premature cardiac aging and heart failure. Circ Res, 93(7), 604-613.

Collado, M., Gil, J., Efeyan, A., Guerra, C., Schuhmacher, A. J., Barradas, M., Benguria, A., Zaballos, A., Flores, J. M., Barbacid, M., Beach, D., & Serrano, M. (2005). Tumour biology: senescence in premalignant tumours. Nature, 436(7051), 642.

Coppe, J. P., Patil, C. K., Rodier, F., Sun, Y., Munoz, D. P., Goldstein, J., Nelson, P. S., Desprez, P. Y., & Campisi, J. (2008). Senescence-associated secretory phenotypes reveal cell-nonautonomous functions of oncogenic RAS and the p53 tumor suppressor. PLoS Biol, 6(12), 2853-2868.

Csiszar, A., Labinskyy, N., Zhao, X., Hu, F., Serpillon, S., Huang, Z., Ballabh, P., Levy, R. J., Hintze, T. H., Wolin, M. S., Austad, S. N., Podlutsky, A., & Ungvari, Z. (2007). Vascular superoxide and hydrogen peroxide production and oxidative stress resistance in two closely related rodent species with disparate longevity. Aging Cell, 6(6), 783-797.

Cui, X., Chen, J., Zacharek, A., Li, Y., Roberts, C., Kapke, A., Savant-Bhonsale, S., & Chopp, M. (2007). Nitric oxide donor upregulation of stromal cell-derived factor-1/chemokine (CXC motif) receptor 4 enhances bone marrow stromal cell migration into ischemic brain after stroke. Stem Cells, 25(11), 2777-2785.

D'Amario, D., Cabral-Da-Silva, M. C., Zheng, H., Fiorini, C., Goichberg, P., Steadman, E., Ferreira-Martins, J., Sanada, F., Piccoli, M., Cappetta, D., D'Alessandro, D. A., Michler, R. E., Hosoda, T., Anastasia, L., Rota, M., Leri, A., Anversa, P., & Kajstura, J. (2011a). Insulin-like growth factor-1 receptor identifies a pool of human cardiac stem cells with superior therapeutic potential for myocardial regeneration. Circ Res, 108(12), 1467-1481.

D'Amario, D., Fiorini, C., Campbell, P. M., Goichberg, P., Sanada, F., Zheng, H., Hosoda, T., Rota, M., Connell, J. M., Gallegos, R. P., Welt, F. G., Givertz, M. M., Mitchell, R. N., Leri, A., Kajstura, J., Pfeffer, M. A., & Anversa, P. (2011b). Functionally competent cardiac stem cells can be isolated from endomyocardial biopsies of patients with advanced cardiomyopathies. Circ Res, 108(7), 857-861.

Danalache, B. A., Paquin, J., Donghao, W., Grygorczyk, R., Moore, J. C., Mummery, C. L., Gutkowska, J., & Jankowski, M. (2007). Nitric oxide signaling in oxytocin-mediated cardiomyogenesis. Stem Cells, 25(3), 679-688.

De Angelis, A., Piegari, E., Cappetta, D., Marino, L., Filippelli, A., Berrino, L., Ferreira-Martins, J., Zheng, H., Hosoda, T., Rota, M., Urbanek, K., Kajstura, J., Leri, A., Rossi, F., & Anversa, P. (2010). Anthracycline cardiomyopathy is mediated by depletion of the cardiac stem cell pool and is rescued by restoration of progenitor cell function. Circulation, 121(2), 276-292.

Deng, Y., Chan, S. S., & Chang, S. (2008). Telomere dysfunction and tumour suppression: the senescence connection. Nat Rev Cancer, 8(6), 450-458.

Di Micco, R., Cicalese, A., Fumagalli, M., Dobreva, M., Verrecchia, A., Pelicci, P. G., & di Fagagna, F. (2008). DNA damage response activation in mouse embryonic fibroblasts undergoing replicative senescence and following spontaneous immortalization. Cell Cycle, 7(22), 3601-3606.

Dimmeler, S., & Leri, A. (2008). Aging and disease as modifiers of efficacy of cell therapy. Circ Res, 102(11), 1319-1330.

Dimri, G. P., Lee, X., Basile, G., Acosta, M., Scott, G., Roskelley, C., Medrano, E. E., Linskens, M., Rubelj, I., Pereira-Smith, O., & et al. (1995). A biomarker that identifies senescent human cells in culture and in aging skin in vivo. Proc Natl Acad Sci U S A, 92(20), 9363-9367.

Farsetti, A., Grasselli, A., Bacchetti, S., Gaetano, C., & Capogrossi, M. C. (2009). The telomerase tale in vascular aging: regulation by estrogens and nitric oxide signaling. J Appl Physiol, 106(1), 333-337.

Fiordaliso, F., Leri, A., Cesselli, D., Limana, F., Safai, B., Nadal-Ginard, B., Anversa, P., & Kajstura, J. (2001). Hyperglycemia activates p53 and p53-regulated genes leading to myocyte cell death. Diabetes, 50(10), 2363-2375.

Freund, A., Patil, C. K., & Campisi, J. (2011). p38MAPK is a novel DNA damage response-independent regulator of the senescence-associated secretory phenotype. EMBO J, 30(8), 1536-1548.

Fukuda, D., & Sata, M. (2008). Role of bone marrow renin-angiotensin system in the pathogenesis of atherosclerosis. Pharmacol Ther, 118(2), 268-276.

Gaur, M., Ritner, C., Sievers, R., Pedersen, A., Prasad, M., Bernstein, H. S., & Yeghiazarians, Y. (2010). Timed inhibition of p38MAPK directs accelerated differentiation of human embryonic stem cells into cardiomyocytes. Cytotherapy, 12(6), 807-817.

Gonzalez, A., Rota, M., Nurzynska, D., Misao, Y., Tillmanns, J., Ojaimi, C., Padin-Iruegas, M. E., Muller, P., Esposito, G., Bearzi, C., Vitale, S., Dawn, B., Sanganalmath, S. K., Baker, M., Hintze, T. H., Bolli, R., Urbanek, K., Hosoda, T., Anversa, P., Kajstura, J., & Leri, A. (2008). Activation of cardiac progenitor cells reverses the failing heart senescent phenotype and prolongs lifespan. Circ Res, 102(5), 597-606.

Gorbunov, N., Petrovski, G., Gurusamy, N., Ray, D., Kim, D. H., & Das, D. K. (2011). Regeneration of Infarcted Myocardium with Resveratrol-Modified Cardiac Stem Cells. J Cell Mol Med.

Hayflick, L., & Moorhead, P. S. (1961). The serial cultivation of human diploid cell strains. Exp Cell Res, 25, 585-621.

Hierlihy, A. M., Seale, P., Lobe, C. G., Rudnicki, M. A., & Megeney, L. A. (2002). The post-natal heart contains a myocardial stem cell population. FEBS Lett, 530(1-3), 239-243.

Hong, E. H., Lee, S. J., Kim, J. S., Lee, K. H., Um, H. D., Kim, J. H., Kim, S. J., Kim, J. I., & Hwang, S. G. (2010). Ionizing radiation induces cellular senescence of articular chondrocytes via negative regulation of SIRT1 by p38 kinase. J Biol Chem, 285(2), 1283-1295.

Hosoda, T., D'Amario, D., Cabral-Da-Silva, M. C., Zheng, H., Padin-Iruegas, M. E., Ogorek, B., Ferreira-Martins, J., Yasuzawa-Amano, S., Amano, K., Ide-Iwata, N., Cheng, W., Rota, M., Urbanek, K., Kajstura, J., Anversa, P., & Leri, A. (2009). Clonality of mouse and human cardiomyogenesis in vivo. Proc Natl Acad Sci U S A, 106(40), 17169-17174.

Hsieh, P. C., Segers, V. F., Davis, M. E., MacGillivray, C., Gannon, J., Molkentin, J. D., Robbins, J., & Lee, R. T. (2007). Evidence from a genetic fate-mapping study that stem cells refresh adult mammalian cardiomyocytes after injury. Nat Med, 13(8), 970-974.

Imanishi, T., Hano, T., & Nishio, I. (2005). Angiotensin II accelerates endothelial progenitor cell senescence through induction of oxidative stress. J Hypertens, 23(1), 97-104.

Ito, K., Hirao, A., Arai, F., Takubo, K., Matsuoka, S., Miyamoto, K., Ohmura, M., Naka, K., Hosokawa, K., Ikeda, Y., & Suda, T. (2006). Reactive oxygen species act through p38 MAPK to limit the lifespan of hematopoietic stem cells. Nat Med, 12(4), 446-451.

Itzhaki-Alfia, A., Leor, J., Raanani, E., Sternik, L., Spiegelstein, D., Netser, S., Holbova, R., Pevsner-Fischer, M., Lavee, J., & Barbash, I. M. (2009). Patient characteristics and cell source determine the number of isolated human cardiac progenitor cells. Circulation, 120(25), 2559-2566.

Jeyapalan, J. C., & Sedivy, J. M. (2008). Cellular senescence and organismal aging. Mech Ageing Dev, 129(7-8), 467-474.

Khan, M., Mohsin, S., Khan, S. N., & Riazuddin, S. (2011). Repair of senescent myocardium by mesenchymal stem cells is dependent on the age of donor mice. J Cell Mol Med, 15(7), 1515-1527.

Kortlever, R. M., Higgins, P. J., & Bernards, R. (2006). Plasminogen activator inhibitor-1 is a critical downstream target of p53 in the induction of replicative senescence. Nat Cell Biol, 8(8), 877-884.

Kosar, M., Bartkova, J., Hubackova, S., Hodny, Z., Lukas, J., & Bartek, J. (2011). Senescence-associated heterochromatin foci are dispensable for cellular senescence, occur in a cell type- and insult-dependent manner and follow expression of p16(ink4a). Cell Cycle, 10(3), 457-468.

Kuilman, T., Michaloglou, C., Vredeveld, L. C., Douma, S., van Doorn, R., Desmet, C. J., Aarden, L. A., Mooi, W. J., & Peeper, D. S. (2008). Oncogene-induced senescence relayed by an interleukin-dependent inflammatory network. Cell, 133(6), 1019-1031.

Kunieda, T., Minamino, T., Nishi, J., Tateno, K., Oyama, T., Katsuno, T., Miyauchi, H., Orimo, M., Okada, S., Takamura, M., Nagai, T., Kaneko, S., & Komuro, I. (2006). Angiotensin II induces premature senescence of vascular smooth muscle cells and accelerates the development of atherosclerosis via a p21-dependent pathway. Circulation, 114(9), 953-960.

Laflamme, M. A., & Murry, C. E. (2011). Heart regeneration. Nature, 473(7347), 326-335.

Lawless, C., Wang, C., Jurk, D., Merz, A., Zglinicki, T., & Passos, J. F. (2010). Quantitative assessment of markers for cell senescence. Exp Gerontol, 45(10), 772-778.

Lee, I. H., Cao, L., Mostoslavsky, R., Lombard, D. B., Liu, J., Bruns, N. E., Tsokos, M., Alt, F. W., & Finkel, T. (2008). A role for the NAD-dependent deacetylase Sirt1 in the regulation of autophagy. Proc Natl Acad Sci U S A, 105(9), 3374-3379.

Li, L., & Clevers, H. (2010). Coexistence of quiescent and active adult stem cells in mammals. Science, 327(5965), 542-545.

Liang, C. C., Park, A. Y., & Guan, J. L. (2007). In vitro scratch assay: a convenient and inexpensive method for analysis of cell migration in vitro. Nat Protoc, 2(2), 329-333.

Martin, C. M., Meeson, A. P., Robertson, S. M., Hawke, T. J., Richardson, J. A., Bates, S., Goetsch, S. C., Gallardo, T. D., & Garry, D. J. (2004). Persistent expression of the ATP-binding cassette transporter, Abcg2, identifies cardiac SP cells in the developing and adult heart. Dev Biol, 265(1), 262-275.

McMurray, J. J., & Pfeffer, M. A. (2005). Heart failure. Lancet, 365(9474), 1877-1889.

Meissner, K., Heydrich, B., Jedlitschky, G., Meyer Zu Schwabedissen, H., Mosyagin, I., Dazert, P., Eckel, L., Vogelgesang, S., Warzok, R. W., Bohm, M., Lehmann, C., Wendt, M., Cascorbi, I., & Kroemer, H. K. (2006). The ATP-binding cassette transporter ABCG2 (BCRP), a marker for side population stem cells, is expressed in human heart. J Histochem Cytochem, 54(2), 215-221.

Messina, E., De Angelis, L., Frati, G., Morrone, S., Chimenti, S., Fiordaliso, F., Salio, M., Battaglia, M., Latronico, M. V., Coletta, M., Vivarelli, E., Frati, L., Cossu, G., & Giacomello, A. (2004). Isolation and expansion of adult cardiac stem cells from human and murine heart. Circ Res, 95(9), 911-921.

Meyer, M. R., Haas, E., Prossnitz, E. R., & Barton, M. (2009). Non-genomic regulation of vascular cell function and growth by estrogen. Mol Cell Endocrinol, 308(1-2), 9-16.

Minamino, T., & Komuro, I. (2008). Vascular aging: insights from studies on cellular senescence, stem cell aging, and progeroid syndromes. Nat Clin Pract Cardiovasc Med, 5(10), 637-648.

Narita, M., Nunez, S., Heard, E., Lin, A. W., Hearn, S. A., Spector, D. L., Hannon, G. J., & Lowe, S. W. (2003). Rb-mediated heterochromatin formation and silencing of E2F target genes during cellular senescence. Cell, 113(6), 703-716.

Orjalo, A. V., Bhaumik, D., Gengler, B. K., Scott, G. K., & Campisi, J. (2009). Cell surface-bound IL-1alpha is an upstream regulator of the senescence-associated IL-6/IL-8 cytokine network. Proc Natl Acad Sci U S A, 106(40), 17031-17036.

Palacios, J. A., Herranz, D., De Bonis, M. L., Velasco, S., Serrano, M., & Blasco, M. A. (2010). SIRT1 contributes to telomere maintenance and augments global homologous recombination. J Cell Biol, 191(7), 1299-1313.

Patschan, S., & Goligorsky, M. S. (2008). Autophagy: The missing link between non-enzymatically glycated proteins inducing apoptosis and premature senescence of endothelial cells? Autophagy, 4(4), 521-523.

Petrovski, G., Gurusamy, N., & Das, D. K. (2011). Resveratrol in cardiovascular health and disease. Ann N Y Acad Sci, 1215, 22-33.

Rahman, S., & Islam, R. (2011). Mammalian Sirt1: insights on its biological functions. Cell Commun Signal, 9, 11.

Ronnebaum, S. M., & Patterson, C. (2010). The FoxO family in cardiac function and dysfunction. Annu Rev Physiol, 72, 81-94.

Rota, M., LeCapitaine, N., Hosoda, T., Boni, A., De Angelis, A., Padin-Iruegas, M. E., Esposito, G., Vitale, S., Urbanek, K., Casarsa, C., Giorgio, M., Luscher, T. F., Pelicci, P. G., Anversa, P., Leri, A., & Kajstura, J. (2006). Diabetes promotes cardiac stem cell aging and heart failure, which are prevented by deletion of the p66shc gene. Circ Res, 99(1), 42-52.

Salih, D. A., & Brunet, A. (2008). FoxO transcription factors in the maintenance of cellular homeostasis during aging. Curr Opin Cell Biol, 20(2), 126-136.

Sharpless, N. E., & DePinho, R. A. (2007). How stem cells age and why this makes us grow old. Nat Rev Mol Cell Biol, 8(9), 703-713.

Shawi, M., & Autexier, C. (2008). Telomerase, senescence and ageing. Mech Ageing Dev, 129(1-2), 3-10.

Smith, R. R., Barile, L., Cho, H. C., Leppo, M. K., Hare, J. M., Messina, E., Giacomello, A., Abraham, M. R., & Marban, E. (2007). Regenerative potential of cardiosphere-derived cells expanded from percutaneous endomyocardial biopsy specimens. Circulation, 115(7), 896-908.

Spallarossa, P., Altieri, P., Barisione, C., Passalacqua, M., Aloi, C., Fugazza, G., Frassoni, F., Podesta, M., Canepa, M., Ghigliotti, G., & Brunelli, C. (2010). p38 MAPK and JNK antagonistically control senescence and cytoplasmic p16INK4A expression in doxorubicin-treated endothelial progenitor cells. PLoS One, 5(12), e15583.

Spallotta, F., Rosati, J., Straino, S., Nanni, S., Grasselli, A., Ambrosino, V., Rotili, D., Valente, S., Farsetti, A., Mai, A., Capogrossi, M. C., Gaetano, C., & Illi, B. (2010). Nitric oxide determines mesodermic differentiation of mouse embryonic stem cells by activating class IIa histone deacetylases: potential therapeutic implications in a mouse model of hindlimb ischemia. Stem Cells, 28(3), 431-442.

Torella, D., Rota, M., Nurzynska, D., Musso, E., Monsen, A., Shiraishi, I., Zias, E., Walsh, K., Rosenzweig, A., Sussman, M. A., Urbanek, K., Nadal-Ginard, B., Kajstura, J., Anversa, P., & Leri, A. (2004). Cardiac stem cell and myocyte aging, heart failure, and insulin-like growth factor-1 overexpression. Circ Res, 94(4), 514-524.

Urbanek, K., Torella, D., Sheikh, F., De Angelis, A., Nurzynska, D., Silvestri, F., Beltrami, C. A., Bussani, R., Beltrami, A. P., Quaini, F., Bolli, R., Leri, A., Kajstura, J., & Anversa, P. (2005). Myocardial regeneration by activation of multipotent cardiac stem cells in ischemic heart failure. Proc Natl Acad Sci U S A, 102(24), 8692-8697.

Vellai, T. (2009). Autophagy genes and ageing. Cell Death Differ, 16(1), 94-102.

von Zglinicki, T., Saretzki, G., Ladhoff, J., d'Adda di Fagagna, F., & Jackson, S. P. (2005). Human cell senescence as a DNA damage response. Mech Ageing Dev, 126(1), 111-117.

Wajapeyee, N., Serra, R. W., Zhu, X., Mahalingam, M., & Green, M. R. (2008). Oncogenic BRAF induces senescence and apoptosis through pathways mediated by the secreted protein IGFBP7. Cell, 132(3), 363-374.

Young, A. R., & Narita, M. (2010). Connecting autophagy to senescence in pathophysiology. Curr Opin Cell Biol, 22(2), 234-240.

10

Age-Related Changes in Human Skin by Confocal Laser Scanning Microscope

Karine Cucumel, Jean Marie Botto, Nouha Domloge and Claude Dal Farra

Ashland Specialty Ingredients – Vincience
France

1. Introduction

Cutaneous aging is a complex biological phenomenon affecting the different constituents of the skin. Two independent processes, clinical and biological, affect the skin during aging. The first is called "chronological aging" and the second one, "extrinsic aging" or "photoaging", which is the result of out-door exposure (such as solar irradiation). In this study, we mainly investigated the features of "chronological aging".

With the advancement of skin research, and the more and more important requests from the cosmetic consumers to have access to technical information on the product efficacy, many non-invasive methods have recently been developed to evaluate and quantify skin-aging parameters. Indeed, numerous available tools allow studying skin-aging as defined by key clinically observed aging parameters such as surface texture, fine lines and wrinkles, skin pigmentation (age spots), firmness, and loss of skin tonicity and elasticity.

In the past decade, laser-scanning confocal microscopy was developed, providing images from *in vivo* human skin without tissue alteration. At the same time, numerous studies comparing and identifying structures imaged by conventional histological sectioning and confocal laser scanning microscope were performed especially by Rajadhyaksha et al., 1995 and Gonzales et al., 2004. Today, the commercially available tool is VivaScope® (Lucid® Inc.), which allows to observe the cellular structure in the whole epidermis, from the *stratum corneum* to the fibrous tissue of superficial dermis, not invasively (without alteration of the tissue), and in real time (without any fixatives). Thanks to the *in vivo* confocal microscopy, a new way to study the signs of age through the epidermis is born. The aim of this chapter was to investigate histometric parameters on several volunteers of different ages in order to better understand the aging process.

2. Focus on VivaScope®

2.1 Parameters of VivaScope® 1500

Confocal laser scanning microscopy opens a "window into your skin" without damaging the skin. Confocal laser scanning microscopy was performed by using the VivaScope® 1500 (Lucid® Inc.). This method allows the observation of the cutaneous micromorphology *in vivo*; thus, for the first time, a real time optical biopsy is possible.

The principle of *in vivo* confocal laser scanning microscopy consists of a laser that emits, with a wavelength of 830 nm, an illumination power inferior to 35 mW and a water immersion objective. The images obtained using this method have a similar resolution to that of classic microscopy: the lateral resolution is 1.25 μm and the vertical one, about 2 μm. The images are black and white, and parallel to the surface of the skin (Curiel-Lewandrowski et al., 2004; Gerger et al., 2006). In this method, the skin imaging is based on different reflection indices of the micro anatomical structures. Melanin offers the strongest contrast; therefore, the cytoplasm of melanocytes appears very bright. Keratin reflects less intensely, so that the cytoplasm of keratinocytes appears darker. Cell nuclei also appear dark, and collagen very bright (Rajadhyaksha et al., 1995).

This skin imaging technique represents a non-invasive and not painful method, and is not tissue-destructive. All these parameters allow imaging the epidermis up to the papillary dermis at a cellular level without any tissue damage. The skin is unaffected during the preparing procedures, thus minimizing visual artifacts. The data collected in real time are rapidly acquired and processed, and the segment of analyzed skin can be re-examined in order to evaluate the dynamic changes. Image stacks can be obtained by compiling images taken every 2 μm, from the horny layer to the dermis.

Confocal laser scanning microscopy can be used for numerous applications and research areas. The images it produces are especially well suited to assist physicians in performing screening examinations, diagnosing skin cancers, etc. This technology can also be applied in cases of burns, dermatitis, and in the cosmetics research industry (Pierard, 1993; Corcuff et al., 1996; Abramovits et al., 2003; Branzan et al., 2007; Ardigo M. & Gill M., 2008).

2.2 VivaScope® 1500 technical performance

Different images, from the *stratum corneum* to the upper part of the dermis, were collected, analyzed, and consecutively transformed into a digital image with different levels of gray. Different measurements can be performed, such as thickness of the *stratum corneum*, of the epidermis, and the number and height of dermal papillae. This paragraph will summarize the methodology to acquire different measurements.

2.2.1 Measurement of objective parameters: *Stratum corneum* and epidermis thicknesses

2.2.1.1 *Stratum corneum* thickness

The *stratum corneum* thickness was calculated measuring the intensity variation on vertical reconstruction of the epidermis with the software ImageJ IJ 43 (Abramoff et al., 2004) giving a plot profile. The process is explained in Figure 1. The plot profile gives the intensity of the luminosity at different depths at the place of the yellow line in the vertical reconstruction. (The yellow line is placed by the operating expert). The *stratum corneum* thickness can then be obtained from the graph but can be more precisely obtained by calculating the derivative of the plot profile using finite differences. The depth separation between the maximum and the minimum of the derivative measures the *stratum corneum* thickness. Five measurements were made on one stack per volunteer.

2.2.1.2 Epidermal thickness

The distance between the surface (*stratum corneum*) and the level showing the dermal tissue measures the thickness of the epidermis. In Figure 2, **A** represents the *stratum corneum* (beginning of the measurement area), **B**, the beginning of the dermal tissue (end of the measurement), and **C**, the measurement done with VivaScope® on H&E. For the epidermis thickness, we performed one measurement per stack, on eight stacks per volunteer. In our experiment, we decided to measure the thickness between interdigitations where the thickness is smaller, i.e. minimal thickness (called Emin).

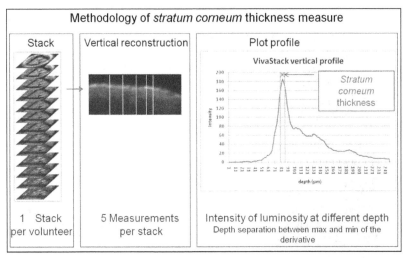

Fig. 1. Methodology of *stratum corneum* thickness measurement: From one volunteer, we collected series of images (stack), then the software provided from the stack a vertical section of the skin, the yellow line was placed by the expert to obtain the measure. The plot profile that gives the intensity at different depth is finally performed with ImageJ software and the calculation of the derivative measurement gave the precise thickness of the *stratum corneum*. Five measurements were made on one stack per volunteer.

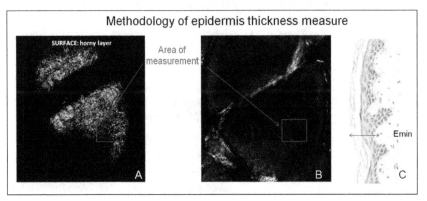

Fig. 2. Methodology of epidermis thickness measurement: The distance between the surface (*stratum corneum*) A and the level showing the first dermal tissue B measures the thickness of the epidermis. C represents the measurement done with the VivaScope® on a H&E biopsy section. For epidermis thickness, we made one measurement per stack, on eight stacks per volunteer.

2.2.2 Measurement of objective parameters: Dermal papillae

2.2.2.1 Measurement of the number of dermal papillae

The number of dermal papillae was evaluated by counting each active/functional papillae showing a lumen. Figure 3 explains how to recognize active/functional papillae from non-active dermal papillae. We made one measurement per stack, on eight stacks per volunteer.

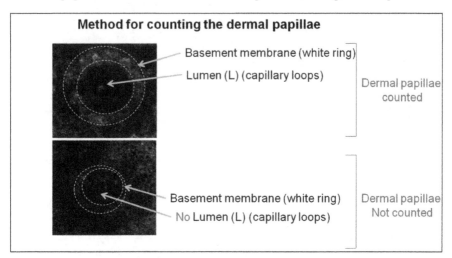

Fig. 3. The number of dermal papillae was evaluated by counting each active/functional papilla showing a lumen. The measurements were performed on eight stacks per volunteer and one measurement per stack.

To determine the height of dermal papillae, we measured the distance between the top of dermal papillae and the level showing no dermal papillae structure at all (Figure 4). We made the measurement on eight stacks per volunteer, and realized one measurement on two dermal papillae per stack. In total, we measured the height of 16 dermal papillae per volunteer.

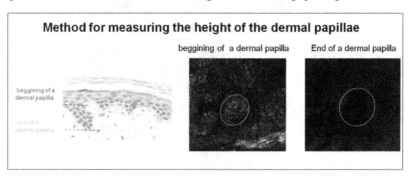

Fig. 4. The height of dermal papillae was measured as the distance between the top of dermal papillae and the level showing no dermal papillae structure at all. The measurements were performed on eight stacks per volunteer, and performed one measurement on two dermal papillae per stack. In total, we measured the height of 16 dermal papillae per volunteer.

2.2.2.2 Observations of reticular collagen

In some cases, we could also observe the reticular dermis, but we must note that we reached the detection limit at this point.

The clinical observations and measurements obtained by VivaScope® microscopy offer the cosmetic industry a great opportunity to trace the changes at different age-stages of the skin; and that will be presented in the second part of this chapter.

3. Skin changes at different age stages by *in vivo* confocal microscopy

3.1 Methodology of the clinical study

To investigate age-related changes, we studied five groups of five healthy volunteers each with Caucasian type of skin:

- The 20-30 group was composed of volunteers aged between 24 and 27 years (mean age: 25.2, 4 females, 1 male)
- The 30-40 group was composed of volunteers aged between 29 and 37 years (mean age: 33.8, 5 females),
- The 40-50 group was composed of volunteers aged between 41 and 46 years (mean age: 44.4, 5 females),
- The 50-60 group was composed of volunteers aged between 51 and 57 (mean age: 54.5, 3 females, 3 males),
- The 60-70 group, the older group, was composed of volunteers aged between 60 and 68 years (mean age: 63.6, 5 females).

Measurements were performed on their volar forearm, over two weeks in May.

For each group, we measured the epidermis and the horny layer thicknesses, the number and height of dermal papillae, and also carefully observed the granular cells, the morphology of dermal papillae, and when possible, the collagen fibers structure.

We compared these results with Hematoxylin-Eosin (H&E) stains on skin biopsies obtained from abdomen plastic surgery of females aged 20 to 79 years. In this part of the study, skin biopsies were fixed in successive baths of formol, alcohol, and xylene, then embedded in paraffin (Excelsior ES, Shandon, UK), and sectioned into 4 µm sections to be used for routine hematoxylin-eosin staining.

In this study, we observed very carefully the morphological features of the skin in the different volunteer groups; for each group, several biopsies were obtained, and for each of them, several skin cross-sections were studied and compared.

3.2 H&E staining observations

The following results are illustrated in Figure 5. In the different age groups, we observed:

- **Group 20-30:** The H&E skin coloration showed lots of dermal papillae as well as numerous invaginations, commonly called "rete pegs". The dermoepidermal junction was highly convoluted, and the basal cells appeared well organized without any damage. The papillary dermis appeared very dense.

- **Group 30-40:** At these ages, the structure of the epidermis was similar to that observed for 20-30 year-old donors. The interface between the epidermis and the dermis appeared highly convoluted due to the presence of rete pegs.
- **Group 40-50:** In this group, we observed a retraction of the rete pegs resulting in a flattened interface between epidermis and dermis. The dermal papillae were less apparent. The basal cells became less organized and the papillary dermis was less dense. It was in this group that we noticed the first morphological signs of age in the skin biopsies.
- **Group 50-60:** Here, the dermoepidermal junction was completely flat with a total absence of dermal papillae and so of rete pegs. The basal cells presented several damage and were disorganized. The papillary dermis decreased in density compared to the group 20-30.
- **Group 60-70:** The skin structure was similar to the one observed in the group 50-60. We found the same flattening of the dermoepidermal junction and the same damage of basal cells and papillary dermis.

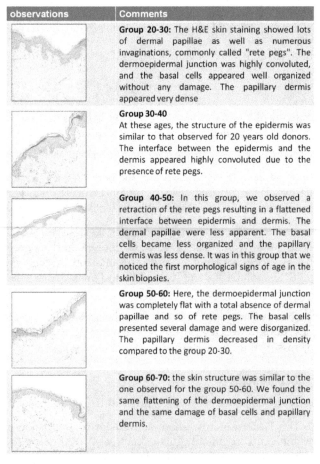

observations	Comments
	Group 20-30: The H&E skin staining showed lots of dermal papillae as well as numerous invaginations, commonly called "rete pegs". The dermoepidermal junction was highly convoluted, and the basal cells appeared well organized without any damage. The papillary dermis appeared very dense
	Group 30-40 At these ages, the structure of the epidermis was similar to that observed for 20 years old donors. The interface between the epidermis and the dermis appeared highly convoluted due to the presence of rete pegs.
	Group 40-50: In this group, we observed a retraction of the rete pegs resulting in a flattened interface between epidermis and dermis. The dermal papillae were less apparent. The basal cells became less organized and the papillary dermis was less dense. It was in this group that we noticed the first morphological signs of age in the skin biopsies.
	Group 50-60: Here, the dermoepidermal junction was completely flat with a total absence of dermal papillae and so of rete pegs. The basal cells presented several damage and were disorganized. The papillary dermis decreased in density compared to the group 20-30.
	Group 60-70: the skin structure was similar to the one observed for the group 50-60. We found the same flattening of the dermoepidermal junction and the same damage of basal cells and papillary dermis.

Fig. 5. H&E staining biopsy observations and comments.

3.2.1 Results of H&E staining

In our light microscopy study, as largely described in the literature, we showed that aged skin revealed a thinner epidermis than young skin. This was primarily due to the retraction of rete pegs resulting in a flattened interface between epidermis and dermis and consequently to a flattened dermoepidermal junction. One of the consequences of this flattening is that aged epidermis becomes less resistant to shearing force, and less vascularized, leading to a bad nutrition of the basal cells. The observation of the *stratum corneum* in light microscopy study, as in other studies, showed that the number of horny cells did not seem to diminish with age, and thus the *stratum corneum* retained its normal thickness (Hull & Warfel, 1983).

In light microscopy, all the studies of the literature are in accordance regarding the evolution of skin thickness with age. However, conflicting results were found in the VivaScope® study and were challenging to interpret, as mentioned and discussed later on.

3.3 Confocal laser scanning microscope (VivaScope® 1500)

3.3.1 Horny layer and epidermis thicknesses

We measured and plotted the mean of horny layer and epidermal thicknesses as a function of age (Figure 6). The mean of horny layer and the epidermal thicknesses became thicker with age. The correlation between the thickening and age was statistically very significant for the *stratum corneum* and highly significant for the epidermal thickness. Table 1 gives the calculated correlation coefficient (r), which measures the correlation between the two variables: X for age and Y for the thickness.

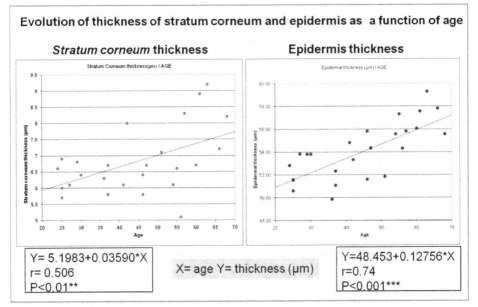

Fig. 6. Measurement and plot of horny layer and epidermis thicknesses as a function of age.

This correlation coefficient was positive, which implied that when age (X) increases, thickness (Y) also increases. The regression line can be used to predict the *stratum corneum* or epidermis thickness by entering the age of an individual.

Correlation test	p<	r=	Regression line	Mean		SD	
				X (age)	Y (thickness)	X	Y
Stratum corneum thickness / age	0.01**	0.506	Y= 5.1983+0.03590*X	44.28	6.788	14.38	1.0199
Epidermal thickness / age	0.001***	0.74	Y=48.453+0.12756*X	44.28	54.102	14.38	2.479

: very significant *: highly significant

Table 1. Parameters of the graph representing the thicknesses of both horny layer and epidermis.

Furthermore, the correlation coefficient of the number and height of dermal papillae being close to 1 suggests that the number and the shape of dermal papillae could be considered as a very good aging indicator. According to the literature, we observed similar results with this technique to the ones that others obtained (Fenske and Lober, 1986; Zaghi et al., 2009).

3.4 Granular cell observation

For each group, we compared the organization and size of the granular cells. For each parameter observed, we assigned a category to each volunteer. Thus, for the granular cell morphology we used four categories: excellent, good, moderate and poor (the data are summarized in Table 2).

Age	excellent	good	moderate	poor
20-30	1	4		
30-40		3	2	
40-50			5	
50-60			2	3
60-70			1	4

Table 2. Granular cell observations: Excellent: very good organization, small and polygonal cells; Good: good organization, small and polygonal cells; Moderate: small cell disorganization and cell spreading, presence of no polygonal cells, vacuoles and dyskeratotic cells; Poor: cell disorganization and cell spreading, lots of cells with irregular shape, vacuoles, and condensed DNA.

- **Group 20-30 and group 30-40:** In these two groups, the granular layer displayed very well organized granular cells. The cells were regular in shape and size and had a good cohesion. We did not notice any cell damage (vacuole, dyskeratotic cells...). The granular layer had a honeycomb pattern as expected in young skin (discussed later in the next paragraph).

- **Group 40-50:** At this age, we observed the beginning of a disorganization of granular cells. The cells lost their polygonal shape and became bigger compared to young granular cells. Dyskeratotic cells appeared in this layer whereas the honeycomb pattern was less apparent.
- **Group 50-60 and group 60-70:** With age, the granular cell organization was completely lost. The cells had no longer a polygonal shape and their size was smaller. Numerous dyskeratotic cells appeared and the granular layer displayed cell spreading with a loss of cohesion.

3.5 Epidermal thickness assessment with VivaScope®

The effect of age on the thickness of the skin is one of the most controversial topics among dermatological researches. Indeed, in our study we observed a statistical correlation between thickening and age in both the horny layer and the epidermis. Focusing on the results of Table 1, we were able to observe a better correlation between epidermal thickness and age than horny layer thickness and age. Actually, this correlation difference pointed out that the *stratum corneum* is more influenced by the environment stress and volunteer's lifetime. Moreover, in the *stratum corneum* we observed that the correlation was not really linear but could look like a curve. This last finding suggested that the horny layer thickness could be constant during the first 50 years of life followed by a thickening in older individuals. More investigations have to be made to confirm these findings.

Controversially with others, we observed with the VivaScope® 1500 that the younger the skin is (and presumably healthier), the thinner the epidermis and the horny layer are. To our opinion, the relative thinness in younger skin (observed by this technique) correlated with what was seen in the granular cells in younger groups; the granular cells were more cohesive, often smaller than in the older groups, and better stacked, thus better organized. Hence, this relative thinness of the epidermis is not correlated with what was seen with light microscopy (i.e. the flattening of dermal junction) but correlated to a better internal organization of the granular cells. Besides, concerning the *stratum corneum*, with age, the cohesion between the corneocytes is lost (Fenske et al., 1986); therefore, they are not well organized and there is more space between each of them, which can explain the relative increase of the thickness.

3.6 Controversial results

Comparing measurements of the skin layer thickness between studies (and also from one individual to another, as well as between assessments) is especially challenging, due to the significant variations in measurements between individuals, between sites within each individual, between seasons, and hormonal differences between individuals.

Indeed, a study using confocal microscopy found that the thickness of living epidermis on the back of the arm decreased with age (Zaghi et al., 1986; Sandly-Møller et al., 2003). Another study using ultrasounds (Gniadecka et al., 1994) found an increase of facial skin thickness between 25 and 90 year-old people. In another study with 61 women with ages ranging from 18 to 94, authors found that the skin thickness increased on the forehead and buttock with age (Pellacani & Seidenari, 1999). In accordance with our results, in another study, Sauermann et al. (2002) considered a relatively old group with a mean age of 72.5 years and reported a significantly larger epidermal thickness in elderly volunteers. In the controversial study of Neerken et al. (2004), the thickness of the *stratum corneum* was found not to change with age, but in the older volunteer group, the minimum thickness of viable epidermis was somewhat larger (but no statistical difference between the investigated groups) and that the maximum thickness of the epidermis statistically significantly decreased with age. In our study, we chose to measure the minimum thickness of epidermis (see Figure 2 for reminder), and this can explain certain discrepancy compared with others.

When we focused on the methodology, we observed that Sauermann et al. measured the epidermal thickness on volar forearm with VivaScope® using the same method as we do, and found an increase in epidermal thickness with age.

Authors (Leveque et al., 1984) measured the total skin thickness on the dorsal and ventral forearm (epidermis + dermis) with Holtain Skinfold Caliper and found that the skin thickness decreased with age.

Takema et al. (1994) measured the skin thickness on the face and the ventral forearm by Dermascan A. They found that, with age, the skin thickness decreased in the area slightly sun-exposed (ventral forearm) whereas it increased in the area markedly sun-exposed.

Gambichler et al. (2006) measured the thickness of the epidermis on the forehead, pectoral area, forearm, buttock, upper back, and calf. They observed no inter-regional variation and a thinner epidermis in their older group.

Finally, Rigal & Leveque (1989) measured the skin thickness with ultrasound images for each decade of life (until 80-90 years) on the volar and dorsal forearms and observed an inter-regional difference. On the volar forearm, skin thickness did not vary significantly between the first and the seventh decade of life, but skin atrophy appeared after the eighth decade. On the dorsal forearm, they observed a phase of maturation (thickness increase) up to 15 years of age, and that atrophy signs began after the seventh decade.

Despite all these disagreements observed in bibliographic data, we can affirm that the keratinocyte turnover slows-down with age, leading to the accumulation and the increase in size of corneocytes in the *stratum corneum* (Leveque et al., 1984; Grove & Kligman; Marks, 1981). In our study, we observed that, with age, the granular cells increased in size, the shape changed, and the honeycomb pattern disappeared (correlating with an observed disorganization). According to Sauermann et al., the correlation between size of cells in the granular layer and age is consistent with a documented increase in corneocytes with age. Moreover, others (Fenske & Lober, 1986), who observed that in supra-basal cell layers the keratinocytes tend to display a decreased vertical height and an increased overall surface area, also reported this effect. Other authors suggested that this decrease in height and increase in irregularity could reflect the decrease in proliferation of the basal cell in aged skin (Sauermann et al., 2002). These changes, coupled with poor corneocytes adhesion, could lead to the increase of the thickness observed in our study.

3.7 Number and height of dermal papillae

Throughout different decades, we observed a statistically highly significant decrease in the number of dermal papillae (p<0.001). Therefore, mathematically, we could determine a negative correlation with r = -0.699 between the number of dermal papillae and age. The measured height of the dermal papillae also statistically highly significantly decreased (p<0.001), with a negative correlation of r = -0.854 (Figure 7).

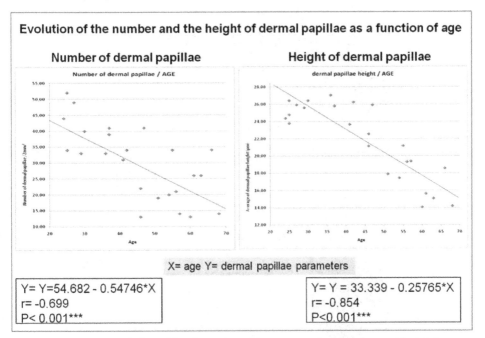

Fig. 7. Measurement and plot of the dermal papilla parameters (Number and height) as a function of age.

Furthermore, the correlation coefficient of the number and height of dermal papillae is close to 1, indicating that the number and shape of dermal papillae could be considered as very good aging indicators. According to the literature, we observed similar results with this technique as the ones that others obtained (Fenske and Lober, 1986; Zaghi et al., 2009).

3.8 Dermal papillae observations

The results of the observations are summarized in Figure 8.

- **Group 20-30:** At these ages, dermal papillae were numerous, round and very well delimited by a "white ring" (intense melanin content), which was constituted by the basal cells. Inside some papillae, we could observe capillary loops. All these observations demonstrated a good vascularization of the epidermis.
- **Group 30-40:** Dermal papillae were still round and pretty well defined although in that case we observed less melanin that interfered in the observation of dermal papillae. Nonetheless, the lumen of capillary loops is still observed, sign of a healthy skin.
- **Group 40-50:** The observed number of dermal papillae decreased in size. The shape was irregular and the dermal papillae were less defined. In this group of age, the capillary loops were hardly observable.

- **Group 50-60 and group 60-70:** The VivaScope® pictures did not display any dermal papillae. In the same manner, cells were disorganized, and the dermoepidermal junction was not observable. Moreover, vascularization was absent suggesting a poor epidermal nutritional status.

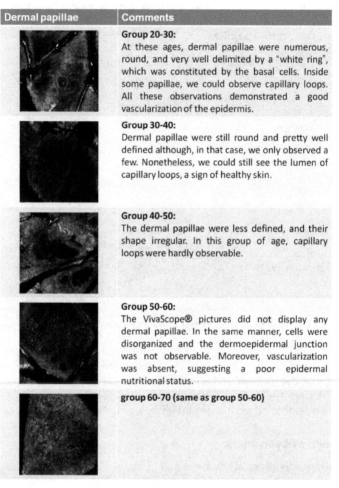

Dermal papillae	Comments
	Group 20-30: At these ages, dermal papillae were numerous, round, and very well delimited by a "white ring", which was constituted by the basal cells. Inside some papillae, we could observe capillary loops. All these observations demonstrated a good vascularization of the epidermis.
	Group 30-40: Dermal papillae were still round and pretty well defined although, in that case, we only observed a few. Nonetheless, we could still see the lumen of capillary loops, a sign of healthy skin.
	Group 40-50: The dermal papillae were less defined, and their shape irregular. In this group of age, capillary loops were hardly observable.
	Group 50-60: The VivaScope® pictures did not display any dermal papillae. In the same manner, cells were disorganized and the dermoepidermal junction was not observable. Moreover, vascularization was absent, suggesting a poor epidermal nutritional status.
	group 60-70 (same as group 50-60)

Fig. 8. Dermal papillae observations and comments.

3.9 Results on dermal papillae

Today, it is well accepted that the number of dermal papillae decreases with age, according to several studies (Fenske and Lober, 1986; Sauermann et al., 2002; Neerken et al., 2004) and we observed the same downward trend. The decrease in the number of dermal papillae with age reflects the flattening of the epidermal-dermal junction (Sauermann et al., 2002). In particular, this flattening demonstrates that not only the height of the dermal papillae decreases with age, but the number of interdigitations also drops with aging. The dermal papillae in the oldest groups were found irregular in shape; this feature could be explained by the reduction of the number of basal cells participating in the cellular cycle. This finding can be supported by the disappearance with age of the white ring surrounding dermal papillae that corresponded to basal cells.

After conducting this study, we conclude that the change in the number and morphology of dermal papillae are the parameters the most correlated with age. These parameters were closely linked to the dermal papilla function in supplying the epidermis with water and nutrients *via* dermal vasculature, and directly to the health of the epidermis. The morphology, number, and shape of dermal papillae could be great indicators of the epidermis health and age.

3.10 Collagen observations

The images obtained by the VivaScope® and the observations are compiled in Figure 9.

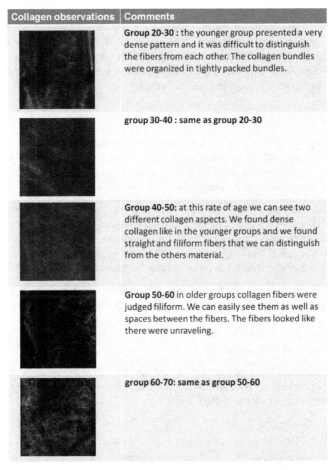

Collagen observations	Comments
	Group 20-30 : the younger group presented a very dense pattern and it was difficult to distinguish the fibers from each other. The collagen bundles were organized in tightly packed bundles.
	group 30-40 : same as group 20-30
	Group 40-50: at this rate of age we can see two different collagen aspects. We found dense collagen like in the younger groups and we found straight and filiform fibers that we can distinguish from the others material.
	Group 50-60 in older groups collagen fibers were judged filiform. We can easily see them as well as spaces between the fibers. The fibers looked like there were unraveling.
	group 60-70: same as group 50-60

Fig. 9. Collagen observations and comments.

- **Group 20-30 and group 30-40:** The younger group presented a very dense pattern of collagen and it was difficult to distinguish the collagen fibers from one another. The collagen bundles appeared organized in tightly packed bundles as discussed in the next paragraph.
- **Group 40-50:** At these ages, we could see two different collagen distribution patterns. We found dense collagen as in the younger groups, and straight and filiform fibers.
- **Group 50-60 and group 60-70:** In this group, collagen fibers were filiform. We could easily observe them, as well as spaces between the fibers, thus the fibers appeared individualized, isolated from one to another; therefore the fibers looked unraveled.

3.11 Results on collagen observations

With a confocal laser scanning microscope, also called VivaScope®, the collagen appearance of young skin was characterized by an organization in tightly packed bundles. At this age range, the collagen is the major dermal matrix component; collagen fiber bundles are compact and dense, which explains why we were not able to distinguish any structure in our observation. In contrast, in older groups, we managed to very well discern some collagen fibers from ground substance. This could be due to the decrease in the rate of collagen synthesis and the thickness of collagen fiber bundles (Fenske and Lober, 1986; Koehler et al., 2008). With age, the matrix becomes thinner, with straight collagen bundle fibers, giving an unraveled appearance to the bundles (Lavker et al., 1987; Zaghi et al., 2009).

4. Conclusion

To analyze the aging process, we investigated histometric parameters on biopsies from several donors of different ages, and performed a clinical study to observe the cutaneous micromorphology *in vivo*, using a confocal laser scanning microscope –VivaScope®1500– on groups of volunteers from 20 to 70 years of age.

The thickness of the epidermis and the horny layer seemed to be still controversial. However, regarding the effect of cosmetic ingredients contributing to an improvement of the skin's parameters, we observed that the healthier is the skin (a better-hydrated appearance), the thinner is the epidermis. The thinness parameter always correlates with an increase in the skin turnover and a decrease in size of granular cells.

As in the bibliographic data, we found that the number and height of dermal papillae decreased with age, and that these effects correlated with the flattening of epidermal-dermal junction seen in H&E staining biopsies. Therefore, after conducting this study, we came to the conclusion that the changes in the number and morphology of dermal papillae are the parameters the most correlated with age.

Finally, *in vivo* confocal laser scanning microscope can undoubtedly be considered as a sensitive and non-invasive tool allowing an easy study of the changes of different parameters of the whole epidermis, at all ages.

5. References

Abramoff, MD., Magalhaes, PJ., & Ram, SJ. (2004). Image Processing with ImageJ. *Biophotonics International*, Vol.11, No.7, pp.36-42.

Abramovits, W., & Stevenson, LC. (2003). Changing paradigms in dermatology: New ways to examine the skin using non-invasive imaging methods. *Clinics in Dermatology*, Vol.21, pp. 353-358.

Ardigo, M., & Gill, M. (2008). Blue nevus. In: *Reflectance confocal microscopy of cutaneous tumors*, Gonzalez S, Gill M, Halpern, AC, (EDs), 146–150, Informa Healthcare, USA.

Branzan, AL., Landthaler, M., & Szeimies, RM. (2007). *In vivo* confocal scanning laser microscopy in dermatology. *Laser in Medical Science*, Vol.22, pp. 73-82.

Corcuff, P., Gonnord, G., Pierard, GE., & Leveque, JL. (1996). *In vivo* confocal microscopy of human skin: a new design for cosmetology and dermatology. *Scanning*, vol.18, pp. 351-355.

Curiel-Lewandrowski, C., Williams, CM., Swindells, KJ,. Tahan, SR., Aster, S., Frankenthaler, RA., & González, S. (2004). Use of *In Vivo* Confocal Microscopy in Malignant Melanoma: an Aid in Diagnosis and Assessment of Surgical and Nonsurgical Therapeutic Approaches. *Archives of Dermatology*, vol.140, No.9, pp. 1127-32.

Fenske, NA., & Lober, CW. (1986). Structural and functional changes of normal aging skin. *Journal of American Academy of Dermatology*, vol.15, pp. 571-585.

Gambichler, T., Matip, R., Moussa, G., Altmeyer, P., & Hoffmann, K. (2006). *In vivo* data of epidermal thickness evaluated by optical coherence tomography: effects of age, gender, skin type, and anatomic site. *Journal of Dermatological Science*, vol.44, No.3, pp. 145-52.

Gerger, A., Koller, S., Weger, W., Richtig, E., Kerl, H., Samonigg, H., Krippl, P., & Smolle, J. (2006). Sensitivity and Specificity of Confocal Laser-Scanning Microscopy for *In Vivo* Diagnosis of Malignant Skin Tumors. *Cancer*, vol. 107, No.1, pp. 193-200.

Gniadecka, M., Gniadecki, R., Serup, J., & Sondergaard, J. (1994). Ultrasound structure and digital image analysis of the subepidermal low echogenic band in aged human skin: diurnal changes and inter individual variability. *Journal of Investigative Dermatology*, vol.102, No.3, pp. 362-365.

Gonzalez, S., Gilaberte-Calzada, Y., Gonzalez-Rodriguez, A., Torres, A., & Mihm, MC. (2004). *In vivo* reflectance-mode confocal scanning laser microscopy in dermatology. *Advances in Dermatology*, vol.20, pp. 371-387.

Grove, GL., & Kligman, AM. (1983). Age-associated changes in human epidermal cell renewal. *Journal of Gerontology*, vol.38, pp. 137-142.

Hull, MT., & Warfel, KA. (1983). Age-related changes in the cutaneous basal lamina: scanning electron microscopic study. *Journal of Investigative Dermatology*, vol.81, pp. 378–380.

Koehler, MJ., Hahn, S., Preller, A., Elsner, P., Ziemer, M., Bauer, A., König, K., Bückle, R., Fluhr, JW., & Kaatz, M. (2008). Morphological skin aging criteria by multiphoton laser scanning tomography: non-invasive *in vivo* scoring of the dermal fiber network. *Experimental Dermatology*, vol.17, pp. 519-523.

Lavker, RM., Zheng, PS., & Dong, G. (1987). Aged skin: A study by light, transmission electron, and scanning electron microscopy. *Journal of Investigative Dermatology*, vol.88, pp. 44s-51s.

Leveque, JL., Corcuff, P., De Rigal, J., & Agache, P. (1984). *In vivo* studies of the evolution of physical properties of human skin with age. *International Journal of Dermatology*, 23: 322-329.

Marks, R. (1981). Measurement of biological aging in human epidermis. *British Journal of Dermatology*, vol.104, pp. 627-633.

Neerken, S., Lucassen, GW., Bisschop, MA., Lenderink, E., & Nuijs, TA. (2004). Characterization of age-related effects in human skin: a comparative study that applies confocal laser scanning microscopy and optical coherence tomography. *Journal of Biomedical Optics*, vol.9, pp. 274-281.

Pellacani, G., & Seidenari, S. (1999). Variation in facial skin thickness and echogenicity with site and age. *Acta Dermato-Venereologica*, vol.79, pp. 366-369

Pierard, GE. (1993) *In vivo* confocal microscopy: a new paradigm in dermatology. *Dermatology*, vol.186, pp. 4-5

Rajadhyaksha, M., Grossman, M., Esterowitz, D., Webb, RH., & Anderson, RR. (1995). *In vivo* confocal scanning laser microscopy of human skin: melanin provides strong contrast. *Journal of Investigative Dermatology*, vol.104, pp. 946–52.

Rigal, J., & Lévêque, JL. (1989). Assessment of aging of the human skin *in vivo* ultrasonic imaging. *Journal of Investigative Dermatology*, Vol.93, pp. 621-625.

Sandby-Møller, J., Poulsen, T., & Wulf, HC. (2003). Epidermal thickness at different bodysites: relationship to age, gender, pigmentation, blood content, skin type and smoking habits. *Acta Dermato-Venereologica*, vol.83, No.6, pp. 410-3.

Sauermann, K., Clemann, S., Jaspers, S., Gambichler, T., Altmeyer, P., Hoffmann, K., & Ennen, J. (2002). Age related changes of human skin investigated with histometric measurements by confocal laser scanning microscopy *in vivo*. *Skin Research and Technology*, vol.8, pp. 52-56.

Takema, Y., Yorimoto, Y., Kawai, M., & Imokawa, G. (1994). Age-related changes in the elastic properties and thickness of human facial skin. *British Journal of Dermatology*. Vol.131, No.5, pp. 641-8.

Zaghi, D., Waller, JM., & Maibach, HI. (2009). The effects of aging on skin. *Nutritional Cosmetics*, pp. 63-77.

11

Multi-Purpose Activities in Ergotherapy

Hulya Yucel

Department of Physiotherapy and Rehabilitation,
Faculty of Health Sciences, Bezmi Alem University, Istanbul,
Turkey

1. Introduction

The elderly population, both in terms of number and relative percentage in society, is increasing throughout the world. Society, relative to previous generations, will be older with a greater burden on health system resources due to chronic ongoing disease and illness management requirements. The demand for services of geriatrics and gerontology in turn increases. Health planners, who define funding levels within government, are required to consider this increasing life expectancy. Long-term care of the unhealthy elderly is a much-debated medical and political issue in developing countries. As the experiences of successful aging increases within society today, attention focusses on what defines successful aging and the factors that promote a healthy aged community (Bowling & Iliffe, 2006). For this reason, World Health Organisation (WHO) described active ageing as improving life expectancy, productivity and quality of life by promoting and maintaining the highest functional capacity of social well-being and physical and mental functioning (WHO, 1998).

Loneliness, a condition relatively common in the elderly, is being increasingly linked to negative quality of life predictors such as chronic diseases, depression and reduced social participation (Alpass&Neville, 2003; Jylha, 2004; Routasalo, 2007). Thoughts of being closer to death, loss of mobility and family bereavement reduce the sense of taking pleasure in life, affecting the extent of community involvement and levels of independence (Alpass&Neville, 2003). This greater isolation from society inhibits effective social behaviour and facilitates passive roles in interactions involving the elderly (Vitkus& Horowitz, 1987). The resulting loneliness and social isolation negatively impacts the psychosocial situation of well-being, the quality of life and cognitive skills (Routasalo et al., 2007).

Researches suggest the interest in and skills associated with activities of daily living are reduced with aging, because of changes in health and social issues commonly experienced in the elderly (Clark&Siebens, 2005; Routasalo et al., 2007; WHO, 1998). Remaining physically, mentally and socially active, things like "doing work", has great importance in maintaining functioning as does the need to avoid excessive levels of the more passive recreations, such as watching television. Indeed, many problems derive from having more unstructured free time. Meaningful leisure time in the elderly is essential to a good quality of life. In Yucel's thesis, the majority of elderly people stated that in their free time, they like to be involved in tasks such as reading and walking. The respondents would prefer not to participate regularly in physical activities. They could not give a concrete reason for this behaviour. One plausible explanation is the elderly may not be conscious enough about the benefits of regular activity programs and the consequences of their relative withdrawal from these (Yucel, 2008).

In a study in Brazil, the lack of adequate financial resources (40.3%) along with fatigue (38.1%) were identified as obstacles to the participation of the elderly in leisure activities (Reichert et al., 2007). Motivation is critical to the success of activity programs designed for the elderly. Many elderly have concerns about participating in activity programs, because they have developed negative behaviors and beliefs throughout their lives associated with activity. Barriers to activity programmes, such as having to park their cars long distances away, along with other factors such as snow and ice, inhibit involvement. Reducing these barriers to exercise is necessary for a more independent and "strong" elderly community (Resnick, 1991; Yoshimoto & Kawata, 1996).

Participating in activities and social integration is one of the important approaches to rehabilitation in the elderly. Ergotherapy programs are created to help to protect life roles in geriatrics. They promote active aging and overall quality of life through participation in activities designed and prescribed to the needs of the individuals (Boswell et al., 1997; Rosalie, 2003). Activity training, which facilitates an active aging process, is an important part of a comprehensive ergotherapy program in the elderly. It aims to care about health, to increase cognitive, emotional and physical capabilities, and ensure the independence of social functions through the choice of different activities in accordance with the individual requirements and needs of the elderly (Donohue et al., 1995; Lachenmayr&Mackenzie, 2004; Nelson, 1997; Vass et al., 2005).

In a survey of 815 elderly participants in Australia, the most desired activities overall were golf, walking, tennis and swimming. The underlying objectives of these activities were to stay healthy, that they were interested in that activity, they wanted to improve their physical capacity and maintain their overall joint mobility (Kolt et al., 2004). The quality of life of older people engaged in activities of their own choice were higher (Duncan-Myers&Huebner, 2000). Conversely, over 80% of elderly people in America spent their free time by visiting friends, watching television and listening to the radio (Lee&King, 2003). The type of activity is less important, because the differences reflect variables such as the individual's health, associated abilities and socio-economic status, than the actual participation itself. Participation is a stronger predictor of quality of life than the type of activity (Ward, 1979).

The maintenance of health and quality of life in older clientele is promoted through the participation in meaningful and purposeful activities (Csikszentmihaly, 1993; Glantz, 1996). The study of Inal et al. showed that life satisfaction scores are significantly higher in elderly people who are interested in a variety of crafts and regular walking (Inal, 2003). In people with life-long activity goals, such as participating in regular physical activity, the normal physiological changes that occur with aging were seen to be delayed or less severe. Elderly people who have defined leisure time activities have a higher quality of life.

Furthermore, Routasalo et al. (2007) showed increased psychological well-being and improved cognitive skills in the elderly by the implementation of activity training. The type of activity prescribed needs to be based on the individual. Studies have showed that leisure / hobbies / social activities are preferred for elderly that have mental health problems rather than physical problems (Mountain, 2005a). Cognitive tests increased significantly in patients with vascular dementia by applied activity treatment (Nagaya et al., 2005). These results suggest that activity training may be protective against the formation of a new dementia. Further investigation into the capacity of multi-purpose activity to inhibit the development of conditions such as depression and dementia in the elderly is necessary.

Whether activity prescribed to the elderly is given on an individual basis or should constitute a group format is open to debate. Researchers following the second world war investigated group behaviour. They concluded people need each other, not just to maintain themselves, but also to feel fulfilled in their lives. The most basic group is family with the members sharing responsibilities and performing required specific tasks. Families expanded roles includes activity such as plays, school activities and the involvement in religious and recreational organizations. All which enhance the integration of life goals and promote the individual's overall quality of life (Matsuo et al., 2003; Royeen & Reistetter, 1996; Yucel et al, 2006a). Some people are social, enjoy spending time with others, while others would prefer to be alone. Some still enjoy making new discoveries, while others want to continue with long-standing interests. Whether the activities are completed as an individual or in a group, whether they are novel or long standing, best reflects the needs and wants of the individual. The elderly should be encouraged to participate in activities appropriate for, and which interest, them. The regular continuity of activities is an important factor in enhancing quality of life.

Future research exploring concepts such as the reintroduction of extinct roles to the elderly, or increasing the diversity of the types of activity undertaken within a specific role, and the implications on quality of life and life satisfaction are necessary. Studies in different ethnic or cultural groups should be encouraged (Ross, 1990). Elderly people need increased diversity of activities to maximize the process of an active older age. Future education and training of health workers in geriatrics is necessary to promote a conciousness of the importance of roles in health outcomes and to provide the skills that facilitate the prescription of optimal activity; activity that best reflects the needs and wants of the individual older person. Activities are as necessary as eating or drinking and to have life; each individual should have regular activity within and outside the home based on their roles and physical and mental health.

This chapter will be issued as below:

Ergoterapy Approaches in Geriatrics, Importance of Leisure Time Activities, Multi-purpose Activities, Activity Training Models, Activity Training, Group Activities.

2. Ergotherapy aproaches

In geriatric rehabilitation, it is important to improve functional capacity and daily living skills of elderly, personal care about areas such as hygiene, rest and nutrition, and to ensure social-emotional support (Lewis&Bottomley, 2002). The goal should be to maintain the independence on functionality of the elderly or to restore if it is decreased.

Considering the following points facilitates to plan appropriate rehabilitation program in geriatrics:

1. There can be the capacity differences among elderly. In training programs which is planned within the scope of rehabilitation for the elderly, the capacity of individuals has to be known. Chosen approach is not important for any activity training for strengthening, but there are some circumstances to be considered peculiar to elderly. For example, late pupil dilation and thickening of the lens with aging mean

environmental clarity and projection can not be tolerated. Therefore three times much light are needed for function of the aged eyes. Additionally, the elderly can not detect the color differentiation which is necessary for driving, Activities of Daily Living (ADL) and ambulation. Such physiological changes affect the functionality of the elderly.

2. The level of activity differs from one aged to the other. For example, a 80 year old can fulfill the physical and cognitive functions whilst the other of the same age may not success.

3. Maximal health is directly associated with the maximal functional ability. Activities that give energy to the elderly to be alive and aim to provide independence in an active life and maintain health should be given to the elderly (Larson et al., 1986; Mountain, 2005a).

Day-care services in geriatric rehabilitation include observation of the elderly by caretakers, caregiver training, daily regular controls of drug intake etc., implementation of treatment services and social / recreational activities. General social services allow older people to maintain their lives in an appropriate environment. And also, they undertake transfer to the hospital and home from the hospital, prevention of diseases and preventive treatment. On the other hand, ergotherapy approaches come into prominence in determining the needs of the elderly, planning/implementation/monitoring of nursing program and revealing of changing needs over time (Mountain, 2005b).

Ergotherapy in geriatric rehabilitation mainly includes the following goals (Yucel, 2006b):

- To maintain basic and enstrumental ADL successfully by increasing physical and/or mental activity performance,
- To restore decreased ability and to improve or maintain quality of life,
- To help to continue on social habits in a society and provide psychosocial support and
- To provide educational support for caregivers.

There are some important cases to be considered in therapy sessions (Lewis&Bottomley, 2002):

1. An therapist who is more patient, relevant, knowledgeable and trustable, reduces the tension of the environment. During therapy session, the most important cause of high anxiety in the elderly is being fumble in front of family members and their own therapists, and fear of humiliation. Therefore, characteristics of therapists are important for the elderly to learn, in terms of providing psychological comfort.

2. Making frequent changes in the curriculum and environment of the elderly should be avoided as much as possible. Since unknown environment will bother elderly, training in their natural ambient is recommended. Once ability is gained, to adapt it to different environments later on will be more suitable.

3. The opportunity of visits to friends and relationships should be given to elderly to be social.

4. It should be put emphasis on family / caregiver training/ support. Family / caregiver training in assessment and treatment of the elderly is important. Because family members and caregivers provide actual physical and emotional support for the elderly. Therefore, their role in the solution of problems should be noted (Larson et al., 1986).

5. Physical and psychological comfort is essential. Ergonomic factors such as noise level, colors, lighting, ventilation, room temperature, comfortance of chair, table height and slipperiness of the floor should be considered. For example, the sounds of water or a computer come from behind should be eliminated.

6. To cope with a feeling of loneliness in the elderly:
   - meaningful relationships are developed,
   - recognizing the names of the people is encouraged to say
   - they are asked to remember the dates of birthdays accurately
   - communicating with plants, animals and children are provided
   - motivation for having something belong to them is provided and
   - alternative occupations are generated.
7. To strengthen the memory in the elderly;
   - audio-visual signals are used to introduce an object to be named permanently
   - principles of vocal motivation are applied to emphasize what is important in their lives
   - they are encouraged to tell their past experiences and to explain and discuss previous achievements
   - they are given sufficient time to remember the events
   - strategical games like chess take part of the programmes.

Fig. 1. Backgammon as a cognitive activity for memory strengthening

In ergotherapy, it is possible to increase skills such as spatial orientation, inductive thinking, fluid intelligence, problem solving and memory flexibility by using different methods. The advanced methods of testing and training are expected to contribute more higher-quality, productive and happy aging by reducing the decline in cognitive skills or by emphasizing cognitive characteristics during this period (Glantz, 1996). Reaching up and bending forward exercises or imagery exercises created from picking up apple from the tree, getting money from ground are effective in the elderly. Imagery exercises develops coordination and cognitive functions. They allow interactive training, because of facilitating a person to format an object mentally (Clark & Siebens, 2005; Nelson, 1997).

Fig. 2. Making puzzle and crosswords contributes to the cognitive health

Ergotherapy programs promote active aging and overall quality of life through participation in activities designed and prescribed to the needs of the individuals. The culture, satisfaction, motivation, interests and role in society of a person are taken into account. Elderly are encouraged to continue their habits and activities such as; gardening, non-strenuous sports, painting, handicrafts, building up a collection, simple repair work, singing and movie watching, which they used to enjoy participation. With some suggestions like "go on vacation, make hobbies and sports" elderly are removed from inactivity and negative psychology.

Old age is generally a period of limited environment. In fact, not only aging, but also an un-well organised environment for elderly restricts their power to live alone (Larson et al., 1986; Yucel et al., 2006a). Many elderly are not aware of being at risk of falling. 85% of falls happens especially on the stairs at home in the bathroom and bedroom. Therefore, this situation reveals that environmental changes and adaptations are needed for elderly to survive independently and self-sufficiently (Yucel et al., 2006a).

Fig. 3. Ergonomic accessibility with proper devices

Assistive aids such as walker should be suggested to increase stability during walking and to relieve stress in painful joints. An ergotherapist is needed to teach the use of assistive aids and joint protection techniques. Safety modifications and family / caregiver / elderly training reduce dangers. And providing adaptive tools which are necessary for age-related changes, positioning, teaching transfers and ambulation, training about health and prevention techniques and home exercise programs to increase the independence are the major topics of work field of a geriatric ergotherapist (Pu&Nelson, 2004).

As a result, service of the targeted rehabilitation to a person is the cornerstone of ergotherapy. Ergotherapy includes ADL, instrumental ADL, psychosocial well being, caregiver training, vocational rehabilitation, social / recreational leisure activities, lifestyle redesign, public health and environmental regulations for performing the roles successfully (Mountain, 2005a).

3. Leisure time activities

During lifelong, towards from young adulthood to middle age, interests and desires increase. Having more free time gives a person the opportunity to involve in an activity. But, leisure time activities in the elderly are more passive and home based. The time spent in outside cultural activities is quite less (Crombie et al., 2004). Older people often spend their times by visiting friends, listening to the radio, watching television and reading at their homes (Lee&King, 2003). Outdoor activities such as; sports, going to the theatre and cinema are activities with less continuity. There are some studies showing that this condition and low levels of recreational activity in the elderly are associated with changes in their body function (e.g. excess body mass index), marital status, low education level, male gender, genetic and metabolic factors (Mc Pherson&Kozlik, 1987; Mouton et al.,2000; Strain et al., 2002; Ross,1990).

Fig. 4. Older people should participate in physical recreational activities

The habits of regular participation in physical activities among the elderly are decreased physiologically (Dipietro, 2001). Activities such as cycling are non-preferred activities, because they may often cause injuries (Gerson& Stevens, 2004). In literature there are some studies showing that male elderly are more active, but the role of women in recreational activities are more than men. Conversely, some studies show that women have less leisure time activity (Bruce & Devine, 2002).

Activity restriction in the elderly may be due to functional limitations in areas, such as vision, hearing and mobility (Cambois et al., 2005; Donohue et al., 1995). In a study in the United States, it has been shown that approximately 10% of the elderly have visual impairments cause depression, social dysfunction and lack of activity (Donohue et al., 1995). In the elderly with severe cognitive problems, some failures in memory, expression, orientation, visual perception and other complex abilities are obstacles that elderly require higher cortical functions to participate in some activities (Adler, 1997).

Elderly's interests and skills to leisure time activities may also be reduced due to changes in health and other social areas with ageing (Clark&Siebens, 2005; Routasalo et al., 2007; World Health Organisation (WHO)). Motivation is critical to the success of activity programs designed for the elderly. Many elderly have concerns about participating in activity programs, because they have developed negative behaviors and beliefs throughout their lives associated with activity. Barriers to activity programmes, such as having to park their cars long distances away, along with other factors such as snow and ice, inhibit involvement. Reducing these barriers to exercise is necessary for a more independent and "robust" elderly community (Resnick, 1991; Yoshimoto & Kawata, 1996).

4. Multi-purpose activities

Multi-purpose activities in ergotherapy programs have a positive impact on the independence of the elderly rather than delays of motor aging process. Scientists in twenty first century, specialized in therapeutic recreational activities, have begun to work to find significant and meaningful activities for the elderly (Cottrell, 1996; Heuvelen et al., 1998).

Snoezelen sensory training spreads over a wide area in clinical practice from learning disorders to dementia in the past decade. In this method, primary visual, hearing, touch, taste and smell senses are stimulated with the effect of light, soothing music, touch and relaxation oils (Chung et al., 2002; Lynch& Aspnes, 2004). Besides that vision and hearing are basic requirements of communication, touch is also an important physical sensation component. These sensory inputs should be taken into account in planning a major activity program for the elderly (Lewis& Bottomley, 2002).

Recreational rehabilitation in occupations such as; board games, handicrafts, playing a musical instrument, playing volleyball with balloon and dancing performs cognitive function activation by increasing the blood flow rate of the prefrontal region. There are some studies showed that having been in a leisure time activity like purposeful cognitive activities, such as reading at least two times a week significantly reduces the risk of dementia (Nagaya et al., 2005; Scarmeas et al., 2001; Yucel et al., 2006b, 2010).

Fig. 5. Dancing performs cognitive function activation

Elderly are the people who are at risk for anxiety and depression. Social participations in activities such as painting, making music and religious meetings protect elderly from these risks (Lynch&Aspnes, 2004). In a study in the UK, it is stated that many activities are not effective as much as participating in religious gatherings that have a significant impact on well-being and quality of life in aged 50-74 (Routasalo et al.; 2007; Warr et al., 2004). Visiting friends and participating in social groups have positive effects on being healthy, having regular physical activities and carrying out ADL independently (Yoshimoto& Kawata, 1996). Reading is recommended in order to organize the behavior of depressive people and remove negative thoughts (Lynch& Aspnes, 2004). Baklien and Carlsson said that visiting a library and borrowing books keep people intellectually active (Baklien & Carlsson, 2000, as cited in Wikstrom, 2004).

One of the primary modalities used to treat depression in the elderly is medicine. However, taking anti-depressants without knowing the underlying reason can cause serious side effects. Therefore, alternative therapies are needed. There are many non-pharmacological treatment methods, such as the real orientation, behavioral therapy, sensory stimulation, music therapy and ergotherapy. Reminiscence therapy is also one of them. It is an effective method to gain self-confidence, socialization, well-being, expression and cognitive function. Reminiscence means discussion with a person or a group about activities, events and experiences done in the past, with the help of photos and / or music archive. The elderly indicate that they feel relaxed when they remember nice memories while looking at photo albums. This method reminds all the elderly of having lived a whole life and it still continues (Royeen & Reistetter, 1996; Stinson & Kirk, 2006; Woods et al., 2005).

Fig. 6. Reminiscence therapy

Activities like looking at photo gallery and dancing are important for successful aging and perform daily activities independently. This kind of activities help the elderly to know that they are prepared for changing conditions, express themselves, change their perspectives about the life. Painting or deal with a music are a visual and auditory experience for them (Wikstrom, 2004). That the music takes place in activities becomes a positive influence on well-being of the elderly, especially who has depression and cognitive problems. Music therapy is a proven, easily accessible and useful method to be able to cope with behavior problems such as stress and anxiety. Activities with musics which old people's own choices are both suitable for the control of agitated behavior and cheap. Carefully selected music tone, type and rythm are important to make activities fun. Light and mid-rythmical music is preferred. Music therapy in different categories, such as orchestral music, piano and jazz, decreases the heart rate and respiratory rate, increases body temperature and also the body relaxation (Hsu& Lai, 2004; Lai, 1999, 2004; Lou, 2001; Sherratt et al., 2004; Sung& Chang, 2005; Hanser & Thompson, 1994). Nevertheless, another study has indicated that music does not have any effect on pain perception in stroke patients having upper extremity exercises (Kim&Koh, 2005). Playing a musical instrument is an effective activity to avoid the elderly from isolation and increase their socialization, and make their free time full (Zelazny, 2001).

Creative activities reduce depression and isolation, and increases the power of decision-making of the elderly. Art is a way of opening people's emotional windows and sensory capacity. This kind of activities allow the elderly to express themselves, permit positive effects of well-being, enable physical, sensory-motor and cognitive therapy and teach appropriate ways to respond to the challenges of passing years. For the aforementioned reasons activities hold an important role in ergotherapy (Callanan, 1994; Hannemann, 2006; Mountain, 2005b).

5. Activity training models

There are some basic models that activity training based on the elderly. According to the activity treatment model developed by Mosey in 1977 in the U.S., people's capabilities which are necessary to survive in a wide range of the community are enhanced. This model has been developed to understand why therapists should make assessment and treatment and they suggest specific activities or plan activities in a specific approach to a person. This activity model lost their validity today, because it could not provide an improvement due to focus on personal development. Other models have been developed based on roles. They allow short-term applications. The Canadian Model of Occupational Performance (CMOP) and The Model of Human Occupation (MOHO) are two of them. In these new models, cognitive and behavioral approaches took part in place of psychodynamic perspectives (Chacksfield, 2006; Forsyth & Kielhofner, 2006; Sumsion & Blank, 2006).

CMOP is focused on how a person is successful in self-care and productivity and how he/she performs the roles in leisure time activities and how much satisfaction gives this to him/her. MOHO emphasized on the personal preferences, habits, roles and performance capacity. For example, an elderly person with dementia has to carry out the activity of making a cup of tea. Talking, willingness, motor / physical / cognitive / mental abilities are required for this activity. Social and physical environment, routine work, past experiences and expectations, etc. are questioned by ergotherapist. Both models are not only in activity training, but nowadays also used frequently in all ergotherapy interventions for all the health problems that can be seen throughout the life.

6. Activity training

Activity training, which began to be more popular in the 1940's, is a part of a comprehensive rehabilitation program in the elderly who want to have active aging. It aimes to keep life healthy, increase cognitive, emotional and physical capabilities, ensure the independence of social functions through purposeful and appropriate activities designed to the desires and needs of the elderly (Donohue et al., 1995; Lachenmayr & Mackenzie, 2004; Maestre Castelblanque & Albert Cunat, 2005; Nelson, 1997; Vass et al., 1995). Older people find an opportunity to apply activities through their own choice for the expectations, that make them have more higher quality of life (Duncan-Myers & Huebner, 2000).

Ergotherapy plays a significant role to develop the skills in leisure time activities (Glantz, 1996). Ergotherapist explaines the meaning of one's activity by revealing age, gender, role performances, cultural values, wishes and preferences of a person. Evaluation of the special functional activity skills is one of the duty of ergotherapist (Mountain, 2005b). Ergotherapist recommends the elderly activities and social relationships to carry out daily activities, continue existing skills for social integration and gain new skills. Advices to continue a quiet and relax life, listening to the songs of the past, talking to tell, watching the beautiful scenery, being sufficient on maintaining self care, go for shopping, cooking and house cleaning, do sports/water exercises and acquisition of new hobbies are effective for the elderly. Accordingly, these activities help elderly to gain and protect abilities in fields such as communication, cognitive functions or hand motor control. Multi-purpose activity approaches aim to improve special functionality, reintegration activities supported by lifestyle / behavioral and family education, sensory stimulation, encourage the elderly to express themselves and ADL training (Wikstrom, 2004; Yucel et al., 2010).

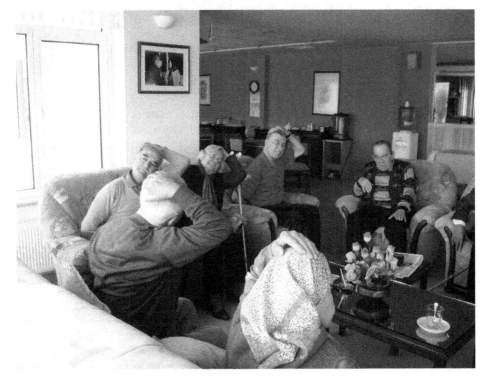

Fig. 7. Group exercise as multi-purpose activities

Fig. 8. Water exercises as an activity training method

The effects of multi-purpose activity training in the elderly:

1. It improves physical/psychosocial health and well-being.
2. It reduces the feeling of loneliness and establishes a close relationship with environment and increases the verbal interaction. Therefore, the elderly in a society peel off a thought of seeing themselves as redundant individuals. It is provided that the society accepts older people as unique individuals.
3. It provides environmental awareness, increases attention and problem-solving ability; reduces orientation distortion, and improves memory.
4. It decreases secondary complications such as decubitus ulcer, urinary tract infection, and it is protective against hypertension, diabetes mellitus, some cancers, osteporosis and depression (Crespo & Keteyian, 1996; Pang et al., 2005).
5. It reduces vital risk factors and contributes to the long-term protection of health status.

7. Group activities

There are many studies in the literature given activities to the elderly as individual or set in groups. However, general opinion is on behalf of effectiveness of group activity sessions. Because it is protective against feeling of loneliness and a lack of hope (Hannemann, 2006). Group activities develop self confidence, cognitive skills and ability of planning. Making a decision becomes easier. The elderly see their own prodecutivity, feel more comfortable and happy, and become more social in group (Landi et al., 1997).

Group activity is a modality which people, selected carefully, need emotional or physical support, are involved into a group by a trained therapist to help each other. Objectives of group activities are to increase awareness and to develop interpersonal and social skills by interaction with the other group members who provide feedback through behaviors. Compared to individual activities, the two main strong points of group activities are that the person can receive immediate feedback from their peers and there is an opportunity for therapist to observe the psychological, emotional and behavioral responses of each person.

Mills offers six different approaches to work with small group of models (Royeen & Reistetter, 1996):

1. Mechanical model: It is a model of mutual interaction with being independent from emotion, norms and believes. Each member controls the behaviors of other members of the group.
2. Organism model: In this model, group looks like a biological organism. Each member of the group has different role and responsibility due to his/her nature.
3. Complex model: This model advocates that independence in the changing needs of people and their obtained resources are limited.
4. Balance model: This model ensures a balance between internal needs and external requirements of the group.
5. Structural functional model: It is a model not only increases its resources, but it is also willing to change the structure and function of the group.
6. Growth model: This model develops depending on the capacity of members of the group and processes information.

Duncombe and Howe have formed ten different ergotherapy groups (Duncombe & Howe, 1985, as cited in Royeen& Reistetter, 1996):

1. Exercise group: Groups are generally formed in such exercises as volleyball, bowling and ping pong in rehabilitation centers and schools.
2. Dining group: This kind of groups are generally being in psychology and rehabilitation programs. Activities in these groups have menu planning, shopping, cooking and eating sections.
3. ADL Group: In these groups, people are prepared to live independently in the community by increasing the required self-care skills.
4. Handicraft group: In this group, art and craft skills are used for psychosocial evaluation and treatment of disabilities.
5. Task group: This group gives social, recreational and educational activities in tasks such as organizing a picnic and publishing newspaper to facilitate communication and socialization.
6. Self-expression group: In this group, interactive work among members is provided to the elderly with pictures, music and self-awareness exercises.
7. Reality orientation discussion group: Role simulation is provided to improve socialization and communication skills in this group.
8. Sensory-motor and sensory integration groups: Thorough these groups, integration of physical skills and sensory development are aimed to increase in the individuals with a wide range of problems like learning disabilities, hearing and visual problems and lack of sense of integrity.

9. Oriented sensation groups: Performing some roles in the style of the game and discussing in poetry and fantasy groups are performed in these groups.
10. Education group: There are groups, individuals and their families receive and discuss information on issues such as drugs.

There are some important factors to consider to determine the suitability of a person in a group (Ozmenler, 2005):

1. The possibility of a high level of a peer anxiety in a person, who has a negative reaction to be in a group, should be considered. Elderly who have destructive relationships with their peers do not want to be in a group, but that will be useful, if they can come to deal with being in a group. On the other hand, those who worry about authority too much, usually with fear of a therapist criticism, may be reluctant to express their thoughts and feelings in a seperate media. So that, they may accept the group as a nice treat and generally prefer group activities. Group media is more comfortable, because of being usually bilateral (one to one) environment (Landi et al., 1997).
2. Determining of impairments of the elderly is important to choose the best activity approach and assess their motivations, capacities and the strengths and weaknesses in their personality. For example, antisocial people do not find a heterogeneous group good and do not accept group standards.
3. It is needed that the therapist gives a detailed depiction of the process to the elderly as possible and to responds every question of the elderly.
4. Group activities are successful in three to fifteen members. Mostly groups with eight to ten members are preferred. The sufficient interactions may not be received in small groups with fewer members. With more than 10 members, the members or the therapist can not follow what is going on. Small groups are the smallest representatives of large communities. It may be difficult to work with large group. So, small groups are preferred (Royeen& Reistetter, 1996).
5. Groups are collected two times per week. One is with therapist and the other one is without therapist. It is important to maintain continuity of sessions. Usually group sessions take 1-2 hours. The time limit should be fixed.
6. Educational planning for the elderly often takes place in ergotherapy training programs. Visual aids and adaptations are used, such as timing and the number used for each person in the group.

The effectiveness of group activities are measured as follows:

How much did the group members reach the objectives?
What is satisfaction level in each individual?
What are quality and quantity of the product?

As a conclusion, good planning should be done in order to encourage the elderly to participate in the activity regularly and continuity. National campaigns are expected to be necessary and effective for the elderly to change perceptions about their levels of physical activity.

8. References

Adler, G. (1997). Driving and dementia: dilemmas and decisions. *Geriatrics*, 52, pp. 26-29, ISSN 0016-867X

Alpass, F.M. & Neville, S. (2003). Loneliness, health and depression in older males. *Aging and Mental Health*, 7 (3), pp. 212-216, ISSN1360-7863

Boswell, R.B.; Dawson, M. & Heininger, R. (1997). Quality of life defined by adults with spinal cord injuries. *Perceptual Motor Skills*, 84 (3), pp. 1140-1149, ISSN 0031-5125

Bowling, A. & Iliffe, S. (2006). Which model of successful aging should be used? Baseline findings from a British longitudinal survey of ageing. *Age Ageing*, 35 (6), pp. 607-14, ISSN 0002-0729

Bruce, D.G. & Devine, A. (2002). Recreational physical activity levels in healthy older women: The importance of fear of falling. *Journal of American Geriatric Society*, 50, pp. 84-89.

Callanan, B.O. (1994). Art therapy with the frail elderly. The *Journal of Long Term Home Health Care. The PRIDE Institute Journal*, 3 (2), pp. 20-23, ISSN 1072-4281

Cambois, E.; Robine, J.M. & Romieu, I. (2005). The influence of functional limitations and various demographic factors on self reported activity restriction at older ages. *Disability and Rehabilitation*, 5; 27 (15), pp. 871- 883, ISSN 0963-8288

Chacksfield, J. (2006). Activities therapy. In: *Foundations for practice in occupational therapy*, E.A. Duncan (Ed.), pp. 143-259, Elsevier Churchill Livingstone, London.

Chung, J.C.; Lai, C.K.; Chung, P.M. & French, H.P. (2002). Snoezelen for dementia. *Cochrane Database of Systematic Reviews*, (4): CD003152.

Clark, G.S. & Siebens, H.C. (2005). Geriatric Rehabilitation, In: *Physical Medicine and Rehabilitation*, J. Lisa (Ed.). pp.1531-1560, Lippincott Williams Wilkins, Philadelphia.

Cottrell, F.R.P. (1996). *Perspectives on purposeful activity: Foundation and future of occupational therapy*. The American Occupational Therapy Assosiation, Inc., USA.

Crespo, C. J. & Keteyian, S. J. (1996). Leisure-time physical activity among US Adults. Results from the third national health nutrition examination survey. *Arcieves of Internal Medicine*, 156, pp. 93-98, ISSN 0003-9926

Crombie, I.K.; Irvine, L. & Williams, B. (2004). Why older people do not participate in leisure time physical activity: asurvey activity levels, beliefs and deterrents. *Age Aging*, 33 (3), pp. 287-292, ISSN 0002-0729.

Csikszentmihaly, M. (1993). Activities and happiness: towards a science of occupation. *Journal of Occupational Science*. 1 (1), pp. 38-42.

Dipietro, L. (2001). Physical activity in aging: Changes in patterns and their relationship to health and function. *Journal of Gerontology*, 56A, pp. 13-22.

Donohue, B.; Acierno, R.; Hersen, M. & van Hasselt, V.B. (1995). Social skills training for depressed, visually impaired older adults. A treatment manual. *Behavioral Medicine*. 19 (4), pp. 379-424, ISSN 0896-4289

Duncan-Myers, A.M. & Huebner, R.A. (2000). Relationship between choice and quality of life among residents in long-term-care facilities. *American Journal of Occupational Therapy*, 54 (5), pp. 504-508, ISSN 0272-9490

Forsyth, K. & Kielhofner, G. (2006). The model of human occupation integrating theory into practice and practice into theory. In: *Foundations for practice in occupational therapy*, E.A. Duncan (Ed.), pp. 80-108, Elsevier Churchill Livingstone, London.

Gerson, L.W. & Stevens, J.A. (2004). Recreational injuries among older Americans. *Injury Prevention*, 10 (3), pp. 134-138, ISSN 1353-8047

Glantz, C.H. (1996). Evaluation and intervention for leisure activities. K.O. Larson (Ed.). *The role of occupational therapy with elderly (ROTE)*. (s. 729-741). The American Occupational Therapy Association,USA.

Hannemann, B.T. (2006). Creativity with dementia patients. Can creativity and art stimulate dementia patients positively? *Gerontology*, 52 (1), pp. 59-65.

Hanser, S.B. & Thompson, L.W. (1994). Effects of a music therapy strategy on depressed older adults. *Journal of Gerontology*, 49 (6), pp. 265-269.

Heuvelen, M.G.; Kempen, G.M.; Ormel, J. & Rispens, P. (1998). Physical fitness related to age physical activity in older persons. *Medicine and Science in Sports Exercise*, 30 (3), pp. 434-441, ISSN 0195-9131

Hsu, W.C. & Lai, H.L. (2004). Effects of music on major depression in psychiatric inpatients. *Archives of Psychiatric Nursing*, 18 (5), pp. 193-199, ISSN 0883-9417

Inal, S.; Subaşı, F.; Ay Mungan, S.; Uzun, S.; Alpkaya, U.; Hayan, O. & Akarçay, V. (2003). Yaşlıların Fiziksel Kapasitelerinin ve Yaşam Kalitelerinin Değerlendirilmesi, *Geriatri* 6 (3), pp. 95-99.

Jylha, M. (2004). Old age and loneliness: cross-sectional and longitudinal analyses in Tampere longitudinal study on aging. *Canadian Journal of Aging*, 23 (2), pp. 157-168, ISSN 0714-9808

Kim, S.J.& Koh, I. (2005). The effects of music on pain perception of stroke patients during upper extremity joint exercises. *Journal of Music Therapy*, 42 (1), pp. 81-92, ISSN 0022-2917

Kolt, G.S; Driver, R.P. & Gilles, L.C. (2004). Why older Australians participate in exercise and sport.. *Journal of Aging and Physical Activity*, 12(2), pp. 185-198, ISSN 1063-8652

Lachenmayr, S. & Mackenzie, G. (2004). Building a foundation systems change: increasing access to physical activity programs for older adults. *Health Promotion and Practice*, 5 (4), pp. 451-458, ISSN 1524-8399

Lai, Y.M. (1999). Effects of music listening on depressed women in Taiwan. *Issues in Mental Health Nursing*, 20 (3), pp. 229-246, ISSN 0161-2840

Lai, H.L. (2004). Music preference and relaxation in Taiwanese elderly people. *Geriatric Nursing*, 25 (5), pp. 286-291, ISSN 0950-0448

Landi, F.; Zuccala, G. & Bernabei, R. (1997). Physiotherapy and occupational therapy, A geriatric experience in the acute care hospital. *American Journal of Physical Medicine and Rehabilitation*, 76, pp. 38-42, ISSN 0894-9115

Larson, R.; Mannell, R. & Zuzanek, J. (1986). Daily well being of older adults with friends and family. *Psychology and Aging*, 1(2), pp. 117-126, ISSN 0882-7974

Lee, R.E & King, A.C. (2003). Discretionary time among older adults: how do physical activity promotion interventions affect sedentary and active behaviors? *Annals of Behavioral Medicine*, 25 (2), pp. 112-119, ISSN 0883-6612

Lewis, C.B. & Bottomley, J.M. (2002). Principles and practice in geriatric rehabilitation, *In: Geriatric Physical Therapy. A Clinical Approach*, pp. 249-287, Prentice Hall, USA.

Lou, M.F. (2001). The use of music to decrease agitated behaviour of the demented elderly: the state of the science. *Scandinavian Journal of Caring Sciences*, 15 (2), pp. 165-173, ISSN 0283-9318

Lynch, T.R &Aspnes, A.K. (2004). Individual and group psychotherapy. In: *Textbook of Geriatric Psyciatry*, D.G. Blazer, D.C. Steffens & E.W. Busse. (Ed.). pp. 443-458, London.

Mac Pherson, B. & Kozlik, C. (1987). Age patterns in leisure participation, In: *Aging in Canada: Social perspectives*, V. Marshall (Ed.), pp. 89-95, Toronto: Fitzhenry and Whiteside.

Maestre Castelblanque, E. & Albert Cunat, V. (2005). Life style activities in older people without intellectual impairment: a population-based study. *Rural and Remote Health*, 5 (1), p. 344, ISSN1445-6354

Matsuo, M.; Nagasawa, J.; Yoshino A.; et.al. (2003). Effects of activity participation of the elderly on quality of life. *Yonago Acta Medica*, 46, pp. 17-24, ISSN 0513-5710

Mountain, G. (2005). Occupational Therapy Interventions, In: *Occupational therapy with older people*. (pp. 160-191). Whurr Publishers, London.

Mountain, G. (2005). Services to meet needs, In: *Occupational therapy with older people*. (pp. 102-135). Whurr Publishers, London.

Mouton, C.P.; Calmbach, W.L; Dhanda, R.; Espino, D.V & Hazuda, H. (2000). Barriers and benefits to leisure time physical activity among older Mexican Americans. *Archives of Family Medicine*, 9 (9), pp. 892-897, ISSN 1063-3987

Nagaya, M.; Endo, H.; Kachi, T. & Ota, T. (2005). Recreational rehabilitation improved cognitive function in vascular dementia. *Journal of American Geriatric Society*, 53 (5), pp. 911-912.

Nelson, D.L. (1997). Why the profession of occupational therapy will flourish in the 21 th century. *American Journal of Occupational Therapy*, 51 (1), pp. 13-24.

Ozmenler, N. (2005). Geriyatrik Psikiyatri, In: *Klinik Psikiatri*, H. Aydın, A. Bozkurt (Ed.), pp. 599-601, Ankara: Kaplan& Benjamin Sadock, Virginia Sadock.

Pang, M.Y.; Eng, J.J.; Dawson, A.S.; Mc Kay, H.A. & Harris, J.E. (2005). A community base fitness and mobility exercise program for older adults with chronic stroke: a randomized, controlled trial. *Journal of American Geriatric Society*, 53 (10), pp. 166-174.

Pu, C.T. & Nelson, M.E. Aging, function and exercise, In: *Exercise in Rehabilitation Medicine*, W.R. Frontera, D.M. Dawson ve D.M. Slovik (Eds.), pp. 391-424, USA: Human Kinetics.

Reichert, F.F.; Barros, A.J.; Domingues, M.R. & Hallal, P.C. (2007). The role of perceived personal barriers to engagement in leisure-time physical activity. *American Journal of Public Health*, 97 (3), pp. 515-519.

Resnick, B.M. (1991). Geriatric motivation: Clinically helping the elderly to comply. *Journal of Gerontological Nursing*, 17 (6), pp. 4-8, ISSN 0098-9134

Rosalie, A.K. (2003). Social assessment of geriatric patients. In: *Brocklehurst's Text book of Geriatric Medicine and Gerontology*, R.C. Tallis & H.M. Fillit (Eds.), p. 187, Churchill Livingstone, Spain.

Ross, R.R. (1990). Time-use in later life. *Journal of Advanced Nursing*, 15, pp. 394-399, ISSN 0309-2402

Routasalo, P.E.; Savikko, N. & Tivils, R.S. (2007). Effectiveness of psychosocial group rehabilitation in relieving loneliness of older people. *Advances in Gerontology*, 20 (3), pp. 24-32, ISSN 1561-9125

Royeen, M. & Reistetter, T.A. (1996). K.O. Larson (Ed.). *The role of occupational therapy with elderly (ROTE)*. pp. 774-804. The American Occupational Therapy Association, USA.

Scarmeas, N.; Levy, G.; Tang, M.; Manly, J. & Stern, Y. (2001). Influence of leisure activity on the incidence of Alzheimer's disease. *Neurology*, 26; 57(12), pp. 2236-2242, ISSN:2090-5513

Sherratt, K.; Thornton, A. & Hatton, C. (2004). Emotional and behavioural responses to music in people with dementia: an observational study. *Aging and Mental Health*, 8 (3), pp. 233-241, ISSN 1360-7863

Stinson, C.K. & Kirk, E. (2006). Structured reminiscence: an intervention to decrease and increade depression and increase self-transcendence in older women. *Journal of Clinical Nursing*, 15 (2), pp. 208-218, ISSN 0962-1067

Strain, L.A.; Grabusic, C.C.; Searle, M.S. & Dunn, N.J. (2002). Continuing and ceasing leisure activities in later life: a longitudinal study. *The Gerontologist*, 42 (2), pp. 217-223, ISSN 0016-9013

Sumsion, T. & Blank, A. (2006). The canadian model of occupational performance (CMOP), *Foundations for practice in occupational therapy*, E.A. Duncan (Ed.), pp. 109-124, Elsevier Churchill Livingstone, London.

Sung, H.C. & Chang, A.M. (2005). Use of preferred music to decrease agitated behaviours in older people with dementia: a review of the literature. *Journal of Clinical Nursing*, 14 (9), pp. 1133-1140, ISSN 0962-1067

Vass, M.; Avlund, K. & Lauridsen, J. (2005). Feasible model for prevention of functional decline in older people: municipality-randomized, controlled trial. *Journal of American Geriatric Society*. 53 (4), pp. 563-568.

Vitkus, J. & Horowitz, L.M. (1987). Poor social performance of lonely people: lacking a skill or adopting a role? *Journal of Perceptual Social Psychology*, 52 (6), pp. 1266-1273.

Ward, R.A. (1979). The meaning of voluntary association participation to older people, *Journal of Gerontology*, 34 (3), pp. 438-445.

Warr, P.; Butcher, V. & Robertson, I. (2004). Activity and psychological well-being in older people. *Aging and Mental Health*, 8 (2), pp. 172-183, ISSN 1360-7863

WHO rapport (1998). A population aged 65 and above, 1997 in An Aging World, 30.10.2007, http://www.who.org/whr/1998/age-97-e.gif.

Wikstrom, B.M. (2004). Older adults and the arts: the importance of aesthetic forms of expression in later life. *Journal of Gerontological Nursing*, 30 (9), pp. 30-36, ISSN 0098-9134

Woods, B.; Spector, A.; Jones, C.; Orrell, M. & Davies, S. (2005). Reminiscence therapy for dementia. *Cochrane Database of Systematic Reviews*, 18 (2): CD001120.

Yoshimoto, T. & Kawata, C. (1996). Negative effect of perceived transportation problems on social activities of elderly people in a small town far from the nearest train station. *Japanese Journal of Geriatrics*, 33 (12), pp. 928-934, ISSN 0300-9173

Yucel, H.; Yıldırım, S.A. & Kayıhan, H. (2006). Assessments in Geriatric Rehabilitation. In: *Geriatri ve Gerontoloji*, S.Arıoğul (Ed.), pp. 183-197, Medikal&Nobel, Ankara.

Yucel, H.; Düger, T.; Uyanık, M. & Kayıhan, H. (2006). Occupational Therapy in Geriatric Rehabilitation, In: *Geriatri ve Gerontoloji*, S. Arıoğul (Ed.), pp. 205-221, Medikal & Nobel, Ankara.

Yucel, H. PhD Thesis, Hacettepe University, Institute of Health Sciences, Ankara, 2008

Yucel, H. & Kayıhan, H. (2010). Effects of multi-purpose activities on cognitive functions in the elderly who reside at home and the nursing home, *Turkish Clinics Medicine Sciences*, 30 (1): pp. 227-232.

Zelazny, C.M. (2001). Therapeutic instrumental music playing in hand rehabilitation for older adults with osteoarthritis: four case studies. *Journal of Music Therapy*, 38 (2), pp. 97-113, ISSN 0022-2917

12

The Level of ROS and DNA Damage Mediate with the Type of Cell Death, Senescence or Apoptosis

Takafumi Inoue and Norio Wake
Department of Obstetrics and Gynecology,
Kyushu University, Fukuoka,
Japan

1. Introduction

Cell senescence, originally defined as the proliferative arrest that occurs in normal cells after a limited number of cell divisions, is now more broadly regarded as a general biological program of terminal growth arrest. Replicative senescence of cells due to telomeric changes exhibits similar features with those seen in DNA damage (Vaziri et al., 1997; von Zglinicki, 2001). Therefore, DNA damage is expected to induce rapid cell growth arrest, which would be phenotypically indistinguishable from replicative senescence (Di Leonardo et al., 1994). This type of accelerated senescence that does not involve telomere shortening is triggered in normal cells by the expression of supraphysiological mitogenic signals (Orr et al., 1994). Not only normal cells, but also cancer cells can be induced readily to undergo senescence by genetic manipulation or by treatment with chemotherapeutic agents, radiation, or differentiating agents.

Reactive oxygen species (ROS), which are byproducts of normal cellular oxidative processes, are involved in senescence (Chen et al., 1998). Senescent cells have higher levels of ROS than normal cells (Hagen et al., 1997), and oxidative stress caused by sublethal doses of H_2O_2 or hyperoxia can force human fibroblasts to arrest in a manner similar to senescence (Dumount et al., 2000). Additionally, both oncogenic Ras and p53 induce senescence in association with increased intracellular ROS (Lee et al., 1999; Macip et al., 2003). p53 induces the accumulation of ROS presumably through a transcriptional influence on pro-oxidant genes (Polyak et al., 1997). Up-regulation of p21 also causes increased ROS levels in both normal and cancer cells (Macip et al., 2002), although the molecular mechanism remains unknown.

Recent evidence has suggested that p21 mediates apoptosis in a p53-independent manner (Roninson, 2003; Hsu et al., 1999), although its role in this apoptotic pathway remains controversial. In view of the possible roles played by ROS in both senescence and apoptosis, and the capacity of p21 to elevate ROS levels, we investigated the involvement of ROS in p21-induced cell death. Additionally, we studied how the status of p21 expression modulated ROS levels to achieve alternative cell fates.

2. Exogenous p21 protein induces senescence and apoptosis induced by p21 mediated through p53

We generated the recombinant adenovirus vector contained either the full-length p21 (Ad-p21). An adenovirus containing an empty vector (mock) was used as a control. We transferred these constructs into two cancer cell lines (LoVo, and HCT116) using an adenovirus infection system. The cell lines were initially selected based on their susceptibility (>80%) to adenovirus. The expression of p21 protein corresponding to the transfected vector was demonstrated by immunoblots (data not shown).

Infection of these cancer cell lines with 20MOI Ad-p21 resulted in growth arrest, which became irreversible after four days. This permanent growth arrest was accompanied by the presence of a senescence-specific marker, SA-β-gal positivity, as well as morphological changes such as a dramatic increase in cell size, and enlarged and prominent nuclei.

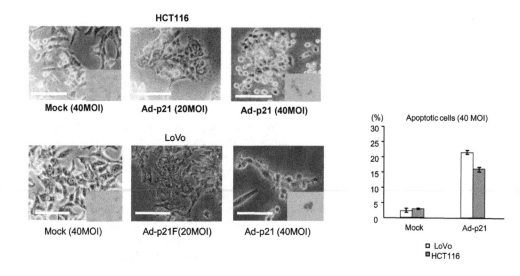

Fig. 1. Induction of senescence and apoptosis by p21 overexpression.
When LoVo and HCT116 cells were transfected with 20 MOI Ad-p21, cells were enlarged, flattened and had increased SA-β-gal positivity. At 40 MOI (LoVo and HCT116), cells were detached from the plate and floating. The small panel on the lower right for 40 MOI shows positive cells with the TUNEL assay. Mock transfected cells showed no change in morphology and were not positive for the TUNEL assays. The populations of apoptotic cells in LoVo and HCT116 infected with 40 MOI Ad-p21 after infection of 48 hrs are shown in right graph as indicated by the significant increased of sub G1 fraction. Data represent the average of three independent experiments and standards deviations are indicated by error bars. Bar=10μm

The apoptosis of cancer cells following DNA damage is p53-dependent and yet p21-independent (Waldman et al., 1996; Deng et al., 1995). However, due to differences in the experimental conditions for investigation the effects, contradictory conclusions have been drawn regarding the relationship between p21 and the induction of apoptosis (Tsao et al., 1999; Kagawa et al., 1999). In order to investigate the basis for the striking differences in the biologic responses to p21 expression, we measured the kinetics of p21 protein increase, as well as its expression levels following Ad-p21 infection by increasing MOI of the viral vector. Whereas 20 MOI Ad-p21 infection resulted in elongated, growth-arrested cells showing the morphological features of senescence, cells infected with 40 MOI Ad-p21 became rounded, contracted, and lost their ability to adhere to the plate with Apo-taq positive staining (Fig. 1). We used flow cytometry to measure the DNA contents of cancer cell lines following treatment Ad-p21. In LoVo and HCT116 cells, a hypoploid peak corresponding to a subG1 population had increased following infection with Ad-p21 (40 MOI) (data not shown). These results are characteristic of apoptosis.

3. The fate of cancer cells as a result of ROS generated by overexpressed p21 from adenoviral transfection

To investigate the roles of ROS in the senescent or apoptotic cell fates triggered by p21 expression in LoVo and HCT116 cells, we measured ROS levels with the fluorescent probe APF (Setsukinai et al., 2003), a marker of changes in the general accumulation of cellular oxidants. FACS analysis of APF-stained LoVo and HCT116 cells revealed a progressive increase in ROS levels following 20 MOI Ad-p21 infection. After three days of infection, when senescent morphological changes were first observed, the ROS levels in the cells were increased more than 2-fold. We next examined whether ROS levels were involved in the decision between senescence and apoptosis in LoVo and HTC116 cells. Both cells infected with 40 MOI Ad-p21 exhibited much higher ROS levels (4-fold) than did the cells infected with 20 MOI Ad-p21 (2-2.5 fold) (Fig. 2a).

We then established whether different cell fate outcomes were due to the levels of induced ROS and p21 protein. We investigated whether the antioxidant N-acetyl-L-cysteine (NAC), could protect cells from senescent phenotypes induced by 20 MOI Ad-p21 infection or the apoptotic phenotype by 40 MOI Ad-p21 infection. Both LoVo and HCT116 cells that harbored the wild type p53 gene were infected with 20MOI Ad-p21 were cultured in the presence of 10mM NAC for three days. Ad-p21 markedly induced SA-β-gal positive cells in the absence of NAC, and cultivation in the presence of NAC significantly suppressed the appearance of SA-β-gal-positive cells (Fig. 2b). Similarly, cultivation of LoVo and HCT116 cells in the presence of NAC for 3 days markedly inhibited apoptosis in response to 40 MOI Ad-p21 infection (Fig. 2c), thus suggesting that the induction of both senescence and apoptosis by p21 occurs via the generation of ROS.

4. Endogenous p21 protein up-regulation by sodium butyrate (NaB) induces cell death in a colon cancer cell line, HCT116

Next, we investigated whether up-regulation of endogenous p21 protein has an effect similar to that of Ad-p21 infection. We have previously demonstrated that NaB induced p21 expression, resulting in growth arrest and cell death in gynecologic cancer cells. [8]

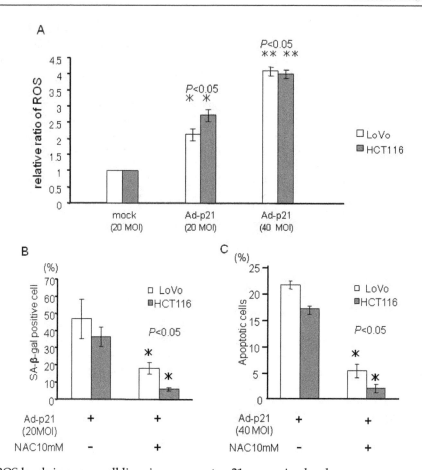

Fig. 2. ROS levels in cancer cell lines in response to p21 expression levels.

(a) ROS levels were evaluated by FACS analysis after staining LoVo and HCT116 cells with the fluorescent probe APF. Relative ratio of the geometric mean that is the average of the logarithm of the linear value for events expressed as the anti log in Ad-p21 (20 MOI or 40 MOI) infected cells as compared to the control (20MOI). Generated ROS levels were significantly higher in the cancer cells infected with 20 MOI Ad-p21 than those in control infected cells. $*P < 0.05$ In LoVo and HCT116 cells, generated ROS levels were significantly higher in the cancer cells infected with 40 MOI Ad-p21 than those in 20 MOI. $**P < 0.05$ Apoptosis was induced in the former and senescence in the latter.

(b) Senescence induced by p21-overexpression was inhibited by NAC (ROS scavenger). LoVo and HCT116 cells were cultured in 10 mM NAC and were infected with 20 MOI Ad-p21. The ratio of SA-β-gal positive cells after 74 hrs of the infection was significantly decreased in the presence of NAC. Results represent mean values of three experiments, and error bar shows the standard deviation.

(c) Induction of apoptosis with 40 MOI Ad-p21 was also inhibited by the addition of NAC in LoVo and HCT116 cells, as indicated by the significant decrease of sub G1 fraction. Results represent mean values of three experiments, and error bars shows the standard deviation.

To evaluate the induction of senescence in HCT116 cells, we analyzed the cell cycle alteration and SA-β gal staining in response to NaB. We measured the DNA contents of cancer cell lines treated with varying concentrations of NaB by flow cytometry. In HCT116, treatment with 0.5 to 1.0 mM NaB resulted in a decrease in the fraction of S phase (21%→4%) and G1 phase cells (57%→22%). Most cells accumulated in G2/M, suggesting arrest at the G2/M checkpoint. A hypoploid peak corresponding to the subG1 population was evident by flow cytometry following treatment with greater than 2.0mM NaB (Fig. 3a). This population corresponded to cells undergoing apoptotic cell death.

A. HCT116

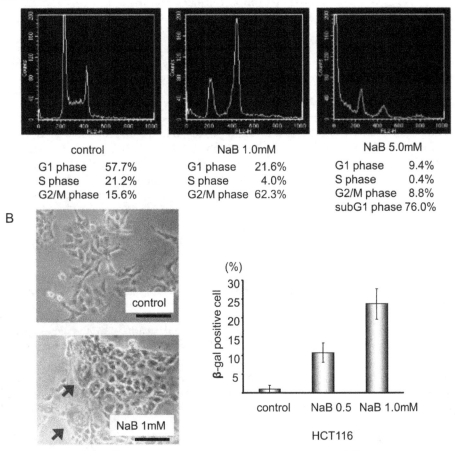

control		NaB 1.0mM		NaB 5.0mM	
G1 phase	57.7%	G1 phase	21.6%	G1 phase	9.4%
S phase	21.2%	S phase	4.0%	S phase	0.4%
G2/M phase	15.6%	G2/M phase	62.3%	G2/M phase	8.8%
				subG1 phase	76.0%

B

Fig. 3. (a) Effect of NaB on cell cycle analysis and the induction of cell death. DNA contents of HeLa and HCT116 cells with or without NaB for 24 hrs were analyzed by flow cytometry. NaB treatment reduced the percentage of cells in S phase and triggered the accumulation of cells in G2/M phase in HCT116 treated with of 1.0 mM NaB.
(b) Treatment with 1.0 mM NaB induced morphological change in HCT116 cells that included enlargement and flattening as well as an increase in the number of SA-β-gal positive cells (arrows). Bar=10μm

Incubation of HCT116 cells with 1.0 mM NaB for five days resulted in morphologic changes. These changes included an enlarged, flattened shape, increased cytoplasmic to nuclear ratio and decreased cell density accompanied by SA-β-gal staining (Fig. 3b).

5. Increased ROS levels in NaB-induced senescence and apoptosis of cancer cells

To investigate the contribution of ROS to senescence or apoptosis of HCT116 cells triggered by the treatment with different concentrations of NaB, we measured ROS levels as described above. FACS analysis of APF-stained cancer cell lines revealed a progressive increase in ROS levels following NaB treatment. The levels of ROS were increased following treatment both with 1 mM of NaB (2-3 fold) that induced senescence in HCT116 cells and with 5 mM of NaB (5-fold) that induced apoptosis compared with no treatment in HCT116 cells (Fig. 4a). The ROS level in apoptotic cells induced by NaB was markedly higher than that in senescent cells.

To further establish whether different cell fate outcomes were due to the induced ROS level, we investigated the effect of the antioxidant N-acetyl-L-cysteine (NAC) on the senescent phenotype. HCT116 cells were treated with 0.5 mM NaB in the presence or absence of 5mM NAC for 5 days. The increase of cell numbers in the G2/M fraction following treatment with 0.5 mM NaB was abrogated by co-treatment with 5mM NAC (Fig. 4b). As shown in Fig. 4c, culture in the presence of NAC significantly suppressed the number of SA-β-gal-positive HCT116 cells (23%→10%). This was accompanied by a decrease in ROS level, which suggested that the induction of senescence by NaB occurred via the generation of ROS. The treatment with NAC, however, could not prevent the apoptotic induction by higher concentrations of NaB in and HCT116 cells (data not shown), though the reason remained unknown.

6. DNA damage response (DDR) signals mediate NaB-induced cancer cell death

To clarify the association of NaB-induced cancer cell death with the DNA damage response, we next assayed for DDR signals including ATM and its downstream signals. One of the first processes initiated by DSB (double strand break) is massive phosphorylation of the tail of the histone variant H2AX (Redon et al., 2002). Foci of phosphorylated H2AX (γH2AX) are rapidly formed at the DSB sites and are thought to be essential for further recruitment of damage response proteins. γH2AX is dependent on the ATM protein and other members of the ATM family. [31] To examine the effect of NaB on DDR signals-related proteins, we analyzed the changes of DSB-related proteins expression levels in response to NaB by immunoblotting. Incubation with 1-5 mM NaB for 48 hrs resulted in the accumulation of γH2AX and ATM in HCT116 cells. The downstream proteins such as p53, phosphorylated p38 MAPK, and p21 were up-regulated by 1-5mM NaB after 48 hrs of incubation (Fig. 5a). In HCT116 cells, the levels of the DSB marker γH2AX were enhanced about 20 times when apoptosis was induced by incubating with 2 mM NaB and about 3.6 times when senescence was induced by 0.5 to 1.0 mM NaB for 24 hrs compared with the control (Fig. 5b).

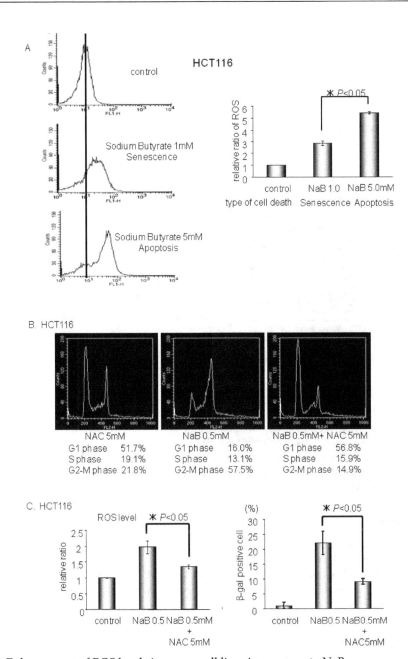

Fig. 4. Enhancement of ROS levels in cancer cell lines in response to NaB.
(a) ROS levels were evaluated by FACS analysis after staining HCT116 cells with the fluorescent probe APF. The ROS levels in apoptotic cells treated with 1.0 mM NaB are significantly higher than those in senescent cells induced by 5.0 mM NaB ($P<0.05$).

Data represent the average of three independent experiments and standards deviations are indicated by error bars.

(b) Treatment of NAC reduced the proportion of cells accumulating in G2/M phase and the ratio of senescent cells following NaB treatment. HCT116 cells were cultured with 0.5 mM NaB in the presence or absence of 0.5 mM NAC for 24 hrs.

(c) ROS level (left graph) and the ratio of senescent cells (right graph) after 96 hrs of treatment significantly decreased in the presence of NAC ($P<0.05$) compared to when NAC was absent. Results represent the mean values of three experiments, and error bars shows the standard deviation.

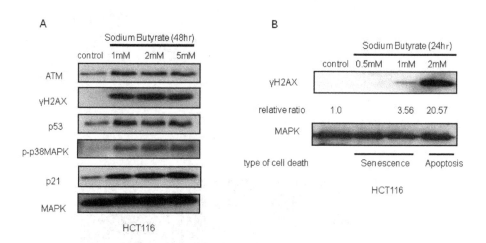

Fig. 5. NaB induced the expression of proteins associated with DSB.

(a) Western blot of ATM, phosphorylated H2AX (γH2AX), p53, phosphorylated p38 MAPK, p21 and MAPK after treatment with NaB treatment for 48hrs.

(b) The level of γH2AX protein in cells treated with 2.0 mM NaB for 24 hrs, was higher than that in cells treated with 1.0 mM NaB, a level which induced senescence. Levels of γH2AX associated with the type determination of cell death.

7. Conclusion

In this study, we obtained the following evidence. Exogenous and endogenous p21 up-regulation (Ad-p21 infection and NaB treatment) are able to induce senescence or apoptosis in cancer cell lines. The magnitude of ROS induced by p21 was critical for the p21-mediated cell fate decision. These findings help account for the differences in the p21-mediated cell fate decisions observed in various studies.

8. References

Chen QM, Bartholomew JC, Campisi J, Acosta M, Reagan JD, and Ames BN. Molecular analysis of H2O2-induced senescent-like growth arrest in normal human fibroblasts: p53 and Rb control G1 arrest but not cell replication. *Biochem J* 1998; 332: 43-50

Di Leonardo A, Linke SP, Clarkin K, and Wahl GM. DNA damage triggers a prolonged p53-dependent G1 arrest and long-term induction of Cip1 in normal fibroblasts. *Genes Dev* 1994; 8: 2540-51

Dumont P. Burton M, Chen QM, Gonos ES, Frippiat C, Mazarati J, Eliaers F, Remacle J, and Toussaint O. Induction of replicative senescence biomarkers by sublethal oxidative stress in normal human fibroblast. *Free Radical Biol Med* 2000; 28: 361-73

Deng C, Zhang P, Harper JW, Elledge SJ, and Leder P. Mice lacking p21CIP1/WAF1 undergoes normal development, but is defective in G1 checkpoint control. *Cell* 1995; 82: 675-84

Hsu SL, Chen MC, Chou YH, Hwang GY, and Yin SC. Induction of p21(CIP1/Waf1) and activation of p34(cdc2) involved in retinoic acid-induced apoptosis in human hepatoma Hep3B cells. *Exp Cell Res* 1999; 248: 87-96

Hagen TM, Yowe DL, Bartholomew JC, Wehr CM, Do KL, Park JY, and Ames BN. Mitochondria decay in hepatocytes from old rats: membrane potential declines, heterogeneity and oxidants increase. *Proc Natl Acad Sci U S A* 1997; 94: 3064-69

Kagawa S, Fujiwara T, Kadowaki Y, Fukazawa T, Sok-Joo R, Jack A, Roth JA, and Tanaka N. Overexpression of the p21[sdi1] gene induces senescence-like state in human cancer cells: implication for senescence-directed molecular therapy for cancer. *Cell Death and Differentiation* 1999; 6: 765-772

Lee AC, Fenster BE, Ito H, Takeda K, Bae NS, Hirai T, Yu ZX, Ferrans VJ, Howard BH, and Finkel T. Ras protein induce senescence by altering the intracellular levels of reactive oxygen spiecies. *J Biol Chem* 1999; 274: 7936-40

Macip S, Igarshi M, Berggren P, Yu J, Lee SW, and Aaroson SA. Influence of induced reactive oxygen species in p53-mediated cell gate decisions. *Mol Cell Biol* 2003; 23: 8576-85

Macip S, Igarshi M, Fang L, Chen A, Pan ZO, Lee SW, and Aaroson S. Inhibirion of p21-mediated ROS accumulation can rescue p21-induced senescence. *EMBO J* 2002; 21: 2180-88

Orr WC, and Sohal RS. Extension of life-span by overexpression of superoxide dismutase and catalase in Drosophila melanogaster. *Science* 1994; 263: 1128-30

Polyak K, Xia Y, Zweier JL, Kinzler KW, and Vogelstein B. A model for p53-induced apoptosis. *Nature* 1997; 389: 300-305

Roninson I. Tumor Cell Senescence in Cancer Treatment. Cancer Research 2003;63:2705-15

Redon C, Pilch D, Rogakou E, Sedelnikova O, Newrock K, Bonner W. Histone H2A variants H2AX and H2AZ. *Curr Opin Genet Dev* 2002; 12: 162-9

Setsukinai K, Urano Y, Kakinuma K, Majima HJ, Nagano T. Development of novel fluorescence probes that can reliably detect reactive oxygen species and distinguish specific species. *J Biol Chem* 2003; 278: 3170-75

Tsao Y, Huang S, Chang J, Hsieh J, Pong R, and Chen S. Adenovirus-mediated p21[(WAF1/SDII/CIP1)] gene transfer induces apoptosis of human cervical cancer cell lines. *Journal of Virology* 1999; 6: 4849-90

Vaziri H, West MD, Allsopp RC, Davison TS, Wu YS, Arrowsmith CH, Poirier GG., and Benchimol S. ATM-depenent telomere loss in aging human diploid fibroblasts and DNA damage lead to the posttranslational activation of p53 protein involving poly (ADP-ribose) polymerase. *EMBO J* 1997; 16: 6018-33 von Zglinicki, T. Telomeres and replicative senescence: Is it only length that counts? *Cancer Lett* 2001; 168: 111-6

Waldman T, Lengauer C, Kinzlar KW, and Vogelstein B. Uncoupling of S phase and mitosis induced by anticancer agent in cells lacking p21. Nature 1996;381:713-6

13

Reviewing the Life Cycle:
Women's Lives in the Light of Social Changes

Anna Freixas, Bárbara Luque and Amalia Reina
Department of Psychology, Faculty of Educational Sciences
University of Cordova, Cordova
Spain

1. Introduction

Our chapter involves both a theoretical revision and a set of reflections that aim to analyse and redefine the significance of the final stages of the female life cycle, which has been revolutionised by two major set of circumstances: the increase in life expectation and the social changes that have taken place in the western world during the second half of the 20th century. In this regard, our reflections deal with the consequences that this spectacular prolongation of life has had for women. We also discuss the new significance that living into old age may have for women who will reach old age during the early decades of this century and who are the beneficiaries of the important social changes of the second half of the past century.

These important social changes have facilitated the virtually universal access of women to basic education, and of a very large proportion of women to higher education, as well as the generalisation and utilisation of new information and communication technologies. The incorporation of women into the labour market and the cash economy has altered relationships and the position of women in the world and has enabled them to renegotiate their intimate relationships, their sexuality and rates of birth-giving, and thus to modify to a great extent their future lives as elderly women during the first half of the 21th century.

2. Feminist critical gerontology

Our theoretical framework is based on critical gerontology, which analyses the extent to which political and socioeconomic factors interact to shape the experience of ageing, and treats age, gender, ethnic background and social class as variables on which the life course of the individual pivots, insofar as it predetermines their position in the social order. Critical gerontology shows a desire to explore the social construction of ageing within a broad sociopolitical and humanitarian context. The field also studies the disparate ways in which individuals grow old, and the social and political disempowerment than often accompany ageing (Minkler & Holstein, 2008: 196).

Feminist gerontology may be regarded as part of a project of development of those epistemologies which, from the perspective of the social sciences, question dominant perceptions of the lives of certain marginalized segments of the population. Feminist gerontological research attempts to document the experiences of elderly women and to promote new interpretations of female ageing, "asking questions about what 'everyone knows', and to examine ideas, positions, theories, and policies from the perspective of the least advantaged" (Minkler & Holstein, 2008: 199).

Critical feminist gerontology has documented the experience of elderly women, encouraging the development of more complete and more complex interpretations of their lives, and has discussed the necessity of studying and understanding their life trajectories in greater detail, revising the lacunae and inconsistencies that a large proportion of current gerontological studies offers, as a victim of the 'ideology of age'. Taking as its point of departure the notion that feminism is "a form of politics which aims to intervene in, and transform, the unequal power relations between men and women" (Hollows, 2001: 3), critical feminist gerontology emerges as a form of study via which we claim to alter the power relationships which, in this case, are mediated via age and gender (Ray, 2006; Freixas, 2008).

3. The new life course

One of the most significant processes of the 20th century has been the gradual ageing of the global population, particularly in the developed world, where life expectation has risen spectacularly. There are two immediate causes of this process in our society; the decline in birth-rates and the increase in life expectancy, which is due to the fall in the death rate at advanced ages. This faces us with a social fact that lacks precedents in human history. We have thus been witnesses to a structural change that has led to *the ageing of old people*, which is to say that the number of nonagerians and centenarians is growing, bringing social, cultural and health-care challenges as well as a duty to study the phenomenon. At this point in time, we can claim that as we approach old age, we still have many 'productive' years ahead of us, time that represents an unprecedented resource in terms of number and potential (Minkler & Holstein, 2008). Today's elderly women were the promoters of one of the most important demographic transformations in history, in that the reduction in their rates of giving birth led to a true demographic transition.

Ageing is not a process that can be viewed solely through the prism of age; it possesses other nuances of great importance, both collective and individual. Growing old for women is not the same as for men, nor does it have the same meaning for members of advanced and developing societies. It is not the same to grow old, having enjoyed a good education, with access to culture and to a health-care system, accompanied by professional activity and emotional and interpersonal relationships, as it is to do so outwith the limits of the system. The fundamental challenge is thus not to live longer, but how to live our extra years in terms of health, financial security, wellbeing, social insertion, and personal, cultural and social significance. Ageing is an achievement, a triumph, not a cataclysmic event (Freixas, 2002). The old vision of age as an inevitable process of loss, illness and decrepitude is no longer valid, as a significant proportion of women and men play important roles as active members of society and enjoy a degree of autonomy and satisfaction to very advanced ages.

The spectacular increase in life expectancy has changed people's psychological position in the life cycle. The old clichés regarding ageing and death once the barrier of the fifties had been passed, have been largely dispelled, and today, we can look forward to a long phase of life to which we must give meaning. Middle age (50 – 65), regarded as a cultural category, has acquired a recognised status as a stage of life distinct from the third age (65 – 80) and the fourth (older than 80). Studies of age need to add a description of each aspect of our state of mind that deals with the life cycle, revealing the fears and assumptions that invade it.

4. Gender and ageing

The characteristics of the lives of women, and their wide individual variations, make it difficult to analyse their experience in terms of the classical stages of development, which are adapted to the masculine model, which is still regarded as the norm. In such studies, theories of adult development have traditionally been based on largely male sample populations, whose experience and perspectives have ignored those of women, while the results obtained on the basis of such samples have been generalised to apply also to women, treating these as deficient when their experience and performance do not correspond to masculine standards. Virtually no studies have attempted to consider the significance and consequences of the differences in socialisation and the life options of women and men in old age.

It is several years since certain female authors first pointed out the necessity for the psychological study of the development of males and females to be separated. In spite of the fact that studies of the psychosocial development of women are still few and far between, we now possess a number of works that illustrate the lives of middle-aged and elderly women from other perspectives (Arber & Ginn, 1995; Bernard et al., 2000; Freixas, 1993; Friedan, 1993; Gannon, 1999; Greer, 1991; Pearsall, 1997). A large proportion of the available studies of the second part of the lives of women have been carried out on sample populations drawn from the middle class; white, heterosexual and with average levels of education, thus leaving in the shade knowledge of the experiences and lives of an important segment of the female population that is the process of becoming old.

A number of female writers (Barnett & Baruch, 1978; Freixas, 1997; Gilligan, 1982) have argued that the words of Erikson and Levinson —who proposed the development of the adult personality through unidirectional, irreversible, hierarchical and universal stages that do not take into account individual differences— does not represent the reality of the situation for women (Erikson, 1950; Levinson, 1978). The life experiences of men are intimately related to their chronological age, as a variable in which the events of their lives are framed, belonging as much to the family as to the occupational sphere. However, this type of model does not function in the life of women, for whom adulthood involves a wide variety of role models that are not based on chronological age, since their lives may offer a large number of combinations in which their occupation, partnership and child-rearing involve several levels of use of their time and commitment that mean that the roles of wife, mother and worker may possess different degrees of importance at different points of their life cycle. This tends not to occur in the lives of men, in whom the unidirectionality of events has usually been clearer. The differences in involvement in the public and private spheres is the cause of completely divergent paths of life, which means that in the development of women, the evolution of relationships frequently exerts greater pressure than does that of chronological age as such (Luque, 2008).

Viewed from a feminine perspective, ageing can be a wide-ranging challenge, insofar as they need to face their personal and social situation which, in many cases has left them in poverty and dependence. Furthermore, they need to uncover certain of the most deeply-rooted sociocultural demands that have anchored them to profoundly restrictive models, related to concepts of beauty and youth that have no respect for the natural processes of human development.

5. Social changes

The important social changes that have taken place in the West in the course of the 20th century have involved new social, political, cultural, sexual, family and financial organisations of such importance that they have transformed the social and private lives of both men and women. A good proportion of the successes achieved by the end of the 20th century originated in the feminist movement and its thinking.

5.1 The new social organisation

In the case of women, the new social organisation has produced such a degree of structural change that their lives will never again be marked by the social conditions that previously constrained them and in which they lived, deprived of education, liberty, financial resources, voting rights, and control of their own bodies and sexuality. Their lives remained at the mercy of men —fathers, husbands, brothers, priests— who were the sole possessors of all rights. These social changes have given them access to education, paid work, social and political participation, as well as to the use of their own property and to their bodies and sexuality, as mentioned above.

In the case of the 'new' elderly women of the 21st century, one of the principal effects of their longevity has been the lack of models of elderly women with meaningful lives. With the aim of filling this blank screen of some 30 years of extra life with content, and in the lack of social models to which we can look, the new elderly women of the first half of the 21st century will need to look to each other if they are to trace out a new route-map. We may assume that this new generation of elderly women will be happier than their grandmothers, given that they have succeeded in dismantling some of the social requirements that previously restricted their lives. Nevertheless, social change never takes place without pain, puzzlement and uncertainty. It is probable that the new life situations that we analyse in this chapter will noticeably improve their sense of satisfaction with life.

This, in the sense of a subjective perception of wellbeing beyond what the objective data might suggest, lies along two axes: that of 'control' of one's own life, and that of 'happiness'. The gender-based division of work that has ruled in industrialised societies has assigned to men the pole of 'control' (access to education, money, work outside the home, power, status) while limiting women to the pole of 'happiness' (relationships, emotions, care). This model of social organisation places men and women in different life spaces; the world of affect, relationships, care-giving, raising children, are all part of the feminine specialisation, while men are assigned the biblical tasks of earning the family's bread, going to war and defending their wives and their flocks, which in practice involves the total management of money and of political and private power. This model has led to the devaluation of all the activities and practices supplied by women, which has meant that the 'feminine space' has come to be regarded as inferior, and thus something to be avoided in the process of masculine identity, which has traditionally been constructed on the basis of denying its feminine side.

However, this division of social roles generated a lack of satisfaction among both parties, distancing men from the life of the emotions and women from power and from control of their own lives. The voices that demanded a more equitable division of these two spaces came from women, as they were more aware than men that this way of locating themselves in the world resulted in deficiencies which had lasting and irremediable consequences, especially in old age. For this reason, from the 1960s onwards, a good number of women progressively and definitively joined the world of paid labour, thus availing themselves of the use and management of their own money and possessions; meanwhile, they were also struggling to access education and the universities, and to have legislation passed that would give them better control of their lives, their bodies and their sexuality, giving them a radical transformation of their everyday life —although they might not have realised that at the time— above all, of the conditions of ageing, assuming the responsibility for their own well-being, not only physically, but also mentally and spiritually. All the above changes modified the relationships and the position of women in the world and enabled them to renegotiate their intimate relationships, their sexuality, birth-giving and, as a result, to modify to a great extent what will be their lives as elderly women during the first half of the 21st century.

We do not know whether these elderly women will be happier than their predecessors, because it is not age as such that is the cause of a lessening of pleasure in growing old, but rather the circumstances associated with life in old age that can determine a greater or lesser feeling of happiness. In this synthesis between the pole of control and the pole of happiness that marks the life of individuals, the old values of the popular song —'there are three things in life: health, money and love'— continue to be basic aims. Family, social and friendship networks, a higher level of education, good health and financial resources are the indicators that sustain us in adequate comfort in our old age.

5.2 Feminism and the life of older women

The great social changes that have marked the second half of the 20th century doubtless affect the pattern of life of men and women of all ages, in such a way that these important social changes will have important consequences for the old age experienced by women. Feminist movement has provided the foundations of the transformation of the public and private life of women —and thus of men— by overcoming the many social and cultural limitations that restricted their lives to the fields of reproduction and the private world.

The 60s and 70s of the previous century produced major social changes in favour of the civil rights of discriminated minorities —particularly those of black people and of women. The activities of the women's liberation movement encouraged the passing of laws that eliminated many of the social barriers that limited the lives of women in fields such as education, civil rights, reproductive and sexual rights, and of rights to work and to hold property, and so on.

Feminism, as a theoretical perspective and a social movement, has illuminated our understanding of power relationships within the family and emotional life, and has unveiled the system of maintenance and reproduction of such concepts. The feminist slogan 'the personal is political' led to a structural change. By claiming that the relationships that rule our private lives are power relationships, it suggests that many of the problems that we regard as 'personal' originate in society and, as such, can only be resolved by social and political change. On this basis, the creation of a collective consciousness mobilised women in their search for objectives and changes in their situation, although the emphasis on equality hid reflections on gender-based differences. Moreover, social recognition of rights has in many cases not been accompanied by true equality and the transformation of social, political, economic or personal life practices, in which women remain at a disadvantage, and which become significantly worse in old age.

The new elderly women of the 21st century to whom we refer in this chapter[1], born in the final third of the 20th century —heiresses of the benefits and discourses of the second wave of feminism, daughters of May'68 and the great social movements propelled by faith in change—, convinced that 'the personal is political' will revise each and every one of the elements of the social contract and of love: they will examine through a magnifying glass the received prescriptions regarding daily life and will denounce the patriarchal agreements that dominate both the personal and political spheres. They will insist on the deconstruction of identity inherited from the feminine 'mystique' (heterosexuality, femininity, passivity, obedience, maternity) across a continuum of crises of identity throughout their lives, until old age which, in the lack of a recognised legacy, offers itself as a blank space, without models and accompanied by many fears and ghosts; with a single strength derived from their links and the bank of arguments of feminist thinking and epistemology.

A good number of these women, who will make up an important fraction of the elderly population of the 21st century, have been characterised as refusing to passively accept the life models left to them by previous generations, have renegotiated the meaning of many received prescriptions and have modified the sense of ageing, taking as their point of departure vital and intellectual positions that are very different from those of their mothers and grandmothers. They have challenged cultural images of the 'little old lady'; asexual, self-sacrificing, lacking opinions, desires and necessities, always available, undervalued and weak, giving way to the model of an elderly woman who is active and sexual, attractive, who utilises her power and her new position in society, in her family, her network and her relationships (Kingsberg, 2002). These women will face old age with experiences of work, finance, family, status and power that are very different from those of their predecessors, and as such, enjoy greater financial, social and intellectual resources. All this has required the redefinition of many of the social roles that they have played up to the present day in terms of partnerships, family, paid employment, money and sexuality, etc. Theories of the life cycle have yet to develop a set of arguments that value the significance of these factors in individual and psychosocial terms.

On the other hand, the reflections produced by feminism at the end of the previous century regarding sexual differences have helped to give value to 'the feminine', and to recover the values that women have historically brought to relationships and the sustainability of life, recognising their civilising efforts (Libreria delle Donne di Milano, 1987). Thanks to these contributions, the elderly women of the 21st century find themselves occupying new fields of

---

[1] The new elderly women to whom we refer represent minor cultural, social, political, economic and personal avant garde. The elderly women of the 21st century also include women who remain rooted in the models of the previous century, for whom social change has not led to any real transformation of their lives, although it may do so for thair daughters, as a result of their efforts.

meaning and presence in which they do not feel a necessity to deny their femininity in occupying spaces that used to be dedicated to men and for which models were in short supply.

6. Elderly women of the 20th century / elderly women of the 21st century

In their youth, a large proportion of the women who are elderly today lived in social, economic and political situations characterised by poverty, deprivation and submission; they laboured within the family unit, in the fields, and in activities that accorded them neither social nor economic recognition. Nevertheless, many of these women were the promoters, whether voluntarily or involuntarily, of many of the social transformations enjoyed by the elderly women of the 21st century. Their role as pioneers in many areas has yet to be recognised. In reality, as Brody (2010) claims, they made up a 'new frontier' for themselves and for new generations. Also, to the extent to which they needed to overcome the many difficulties and deficiencies from which they emerged more than successfully, we may feel that they have demonstrated an admirable resilience. They have understood how to adapt to losses, to the new environments of life and work, to social and familial change, in a society that is profoundly changing and in which changes they have participated through their questioning. In this too, they were pioneers.

The demographic aspects are the basis of some of the great changes in the lives of elderly women. Thus, the increase in life expectancy will produce one of the most important transformations: while at the age of 60, or even earlier, life for both men and women had more or less come to an end during the first half of the 20th century, for the new elderly, the same age involves launching new initiatives; experiences and lives unlived until the moment at which they emerge as new expectations. To a great extent, this is because the increase in life expectancy is taking place in an epoch in which potent social changes have given the female population a wide range of new resources, both economic and intellectual as well as in the sphere of health. Furthermore, the revision carried out by women who were socialised under the model of submission to the patriarchy —feminists *avant la lettre*— will provide future generations with the ability to choose and to make decisions regarding their own lives that hitherto have been unthinkable.

On the other hand, the social and personal advances encouraged by feminism and the social movements will be concretised in a re-evaluation of age, and thus of the visibility and occupation of public space on the part of elderly people. Elderly women and men will enjoy a social and vital space and recognition that they have lacked until now.

6.1 Education, culture, freedom

One of the most important tasks in the process of ageing consists of 'assigning meaning to life itself', a task that demands a conjunction between reminiscence —giving meaning to one's own past life— and premonition —planning the future. To be able to identify a personal pathway to graceful ageing, accepting one's past and designing one's future, the new elderly women possess highly valuable elements, of which education is probably the most important. This group, born around the middle of the 20th century, does not include illiterate persons. The virtual universal access of women to basic general education and of an important fraction of the female population to higher education, as well as to the spread in the use and knowledge of the new information and communication technologies will offer them an extremely interesting panorama.

The new elderly women will communicate with their daughters, grandchildren and friends by email, and buy their airline and theatre tickets via the Internet, on which they will also check the weather and read the papers in their own homes.

Education allows access to information, and that in turn to liberty. The elderly women of the 21st century, given that they will have had an education, will be able to take up paid work (their own money) and will have had their own experiences of management and access to various forms of power and control. In no way do they resemble their grandmothers who, to the extent that they came from a society in which women were regarded as being less intellectually gifted than men, had very restricted access to education, which was reduced to a rudimentary level. This limited them to the role of housewife (wives, mothers, grandmothers) as the only possible occupation, leaving them in old age without money of their own and with very limited access to cultural and intellectual resources.

6.2 The work cycle of women

The most obvious consequence of the access of women to education is their incorporation into the labour market, which has markedly changed —although still with important limitations— domestic life, financial relationships within the family and, in consequence, the power relationships within the family and in social life. The model of social organisation of our culture is profoundly androcentric and not only ignores the peculiarities of the life cycle of women, but even punishes them for not running with the same rhythm as their male companions and being 'distracted' from their professional career by their child-care duties, performing tasks of care-giving and emotional support that are 'not the concern' of their male companions. All in all, the years dedicated to these tasks of sustaining life do not count in their curricula, thus giving men an advantage through their lack of solidarity in the tasks involved in civilising the world. A similar problem arise in the area of pensions, which are calculated according to a model of working life derived from the division of labour between the sexes, as a result of which the typical variability in the trajectory of the feminine career obviously operates to the disadvantage of the financial level of women in old age.

The financial insecurity which women in previous generations faced in old age is related to the fact that in many cases they entered and left the labour market as a function of the financial requirements of the family and the demands for care of their husbands and children. These are women who have abandoned interesting jobs in order to follow their husbands in his work, who have been unable to rise in their professional career for fear of injuring masculine honour, and who have left the house to work 'at anything' when their male's wage has not come in. All in all, the heterosexual definition currently defines the age of women who invested their capital in marriage, with the idea that this would supply their financial necessities in their old age, but when the hour of truth arrives find themselves in misery. Moreover, in a society in which women's most important value lies in their reproductive capacity, having an intellectual occupation was often seen in a negative and suspicious light, since intellectual work for women was considered to have a negative impact on childbirth (Hirdman, 1994).

We may think that some of the new elderly women —insofar as they have had a history of work that is more continuous and of higher status than that of their predecessors— will enjoy old-age pensions that will permit them to live the long later years of their life in a better financial situation than their mothers who, in accordance with the model of gender-based division of labour, put their efforts into unpaid domestic work, and thus found themselves consigned to poverty and financial dependence in their old age.

The fact that feminism has emphasised the necessity for women to have a 'purse of one's own' (Woolf, 1938) —a financially independent life, a career— should not be interpreted as a desire to adopt the masculine model of work. In fact, this is not the case, although some women are obliged to assimilate themselves to this model if they wish to advance in their professional careers. Many women of recent generations, in spite of their definitive incorporation into the world of work, continue to structure their working life around the requirements of the family and their duties of care. The flexibility in work demonstrated by women is not usually a matter of choice. They need flexible working conditions in order to be able to reconcile this aspect of life with the demands of the family.

Women want an equilibrium between paid work and their other activities, including leisure, care and voluntary work. One of the challenges of contemporary life will be to bear in mind the new, creative forms of life that women are putting into practice —most of them as solitary adventurers— given the difficult conditions that the traditional forms of relationship, derived from heterosexual romantic love, place on them.

6.3 Rethinking the model of retirement

The androcentric model of retirement, in which one changes from paid work to 'not working' from one day to the next, is probably in need of revision, attributing value to the work cycle of women who pursue extremely diverse career paths which, besides paid work *sensu strictu*, include care-giving, providing emotional support, voluntary work and a range of community and social activities of great importance and enormous social value, such as the irreplaceable tasks of making life more supportable and humane.

Just as diversity is the norm when we speak of the family types of the 21st century, so is it too when we observe the wide range of situations that have defined the social life of women since the late 20th century. The reflections aroused by a number of studies suggest a pre-retirement model of work that takes the form of a transition between working and retirement that could bring about an improvement in the experience of retirement, ameliorating the crisis of meaning and identity that can result from a single life-model, such as we often find among men (Everingham et al., 2007). In fact, some companies already ease the transition to retirement of both men and women by allowing them to reduce the number of hours worked and the range of tasks they undertake. For example, some universities reduce teaching hours, while maintaining the amount of time available for research, thus making best use of the intellectual capacity of their academic staff. Such a reduction in hours worked as retirement approaches could be adapted to a great extent to the range of women's careers.

It might be useful on financial grounds, given the erratic work-life histories of many women, with their serious consequences for retirement pensions, to lengthen their working lives, albeit in a partial fashion, allowing them to include periods of paid work alongside periods of other activities of interest, even if unpaid. This is not to forget the structural changes that are a result of our present longevity: for earlier generations, retirement coincided with the beginning of 'old age', the loss of one's faculties, senescence. Today, at 65, most of us enjoy an enviable state of health and can boast of knowledge and skills that make us useful in many spheres of work, social and community life. All of this favours the idea of creating à la carte retirement policies that would allow those who wish to do so, to compensate for some of the difficulties that they are liable to meet at the moment of compulsory retirement.

A unitary model of retirement does not correspond to the different implications of the work and family world of women. In fact, the new forms of retirement that society ought to be organising emerge from this plural reality and also, in these times of crisis, from that of the new generations of women and men for whom work will be no more than a matter of security, as it has long been for men, both historically and up until the present day. We might say that we find ourselves facing a society in which work insecurity is drawing both women and men down to the same level, and that the time has come to revise the forms of retirement, in order to be able to take new situations into account. This demand has been made time and time again by women. Although the different way in which women have interacted with the labour market has historically been regarded as 'second-best', the current crisis appears to be minimising gender differences in the world of work (Everingham et al., 2007).

The fear of financial insecurity will continue to be a prevalent feeling in the women of the coming generations who have worked intermittently, particularly for those who lack a partner but have family obligations. Such insecurity is based on the practice of part-time working, in the need to combine a number of obligations, and in the culture of flexible working that has left women in a financially vulnerable situation by denying them a secure income and a continuity of employment that would guarantee them a successful retirement. Thanks to a number of ideas derived from feminist thinking, the new generation of elderly women will probably take more seriously the topic of continuity of employment and pensions, which ought to reduce the numbers of elderly women in poverty.

6.4 Health and paid work

The health of elderly women in the 21st century is benefitting from the thinking derived from women's health-oriented networks which, since the end of the nineties, have produced interesting studies of differences in morbidity between men and women and have questioned diagnoses, treatments and medical practices regarding women's health (Valls et al., 2008). To date, little research has been done on the topic of the benefits of paid work on women's health, while the few studies that do exist suggest that social and relationships resulting from work outside the home protect women from mental and physical illness, by enabling them to raise their self-esteem and sense of security in decision-making, while offering them social support as well as a greater feeling of satisfaction with life (Sorensen & Verbrugge, 1987).

Paid work also offers other health advantages; it enables people to structure their own time and provides financial benefits, social contacts and professional identity. Where women are concerned, participation in the world of work improves their health insofar as it offers social status and power in addition to financial independence and self-esteem. The social support provided by paid work is valued by women as the most important element in keeping them there, quite apart from the potential financial necessity. The participants in Forssén and Carlstedt's study of health and paid work emphasised that this enabled them to control and relieve their illnesses and that it offered a number of health benefits, of which they mentioned the importance of enjoying a meaningful life, feeling competent and needed, and being recognised and enjoy a good mood, and that it helped to structure their days (Forssén & Carlstedt, 2007).

A topic of great importance in the lives of elderly females in the future will be the relationship that emerges between the massive incorporation of women into paid employment and their experiences of health. To what extent has the participation in the labour market been a source of physical and psychological health —even when such work has been hard, poorly paid and little recognised— in comparison with that of their predecessors as housewives or unpaid labour? Future studies ought to take into account the new relationships between women, their bodies, health and attractiveness, topics of great importance in the ageing process, while bearing in mind the double standard of ageing denounced by Susan Sontag (1972).

6.5 Body and beauty

Growing old is not easy in a society such as ours, in which the concept beauty is based on two elements that are difficult to maintain as we grow older: youth and slimness. Staying young when we have passed 60 is an oxymoron: we cannot be both old and young, while the need to remain slim, which is derived from an inadequate history of nutrition and an upbringing under an aesthetic model which itself is static, is no easy task in old age. The cultural change concerning the image of the female body has basically taken place since the early 20th century. Naomi Wolf located the start of our preoccupation with diet and slimness in the 1920s, when western women started to obtain the vote and legal emancipation, with certain swings that were functions of the greater or lesser exaltation of maternity as the destiny of women. However, it was most clearly after 1965, with the emergence of the skeletal model Twiggy, that women began to slim seriously and to suffer for the weight that they always regard as excessive. In spite of the fact that women have made advances in terms of rights, status and power, which ought to have brought them a greater sense of self-esteem and of competence and value, their obsession with weight has led them as a rule to feel unhappy, in spite of the advances that ought to have been concretised in the very opposite perception (Fredrickson & Roberts, 1997; Wolf, 1991).

Some of the sociocultural requirements regarding attractiveness that have such a great effect on the life of women while they are still young continue to place limits on their feeling of satisfaction with life and of wellbeing in their old age. Cultural requirements have historically impelled women to involve themselves in 'disciplinary practices' at a high physical, financial and psychological cost in order to maintain their appearance; practices that imply global strategies of control; in this case a biopolitics of control of the female body that leads women into a continual 'must do it', because otherwise they would not exist. Naomi Wolf put it succinctly as follows: "The real problem is the lack of choice." (Wolf, 1991: 354). Faced with this situation, the old women of the future will start to make decisions regarding how they wish to dress, make up and display themselves. They will develop a standard of personal care that is no longer a matter of pain or obligation, an imperative that makes them suffer because if they do not live up to it they will be excluded, but rather as an element of pleasure or enjoyment, of personal identity and acceptance of what nature has given to each of them, of liberty, of beauty, in order to feel good.

The new elderly women have managed to reflect individually and collectively about the messages and mandates they have received regarding attractiveness, and have been able to construct their own modes of thinking on this topic. Today, ideas of beauty have changed, and we can look to a new concept of beauty that integrates and fulfils them as individuals, that does not demand a particular external appearance, but rather looks into the interior of each individual being, taking self-esteem as its point of departure. Given that the conventional images of women older than 60 with which we have grown up no longer have anything to do with current reality, we will have to construct new patterns for them. The crucial topic here as far as new elderly women are concerned lies in the search for a model of beauty that moves them from the image of a wrinkled little old lady, dressed all in black, whose involvement in her own body image shines by its very absence, to a typology in which there is room for diversity and enjoyment rather than merely the obligation to wear mascara in the process of hiding one's age. All in all, it is a matter of questioning how far elderly women are prepared to go in identifying themselves with what is regarded as attractive in our society: how far they are willing to conform to a model in order to achieve a 'correct' image of growing old.

6.6 New family life

The demographic fact of greater longevity, allied to the lowering of the birth-rate, has altered the structure of the family; this has changed from that of the extended family with many children and few generations to one made up of few children and several generations. This situation has led to a change in the pattern of relationships, which are no longer horizontal (between brothers and sisters), to vertical links (between generations). However, new forms of social organisation have encouraged the appearance of new modes of family life that have already become the norm in the 21st century. Diversity characterises emotional life and the relationships of women who are currently in the process of becoming old, though not without some pain. In spite of everything, the family is, and will probably continue to be, an important aspect of the life of women for all time.

If anything does define the life of the old women of the 21st century, it is the fragmentation of their emotional and work careers. The concept of 'definitive' or 'for ever' under which their grandmothers were socialised, has disappeared at the stroke of a pen. We now live in what Ulrich Beck has called the "risk society", in which we have to be prepared for change, for the ephemeral, for breaks in long-term family and professional careers (Beck, 1992). The normal 'chaos of love' that has characterised our society since May'68 has fragmented our emotional lives (Beck & Beck-Gernsheim, 1995). No longer is anything 'for ever', while to introduce this concept into our emotional programme would not be a simple matter, particularly after having put so much effort into the creation of spaces of relationships and connections that were believed to be lasting.

The ideology of the traditional family has permeated the lives of women of all ages, in spite of the fact that for almost half a century new forms of family life have been emerging. The traditional family model (regarded as a universal model, in spite of the fact that it is basically the result of the gender-determined division of labour dating from the 19th century), which is characterised by the financial dependence of the woman and the lack of male involvement in care-giving and domestic tasks, has been followed by other family models derived from the incorporation of women into the labour market and feminist demands for fairness in the division of responsibilities. These new models have deconstructed the familiar myth, revealing a wide variety of models that have achieved similar degrees of legal, social and personal validity, among which we find the egalitarian heterosexual family, the female single-parent family as well as the male ditto, the homosexual family —both lesbian and gay—, and the recent concept of 'families of choice'. All of these models are derived from the social changes of the 20th century and from the theoretical thinking of the feminist movement (Fortin, 2005).

Heterosexual marriage continues to be a goal in the life of the new generation of elderly women, albeit to a lesser extent than before. What is certain is that in their time, many of these women married for love, rather than to obtain financial security as had previous generations, marrying men who, while they supported the discourse of equality in theory, continued in practice to behave like their fathers. Cohabitation with these husbands whose theoretical discourses displayed a social sensitivity and democratic framework, but who refused to renounce the privileges of their sex, which they assumed to be something 'natural', often generated relational conflicts which in turn raised the rates of separation and divorce (Coria, 2001). Women of recent generations who have wanted to maintain relationships with their partners on a basis of equality have had to deploy a range of strategic discourses and take part in various practices in order to control the context of daily life (Elizabeth, 2003). The inequalities in intimate relationships are not only a product of interpersonal relationships, but also the result of the limiting effects of cultural norms and other socially significant spaces, such as the family of origin.

6.7 New life-styles

Today, more and more people are adopting life-styles that combine intimacy, physical contact, emotional relationships and company, even though they live apart. Sharing interests, intimacy and social and personal activities need not involve sharing a home or a residence, which resolves some of the problems of cohabitation in later life with persons with whom one has not earlier come to agreement concerning everyday life together. It is thus clear that the structural changes of recent times involve a wide range of intimate relationships which, for the new generation of elderly women, involve moving on from a monogamous matrimonial relationship —which essentially means emotional and financial dependence— to a relationship that Anthony Giddens (1992) called a 'pure relationship' which is kept up only as far as both parties find that it gives them sufficient satisfaction to remain in it. Couples involved in such relationships lack models to follow, and in theory are more egalitarian, autonomous and happy than those in the classical model, preferring to replace marriage with some form of mutual commitment (Gross & Simmons, 2002).

In the new generation of elderly women we already find different 'trials' of relationships in which various alternatives to cohabitation within the framework of classical heterosexual marriage are practised. Some of these go in for cohabitation —i.e. living together with another person outwith a marriage contract— as an alternative to marriage as such. This mode tends to involve a more egalitarian relationship and may be more satisfactory for women. Another formula that has been tried out with some success is to be a couple, each of whose members live in their own house (LAT; living apart together), a formula that involves commitment, sexual relations and social recognition while satisfying the necessities for intimacy and autonomy, company and independence desired by many people in later life —particularly women. This formula, which is very similar to the 'pure relationship' described by Giddens (1992), is less institutionalised and requires a high level of negotiation. Meanwhile, a relationship with another person of the same sex is an option chosen by some mature women, who find areas of mutual understanding with their equals that they had not enjoyed in their heterosexual relationships. Recent legislation that recognises homosexual marriage has offered a certificate of legitimacy to this option, and has contributed to the elimination of social, cultural and family homophobia (Connidis, 2006).

6.8 Links and networks

Enjoying relationships in old age —whatever their configuration— has positive domestic, psychological, social and even financial benefits, insofar as they offer social and emotional support and physical and sexual contact, while also enabling care-giving and domestic tasks to be shared. Furthermore, networks allow different strategies for life in old age to be shared. This would appear to be a positive programme, but to make a success of it demands a good dose of internal freedom that no-one has ever said cannot be exploited, although women may not allow themselves to do so.

Women have always been creative experts in relationships of friendship which, as they grow older, display their inestimable value in the maintenance of wellbeing and a positive ageing process. For many reasons, the depositories of these links of friendship tend to be other women, whether younger, older or of similar ages, shading out differences in age, and giving and receiving, as appropriate, in rich intergenerational exchanges. Carolyn Heilbrun put it thus: "[since the 50s]...after a lifetime of solitude and few close and constant companions, women friends and colleagues, themselves now mature adults, whose intimacy helped to make the sixties my happiest decade" (Heilbrun, 1997: 4).

Too often is age associated with isolation; however, women of all ages form powerful networks of social and emotional support and set up instrumental and emotional links at time when life seems to be losing its balance, and these turn out to be extremely solid structures. Women have ample experience of this; since it is they who establish and maintain relationships within and beyond the family they are very well prepared to develop and maintain new social links when they find themselves facing the vicissitudes of old age.

Social networks are closely related to the quality of life, to the ability to deal with the stresses of everyday life, and to health and longevity. The greater life expectancy of women, their solid social networks and personal capacity for dealing successfully with the changing circumstances of life, ought to be recognised as closely interrelated feminine strengths.

6.9 Finally alone

Solitude is one of the great discoveries of maturity, at least for those who possess sufficient health and emotional support networks to enjoy it. Dreaded when we are young as a symbol of abandonment and lack of social participation, with the passing of the years it turns into a challenge, a personal space of liberty and happiness. In the first stage of adulthood, women are often so caught up in various central aspects of their own and others' lives that there comes a point in their later lives when solitude ceases to be something feared and turns into a happy meeting with oneself. Solitude is a temptation for persons who have lived with too much company and thus have scarcely enjoyed time and space to themselves. Heilbrun (1997) describes it as a pleasure for those who have managed to give meaning to their lives, as an opportunity to live in the present, as a gift that they do not wish to allow to escape. Solitude enables us to hold the reins of our daily lives, to organise our time.

Living alone is not usually a problem for most women. Used as they are to organising their own and others' lives, they suddenly find themselves on their own stage. Living alone brings a meeting with long-postponed desires that had become unrecognisable. Now they have all the time and space they need to organise their environment as they wish. All in all, they can embrace personal freedom and make room for their own desires.

The trends in gerontology that emphasise activity as a tool in successful ageing assign a value to 'doing' over to 'being', leaving little room for more quiet life choices, and obliging elderly persons to maintain family, leisure and care-giving activities and responsibilities that they have not chosen themselves. In the lives of elderly women, active ageing frequently implies yet other obligations: to remain active, to go out, participate, provide care, to show that one is bursting with life and activity. In reality, so-called productive and successful ageing imposes totalising ideals about the meaning of a 'good old age' (Minkler & Holstein, 2008).

6.10 The 'single'culture

If their predecessors were wives, mothers, daughters and neighbors, the new elderly women of the 21st century are partners, lovers, mothers and step-mothers, sisters, colleagues, cyber-girlfriends and, above all, divorcees. The elderly women of the future, i.e. women who are currently in their fifties, are twice as likely to be divorced as the elderly of today (Thomas & Fogg, 2000), to the extent that the current generation of the 'emerging elderly' has launched the practice of the 'single' life. On the one hand, they divorce and separate when the contradictions of equality in daily life become too obvious, while on the other, many of them opt much earlier to live a life that does not have room for heterosexual marriage.

It was this generation that deconstructed the old concept of 'spinster' that tormented and influenced the options of its mothers and grandmothers. The old women of the 21st century have continued to live with the specter of spinsterhood, although now only half-voiced, within a society that no believed in it. They do not wish to burden themselves with conventional marriage, while the convent is no longer an interesting option, so they have launched themselves on the paths of a profession and the control of their emotional life under new parameters. Whatever the reason, they have learned to live alone much earlier than their mothers, who only did so with the blessing of society when they had become widows, although the group of divorcees who had not made provision for the state of separation tends to find themselves in a poorer financial position in old age. In this way they can control all the threads of their life (economics, sexuality, independence), albeit not without difficulty, given the lack of models by which to view themselves.

In the face of the transitoriness of emotional relationships and the fragility of the ties of love, the new generations of elderly women have begun to practise forms of relationship and support that we might call 'families of choice': usually networks of women, spiced up with a few men, that make up a powerful support organisation and offer an antidote to solitude and isolation. Sources of cultural knowledge, of social support, of exchange of knowledge, of connections and emotional security, these new forms of family life enable elderly women to enjoy life alone in the security of knowing that nothing bad can happen to them, thanks to the efficient functioning of the structural networks of such 'families of choice'.

6.11 Habitat

One of the important conquests in the lives of elderly women is the possibility of deciding where and how to live, as much in terms of space as in forms of living and relationships. Present-day elderly women often find themselves living in old people's homes or in the homes of their children; perhaps even subjected to a peripatetic life, moving from the house of one child to the next, deprived of all intimacy, memories and mementos. The new generation of elderly women have considered how they are to be lodged in old age, about the design and features of the space in which they want to live, looking at it from their own perspective, and including in the balance the necessities that they may have in the future.

One's own house is the space that we have in which to live, to relate to the persons with whom we live, to receive our friends. It is also one's own personal and intimate space, in which we can enjoy freedom, but it is also the physical space in the city that enables us to participate in the community, maintain relationships and connections beyond the domestic circle, as active participants in the neighbourhood. The new generation of elderly women thus realise that they need a sufficiently intimate and private space that also provides for contact with the community and allows for relationships, avoiding dependence on the goodwill of other people.

In old age, we have to live in proximity with women and men of all ages; young people must not be deprived of the experience of relationships with elderly people, but nor can we grow old without participating in the interests and projects of younger generations. We are part of a community of care, with obvious mutual benefits (Tronto, 2000). If old people could maintain good connections with younger generations through participation in community life, they would retain an idea of themselves that was not fragmented by age and would contribute to a stimulating intergenerational continuity, in which reciprocity and interdependence would create a mutually enriching style of exchange. There would be fewer complaints, fewer aggravations, fewer misunderstandings. We would function as tribal chiefs, standing up for the interests of the future and preserve the valuable continuity between generations (Woodward, 1999).

6.12 A community of care

Caring is a crucial activity in human development; it configures us as emotional, empathetic beings, sensitive to the needs of our congeners. Caring activities comprise everything that we do to preserve life and wellbeing; our bodies, our souls and the environment, i.e. everything that enables us to sustain life on earth and makes us complete human beings. Caring is not a simple task. It produces internal satisfaction and peace of mind, but it costs us effort, and requires us to renounce alternatives. It generates internal contradictions and with them, feelings of guilt and frustration; it can also make us angry.

The ethics of care, as discussed by Carol Gilligan, assumes a moral virtue that goes beyond the simple assumption of responsibility like an obligation or a routine (Gilligan, 1982). It involves a personal commitment, internal and freely assumed, regarding the wellbeing of other persons. And precisely there lies one of its fundamental problems, whose crux is rooted in the differential traditional socialisation that has exempted men from this moral responsibility, to such an extent that, as Tronto (2000) claims, they have enjoyed the 'privilege of irresponsibility' in the care both of themselves and of others.

The unrewarded efforts of women on behalf of their children, partners and/or older members of the family is a public and social good that permits the economic development of society, thanks to the savings made by the state and the practical and emotional benefits that fall to individuals. Women find it difficult to set limits on the care they provide, just like the historical problems that they have experienced in offering their time gratis; in spite of this it is still the subject of contradictory feelings.

We live in a society in which the fall in the number of children per family reduces the proportion of the members of the group available to look after old people. Furthermore, caring for an old, dependent person does not offer the same pleasure as the care of a baby, whose day-by-day progress is obvious and stimulating. Care of the elderly is not a highly regarded activity, because the persons who receive such care are little valued by society (Calasanti, 2006). On the other hand, the realisation of dependence and the loss of capacities of a loved one force us to confront our own existence, precisely at the moment of our life cycle at which we are performing our own personal evaluation. This leads to clear physical and emotional wear and tear and generates many personal, partnership and intergenerational conflicts.

The new generation of elderly women may regard care-giving as an ethical and emotional opportunity that many of them assume, beyond the stress involved and the call of duty, which mixes feelings such as the need to protect the dignity of the loved one, company, and the desire to help the person involved to maintain a sense of self that protects their integrity. On the other hand, the experience of caring for a loved one offers an opportunity for emotional exchanges, for pardon, compassion, and is accompanied by an interesting and necessary reflection on dependence relationships. These also include those who perceive the necessity to liberate themselves from the oppression of care as a social demand that falls upon women, maintaining the right not to care for others and not to receive care themselves in the future, thus liberating the family from this responsibility and delegating it to the social services.

6.13 A social movement toward visibility

Finally, after so many years of playing an externally dictated role, the new generation of elderly women feel that they are entering a period of authenticity. In the second half of their lives, they may become what they have been building up in the course of time and they possess a full repertoire of knowhow that they can validate via participation social, neighbourly, political, cultural and leisure activities. The wisdom that accompanies the process of ageing enables them to distance from many of the preoccupations that in other periods had dominated their lives, and now they can create a space for the development of other persons; they have more time available, they enjoy collective activities as though they were a personal project and in this context can offer their experience, knowhow, and tricks and strategies learned in the course of their previous everyday lives.

The participation of middle-aged and elderly women in cultural, political and social organisations, in NGOs and women's associations has become a transformating element of great importance, for the psychological wellbeing of both the women themselves and for community, which benefits from the free, disinterested and wise richness of resources that the voluntary efforts bring to it.

Social participation and an active lifestyle are important elements of personal satisfaction throughout our lives. They generate pleasure, raise our self-esteem, and help to blunt the stressful and traumatic events that occur in the course of our lives. Nevertheless, not everyone feels the same necessity for interaction and social and community participation and, now and again, silence and solitude are essential, beyond the well-meaning prescriptions that would bring the elderly out on the streets at all times. Many women wish to combine activity with enjoyment of serenity and silence. To live occasionally parked by the wayside may also be a source of happiness; a necessity.

6.14 Citizens and pioneers

The passage from a fundamentally private life to the participation of this generation of women in public life is one of the key changes in the configuration of the ageing process in the women of the 21st century. Their personal position as 'citizens' with the experiences resulting from this concept and their refusal to be excluded from the practices of citizenship for which they have sacrificed so much as a generation demands a serious discussion. Citizens, women who participate in public life which, in the case of our future old women, does not usually begin in old age. Frequently, in earlier time, they were involved in other fields of neighbourhood, social and/or political activity. Some of them came from careers of social involvement in the public sphere and it is impossible to say whether their participation derives from their new situation as retirees or because they now have more free time. They are activist women who have grown old. In fact, such continuity in social involvement and participation is linked to high levels of education, independent of age or sex (Milan, 2005). We can say that we are faced with a 'culture of militancy' or of participation in its various forms, in elderly women.

Elderly women are active in many sectors of social life, as well as playing important roles in family care and voluntary work and in democratic and political life (Magarian, 2003). In their eagerness to combine different worlds; family, work, community, they have maintained a delicate balance between these public spheres and their family life, in which negotiations in terms of time and desires have often not been easy (Charpentier et al., 2008). The family and the needs of their loved ones have shaped the social commitments of these women, who have demonstrated a sustained willingness to reconcile their social, political and neighborhood connections with the needs of their families. This generation has left the house for the city; it has ventured to occupy new spaces, assumed new social roles and created new models of women in the public world, of committed citizens, beyond the frontiers of age. In the course of their lives, they have gained civil rights (the vote), personal and social rights (education, paid work, money of their own), and are convinced that 'the personal is political' (abortion, divorce, birth control, control of their bodies and of their sexuality).

They have been pioneers in many fields: in politics, in the universities, in trade unionism. They have fought for coeducational schools; they were the precursors of all the legislative changes that permit the access of women to the control of their sexuality, have gained contraception and the depenalisation of abortion. They were the first university women, the first lawyers, doctors, architects, scientists, philosophers, educators, the first female politicians to achieve power. Their valour has helped to ensure that women of subsequent generations enjoy better and easier access to civil rights, to their own bodies, to culture and the labour market, thanks to their participation in the tasks of political representation, the defence of their rights and in social activities and voluntary work. Although many of them do not define themselves as feminists, their lives and the personal and social victories for which they have struggled have clearly contributed to the improvements in women's lives.

All of the above they have done in the conviction that there has been a path not all of whose turnings they knew, and which lacked images that would show them alternative futures. There has been a debate regarding the necessity to build models of elderly women on which to base themselves at a distance of 15, 20 or 30 years, versus the wide range of possibilities that precisely the lack of such models open up for them, with the liberty that this situation offers to devise new forms; diverse, plural, contradictory, which would destroy the homogeneity of the images of the older person that has been available to social and family settings, and in which most of them do not recognise.

The new generation of elderly women requires personal models to live up to. Just now, they are the protagonists in a historical and demographic situation that lacks precedents. Never before so many women have lived for so long, in possession of so much freedom, knowledge, culture, financial independence and good health. The result is that life presents itself as an adventure in which it is possible to discover new territories that guide one's own path and that of coming generations, who will be able to feel that to grow old is not so bad after all. The new generation of elderly women now enjoys a consciousness that enables them to design their own future.

6.15 Useful or exploited?

The work done by women through all the ages, for low or zero remuneration and in many cases, as an obligation resulting from personal or family circumstances, must not be confused with civic commitment or citizen participation performed on a voluntary basis. The participation of elderly women in the life of the community can be regarded as a voluntary effort, although in many cases it has not been chosen by the women themselves. Often, civic commitments are "targeted at the privileged few who have the time, good health, resources and prior experience that allow them to engage in significant volunteer activities" (Minkler & Holstein, 2008: 199).

Too often do we regard elderly people in terms of their potential for voluntary work, taking care of a series of social necessities that no-one would otherwise cover. Elderly women have been trained in renouncing their own free time and their own desires, which makes it difficult for them to say 'no' and to set limits. They wish to be involved in the community and in society, but they occasionally feel that they are being exploited. Through their participation, they want to ensure continuity of the causes that they had earlier adopted (feminism, citizenship, society, family, etc.), and are proud of having paved the way for new generations. They are generative (Erikson, 1950), and wish to pass on their knowledge and their efforts to younger generations. They like to be regarded as active persons, forever learning, open to new trends and trying to change the world for the better (Charpentier et al., 2008).

However, the growing tendency to discuss elderly people in utilitarian terms also implies a form of ageism that is beginning to be questioned by the elderly themselves, as some of the traps that lie within the culture of participation are uncovered. Many societies have regarded old people as a vast and largely underutilized resource for meeting the needs of the community. On the other hand, certain feminist thinkers also demand the right of women to be 'non-productive' in their old age, to use their time for pure enjoyment or leisure activities (Minkler & Holstein, 2008).

7. Old women of today and tomorrow

Many of the vital, professional and relational transformations that women have achieved in the course of the 20th century were noted by the visionary writers of the late 19th and early 20th century; Edith Wharton, Charlotte Perkins Gilman, Kate Chopin, Willa Cather, Virginia Woolf, Katherine Mansfield, etc., who had the historical acuity to write stories and novels in which they advocated a new epoch of relationships and traced the paths of these great changes, proposing models of women who would be professionally, politically and emotionally independent. These are authors whose works have been a lighthouse that has illuminated the life of women, orienting the new life cycle of the women of the future.

If gerontological feminist research wishes to encourage interpretations of ageing that display the variety and complexity of lives and realities, if it wishes to suggest and invent new ways of ageing, overcoming the traditional ideas that restrict, limit and circumscribe the lives of old women, that is, if it wishes to alter the reality facing elderly women, it will need to be capable of generating an idea that, used as a motor, will result in an adequate explanatory framework. An idea that will enable old age to imagine and create and, what is more important, an idea that will help to destroy other beliefs that are currently being involuntarily sustained. For this, it will be necessary to recognise and name the changing contexts in which the women of today live their lives, which will in no way resemble those of their mothers and probably not those of their daughters either. The fact is that everything is simpler when it enjoys the support of a close community of empathetic beings that will allow the resolution of the dissonance that can be perceived between the manner in which old women perceive themselves and their image in society

The social changes that have transformed the lives of women and men in the course of the 20th century will require us to implement highly creative strategies aimed at living happily and peacefully during the last years of a long life. Although the tide of social change will carry the new generation of elderly women to different beaches than those which hosted their predecessors, they have not been brought there by circumstances alone. For the first time in history, these mature women have chosen their own route and navigated by their own charts. Obviously this will not eliminate shipwrecks, but it will make the voyage more interesting and more significant.

8. Conclusion

Although it is clear that the coming generations of old women are not going to resemble their grandmothers in almost anyway, they are not there yet. Growing old is a good time for evaluating the past, of successes achieved and of tasks still to be tackled. The difference between successes and the hopes that were their starting point may be a source of dissatisfaction and uneasiness from which the new elderly may have liberated themselves, insofar as they have managed to reject the model by which their grandmothers had been socialised. It is not a simple task to identify one's own desires, validate them and put them into practice, without feeling a certain dissonance.

Change is a characteristic of individuals of all ages, including the oldest of us. The characteristics of the elderly are also in constant flux due to changes in sociocultural, economic and health factors, new ideas, beliefs and social trends. Nothing is static, which means that gerontological research possesses an unending source of renovation and a huge field ahead of itself. New ideas that need to be studied in depth are constantly emerging, and to do so we need to listen to what older people are really saying... nor only to the words but to cries, whispers and silences (Brody, 1985). Although old people are more visible than ever, much research remains to be done in this respect. Evolutionary psychology and critical gerontology face an important task of explaining and understanding these new ages, beyond the catastrophic vision that dominates the theory in current use.

Topics still to be dealt with that are of decisive importance in the configuration of lives and professional careers include the essential redefinition of the central role played by the family in women's lives, in that life options appear to be less marked by the concept of romantic love. A new evaluation of the roles of men and women in the constitution and harmonious functioning of the family unit could ease the integration of women into the labour market, as well as a fairer share of the tasks of sustaining life, placing care-giving at the centre of the organisation of society and sharing responsibility for this task between both sexes (Carrasco, 2003; Luque & Freixas, 2008). All this can be made concrete in the course of time in a healthier and more comfortable old age for women.

Socialised as they are as 'beings in the service of others' it is difficult for them to identify the path of individuality, achievement of individual identity, and balancing the value of relationships with the necessity for silence and autonomy. There are still a number of extremely important topics that wear out the lives of women of all ages, and these topics have still to be dealt with through research and reflection: the definition of beauty on the part of women themselves; beauty at all ages; the redefinition of personal identity, beyond that given by domestic tasks; balancing the weight of love in the course of life; not being perennially available, trying to respect our way of thinking and our pathways; relationships beyond gender-related violence.

9. References

Arber, S. & Ginn, J. (Eds.) (1995). *Connecting gender and ageing: a sociological approach*, Open University Press, ISBN 0-335-19471-0, Maidenhead, Berkshire, UK

Barnett, R.C. & Baruch, G.K. (1978). Women in the middle years: A critique of research and theory. *Psychology of Women Quarterly*, No.3, pp.187-197, ISSN 0192-513X

Beck, U. (1992). *Risk Society: Towards a New Modernity*, Sage, ISBN 0-8039-8345-X, London, UK

Beck, U. & Beck-Gernsheim, E. (1995). *The Normal Chaos of Love*, Polity Press, ISBN 0-7456-1071-4, Cambridge, UK

Bernard, M.; Phillips, J.; Machin, L. & Davies, V.H. (Eds.) (2000). *Women Ageing. Changing Identities, Challenging Myths*, Routledge, ISBN 0-415-18944-6, London, UK

Brody, E.M. (1985). *Mental and physical health practices of older people*, Springer, ISBN 0-8261-4870-0, New York, USA

Brody, E.M. (2010). On being very, very old: An insider's perspective. *The Gerontologist*, Vol.50, No.1, pp. 2-10, ISSN 0016-9013

Calasanti, T.M. (2006). Gender and Old Age. Lessons from Spousal Care Work, In: *Age Matters. Realigning Feminist Thinking*, T.M. Calasanti & K.F. Slevin, (Eds.), 269-294, Routledge, ISBN 978-0415952248, New York, USA

Carrasco, C. (2003). *Tiempos, trabajos y flexibilidad: una cuestión de género*, Instituto de la Mujer, ISBN 207-03-074-X, Madrid, Spain

Charpentier, M.; Quéniart, A. & Jacques, J. (2008). Activism among older women in Quebec, Canada: Changing the world after age 65. *Journal of Women & Aging*, Vol.20, No. 3-4, pp. 343-360, ISSN 0895-2841

Connidis, I.A. (2006). Intimate relationships. Learning from later life experience, In: *Age Matters. Realigning Feminist Thinking*, T.M. Calasanti & K.F. Slevin, (Eds.), 123-153, Routledge, ISBN 978-0415952248, New York, USA

Coria, C. (2001). *El amor no es como nos contaron... ni como lo inventamos*, Paidós, ISBN 978-950-12-2683-6, Buenos Aires, Argentina

Elizabeth, V. (2003). To Marry, or Not to Marry: That is the Question. *Feminism & Psychology*, Vol.13, No.4, pp. 426-431, ISSN 0959-3535

Erikson, E.H. (1950). *Childhood and Society*, Norton, ISBN 0-393-31068-X, New York, USA

Everingham, C.; Warner-Smith, P. & Byles, J. (2007). Transforming retirement: Re-thinking models of retirement to accommodate the experiences of women. *Women's Studies International Forum*, Vol.30, No.6, pp. 512-522, ISSN 0277-5395

Forssén, A.S.K. & Carlstedt, G. (2007). Health-Promoting Aspects of a Paid Job: Findings in a Qualitative Interview Study With Elderly Women in Sweden. *Health Care for Women International*, Vol.28, pp. 909-929, ISSN 0739-9332

Fortin, P. (2005). Un regard féministe sur les modèles de famille. *Atlantis*, Vol.30, No.1, pp. 80-91, ISSN 0210-6124

Fredrickson, B.L. & Roberts, T.A. (1997). Objectification theory: Toward understanding women's lived experiences and mental health risks. *Psychology of Women Quarterly*, Vol.21, No.2, pp. 173-206, ISSN 0361-6843

Freixas, A. (1993). *Mujer y envejecimiento. Aspectos psicosociales*, laCaixa, ISBN 84-7664-399-3, Barcelona, Spain

Freixas, A. (1997). Envejecimiento y género: otras perspectivas necesarias. *Anuario de Psicología*, No.75, pp. 31-42, ISSN 0066-5126

Freixas, A. (2002). Dones i envelliment. Apunts per a una agenda. *DCIDOB*, No.82, pp. 36-39, ISSN 1133-6595

Freixas, A. (2008). La vida de las mujeres mayores a la luz de la investigación gerontológica feminista. *Anuario de Psicología*, Vol.39, No.1, pp. 41-57, ISSN 0066-5126

Friedan, B. (1993). *The fountain of age*, Simon & Schuster, ISBN 9780743299879, N. York, USA

Gannon, L. (1999). *Women and Aging. Transcending the Myths*, Routledge, ISBN 978-0415169103, London, UK

Giddens, A. (1992). *The transformation of intimacy. Sexuality, love and eroticism in modern societies*, Polity Press, ISBN 978-0804722148, Cambridge, UK

Gilligan, C. (1982). *In a Different Voice: Psychological Theory and Women's Development*, Harvard University Press, ISBN 0674445449, Cambridge, Ma, USA

Greer, G. (1991). *The Change. Women, Ageing and the Menopause*, Hamish Hamilton, ISBN 978-0449908532, London, UK

Gross , N. & Simmons, S. (2002). Intimacy as a double-edged phenomenon? An empirical test of Giddens. *Social Forces*, Vol. 81, No 2, pp. 531-555, ISSN 0037-7732

Heilbrun, C.G. (1997). *The last gift of time. Life beyond sixty*, Ballantine, ISBN 0-345-42295-3, New York, USA

Hirdman, Y. (1994). *Women from possibility to problem?*, Arbetsliuscentrum, ISBN 0-415-15350-6, Stockholm, Sweden

Hollows, J. (2001). *Feminism, Feminity and Popular Culture*, Manchester University Press, ISBN 9780719043956, Manchester, UK

Kingsberg, S.A. (2002). The impact of aging on sexual function in women and their partners. *Archives of Sexual Behavior*, Vol.31, No.5, pp. 431-437, ISSN 00040002

Libreria delle Donne di Milano (1987). *Non credere di avere diritti*, Rosenberg & Sellier, ISBN 8870112756, Turin, Italy

Levinson, D.J. (1978). *The seasons of a man's life*, Knopf, ISBN 0345339010, New York, USA

Luque, B. (2008). El itinerario profesional de las mujeres jóvenes: una carrera de obstáculos. *Anuario de Psicología*, Vol.39, No.1, 101-107, ISSN 0066-5126.

Luque, B. & Freixas, A. (2008). Itinerarios vitales y profesionales de mujeres y hombres y su relación con los proyectos iniciales. *Estudios de Psicología*, Vol.29, No2, pp. 197-207, ISSN 0210-9395

Magarian, A. (2003). Les mouvements associatifs. *Gérontologie et Societé*, No.106, pp. 249-261, ISSN 0151-0193

Milan, A. (2005). Volonté de participer. L'engagement politique chez les jeunes adultes. *Tendances sociales canadiennes*, (Statistique Canada, 1-008), pp. 2-7, ISSN 0831-5701

Minkler, M. & Holstein, M.B. (2008). From civil rights to... civic engagement? Concerns of two older critical gerontologist about a 'new social movement' and what it portends. *Journal of Aging Studies*, Vol.22, pp.196-204, ISSN 08904065

Moreno, H. (2011). Una buena encarnación. *Debate Feminista*,Vol.21, No.42, pp. 111-135, ISSN: 0188-9478

Pearsall, M. (Ed.). (1997). *The other within us. Feminist explorations of women and aging*. Westview Press, ISBN 978-0813381633, Boulder, Co, USA

Ray, R. E. (2006). The personal as political. The legacy of Betty Friedan. In: *Age Matters. Realigning Feminist Thinking*,T.M. Calsanti & K.F. Slevin, (Eds), 21-45, Routledge, ISBN 978-0415952248, New York, USA

Sontag, S. (1972, 23 sept.). The double standard of aging. *Saturday Review*, pp. 29-38

Sorensen, G. & Verbrugge, L.M. (1987). Women, work, and health. *Annual Review of Public Health*, Vol.8, pp. 235-251, ISSN 01637525

Thomas, G. & Fogg, S. (2000). *The best times, the worst of times. Older woman's retirement experience. Messages for future older women (Report)*, Older Women's Network, ISBN 1-900578-16-6, Sydney, Australia

Tronto, J. C. (2000). Age-Segregated Housing As a Moral Problem: An Exercise in Rethinking Ethics. In: *Mother Time. Women, Aging and Ethics*, M.U. Walker, (Ed.), 261-277, Rowman & Littlefield, ISBN 978-0847692606, Lanham, UK

Valls, C.; Banqué, M.; Fuentes, M. & Ojuel, J. (2008). Morbilidad diferencial entre mujeres y hombres. *Anuario de Psicología*, Vol. 39, No.1, pp. 9-22, ISSN 0066-5126

Wolf, N. (1991). *The Beauty Mythe*, Morrow, ISBN 978-0060512187, New York, USA

Woodward, K. (Ed.). (1999). *Figuring Age. Women, Bodies, Generations*, Indiana University Press, ISBN 0-253-21236-7, Bloomington, In, USA

Woolf, V. (1938). *Three Guineas*, ISBN 0-7012-0949-6, Hogarth Press, London, UK

Permissions

The contributors of this book come from diverse backgrounds, making this book a truly international effort. This book will bring forth new frontiers with its revolutionizing research information and detailed analysis of the nascent developments around the world.

We would like to thank Dr. Tetsuji Nagata, for lending his expertise to make the book truly unique. He has played a crucial role in the development of this book. Without his invaluable contribution this book wouldn't have been possible. He has made vital efforts to compile up to date information on the varied aspects of this subject to make this book a valuable addition to the collection of many professionals and students.

This book was conceptualized with the vision of imparting up-to-date information and advanced data in this field. To ensure the same, a matchless editorial board was set up. Every individual on the board went through rigorous rounds of assessment to prove their worth. After which they invested a large part of their time researching and compiling the most relevant data for our readers. Conferences and sessions were held from time to time between the editorial board and the contributing authors to present the data in the most comprehensible form. The editorial team has worked tirelessly to provide valuable and valid information to help people across the globe.

Every chapter published in this book has been scrutinized by our experts. Their significance has been extensively debated. The topics covered herein carry significant findings which will fuel the growth of the discipline. They may even be implemented as practical applications or may be referred to as a beginning point for another development. Chapters in this book were first published by InTech; hereby published with permission under the Creative Commons Attribution License or equivalent.

The editorial board has been involved in producing this book since its inception. They have spent rigorous hours researching and exploring the diverse topics which have resulted in the successful publishing of this book. They have passed on their knowledge of decades through this book. To expedite this challenging task, the publisher supported the team at every step. A small team of assistant editors was also appointed to further simplify the editing procedure and attain best results for the readers.

Our editorial team has been hand-picked from every corner of the world. Their multi-ethnicity adds dynamic inputs to the discussions which result in innovative outcomes. These outcomes are then further discussed with the researchers and contributors who give their valuable feedback and opinion regarding the same. The feedback is then collaborated with the researches and they are edited in a comprehensive manner to aid the understanding of the subject.

Apart from the editorial board, the designing team has also invested a significant amount of their time in understanding the subject and creating the most relevant covers. They scrutinized every image to scout for the most suitable representation of the subject and create an appropriate cover for the book.

The publishing team has been involved in this book since its early stages. They were actively engaged in every process, be it collecting the data, connecting with the contributors or procuring relevant information. The team has been an ardent support to the editorial, designing and production team. Their endless efforts to recruit the best for this project, has resulted in the accomplishment of this book. They are a veteran in the field of academics and their pool of knowledge is as vast as their experience in printing. Their expertise and guidance has proved useful at every step. Their uncompromising quality standards have made this book an exceptional effort. Their encouragement from time to time has been an inspiration for everyone.

The publisher and the editorial board hope that this book will prove to be a valuable piece of knowledge for researchers, students, practitioners and scholars across the globe.

List of Contributors

Jing-ye Zhou, Yong Yu, Xian-Lun Zhu, Chi-Ping Ng, Gang Lu and Wai-Sang Poon
Division of Neurosurgery, Department of Surgery, The Chinese University of Hong Kong, Hong Kong, China

Teimuraz A. Lezhava, Tinatin A. Jokhadze and Jamlet R. Monaselidze
Department of Genetics, Iv.Javakhishvili Tbilisi State University, Tbilisi, Georgia

Susana Novella, Gloria Segarra, Carlos Hermenegildo and Pascual Medina
Departamento de Fisiología, Universitat de València, Instituto de Investigación Sanitaria INCLIVA, Hospital Clínico Universitario, Valencia,

Ana Paula Dantas
Institut d'Investigacions Biomèdiques August Pi i Sunyer (IDIBAPS), Institut Clinic de Tòrax, Hospital Clinic Barcelona, Spain

Sergio Davinelli and Giovanni Scapagnini
Department of Health Sciences, University of Molise, Campobasso, Italy

Sonya Vasto and Calogero Caruso
Department of Pathobiology and Biomedical Methodologies, Immunosenescence Unit, University of Palermo, Palermo, Italy

Davide Zella
Department of Biochemistry and Molecular Biology, Institute of Human Virology-School of Medicine, University of Maryland, Baltimore, MD, USA

Patrizia d'Alessio
University Paris Sud-11 ESTEAM Stem Cell Core Facility and AISA Therapeutics, Evry, France

Annelise Bennaceur-Griscelli
University Paris Sud-11 ESTEAM Stem Cell Core Facility and Inserm U935, Villejuif, France

Rita Ostan and Claudio Franceschi
Department of Experimental Pathology, University of Bologna, Bologna, Italy

Kayoko Maehara
Department of Maternal-Fetal Biology, National Research Institute for Child Health and Development, Tokyo, Japan

Krisztian Kvell and Judit E. Pongracz
Department of Medical Biotechnology, University of Pecs, Hungary

Maria Crisan, Radu Badea and Diana Crisan
University of Medicine and Pharmacy "Iuliu Hatieganu", Romania

Carlo Cattani
University of Salerno, Italy

Daniela Cesselli, Angela Caragnano, Natascha Bergamin, Veronica Zanon, Nicoletta Finato, Carlo Alberto Beltrami and Antonio Paolo Beltrami
Department of Medical and Biological Sciences, University of Udine, Udine, Italy

Ugolino Livi
Department of Experimental and Clinical Medical Sciences, University of Udine, Udine, Italy

Karine Cucumel, Jean Marie Botto, Nouha Domloge and Claude Dal Farra
Ashland Specialty Ingredients – Vincience, France

Hulya Yucel
Department of Physiotherapy and Rehabilitation, Faculty of Health Sciences, Bezmi Alem University, Istanbul, Turkey

Takafumi Inoue and Norio Wake
Department of Obstetrics and Gynecology, Kyushu University, Fukuoka, Japan

Anna Freixas, Bárbara Luque and Amalia Reina
Department of Psychology, Faculty of Educational Sciences, University of Cordova, Cordova, Spain